U0211621

Springer Handbook of Electronic and Photonic Materials

Safa Kasap, Peter Capper (Eds.)

Springer Handbook of Electronic and Photonic Materials
Organization of the Handbook

Each chapter has a concise summary that provides a general overview of the subject in the chapter in a clear language. The chapters begin at fundamentals and build up towards advanced concepts and applications. Emphasis is on physical concepts rather than extensive mathematical derivations. Each chapter is full of clear color illustrations that convey the concepts and make the subject matter enjoyable to read and understand. Examples in the chapters have practical applications. Chapters also have numerous extremely useful tables that summarize equations, experimental techniques, and most importantly, properties of various materials. The chapters have been divided into five parts. Each part has chapters that form a coherent treatment of a given area. For example,

Part A contains chapters starting from basic concepts and build up to up-to-date knowledge in a logical easy to follow sequence. Part A would be equivalent to a graduate level treatise that starts from basic structural properties to go onto electrical, dielectric, optical, and magnetic properties. Each chapter starts by assuming someone who has completed a degree in physics, chemistry, engineering, or materials science.

Part A Fundamental Properties

Part B provides a clear overview of bulk and single-crystal growth, growth techniques (epitaxial crystal growth: LPE, MOVPE, MBE), and the structural, chemical, electrical and thermal characterization of materials. Silicon and II–VI compounds and semiconductors are especially emphasized.

Part B Growth and Characterization

Part C covers specific materials such as crystalline Si, microcrystalline Si, GaAs, high-temperature semiconductors, amorphous semiconductors, ferroelectric materials, and thin and thick films.

Part C Materials for Electronics

Part D examines materials that have applications in optoelectronics and photonics. It covers some of the state-of-the-art developments in optoelectronic materials, and covers III–V Ternaries, III–Nitrides, II–VI compounds, quantum wells, photonic crystals, glasses for photonics, nonlinear photonic glasses, nonlinear organic, and luminescent materials.

Part D Materials for Optoelectronics and Photonics

Part E provides a survey on novel materials and applications such as information recording devices (CD, video, DVD) as well as phase-change optical recording. The chapters also include applications such as solar cells, sensors, photoconductors, and carbon nanotubes. Both ends of the spectrum from research to applications are represented in chapters on molecular electronics and packaging materials.

Part E Novel Materials and Selected Applications

Glossary of Defining Terms There is a glossary of *Defining Terms* at the end of the handbook that covers important terms that are used throughout the handbook. The terms have been defined to be clear and understandable by an average reader not directly working in the field.

使用说明

1.《电子与光子材料手册》原版为一册，分为A、B、C、D、E五部分。考虑到各部分内容相对独立完整，为使用方便，影印版按部分分为5册。

2.各册在页脚重新编排页码，该页码对应中文目录。保留了原书页眉及页码，其页码对应原书目录及主题索引。

3.各册均有完整5册书的内容简介。

4.作者及其联系方式、缩略语表各册均完整呈现。

5.名词术语表、主题索引安排在第5册。

6.文前页基本采用中英文对照形式，方便读者快速浏览。

材料科学与工程图书工作室

联系电话　0451-86412421

　　　　　0451-86414559

邮　　箱　yh_bj@yahoo.com.cn

　　　　　xuyaying81823@gmail.com

　　　　　zhxh6414559@yahoo.com.cn

Springer手册精选系列

电子与光子材料手册

电子与光子材料的制备和特性

【第2册】

Springer Handbook of *Electronic and Photonic Materials*

〔加拿大〕Safa Kasap
〔英　国〕Peter Capper
主编

（影印版）

哈爾濱工業大學出版社
HARBIN INSTITUTE OF TECHNOLOGY PRESS

黑版贸审字08-2012-031号

图书在版编目（CIP）数据

电子与光子材料手册. 第2册，电子与光子材料的制备和特性=Handbook of Electronic and Photonic Materials Ⅱ Growth and Characterization：英文／（加）卡萨普（Kasap S.），（英）卡珀（Capper P.）主编. —影印本. —哈尔滨：哈尔滨工业大学出版社，2013.1

（Springer手册精选系列）

ISBN 978-7-5603-3761-6

Ⅰ. ①电… Ⅱ. ①卡…②卡… Ⅲ. ①电子材料-手册-英文②光学材料-手册-英文Ⅳ. ①TN04-62②TB34-62

中国版本图书馆CIP数据核字（2012）第189774号

材料科学与工程
图书工作室

责任编辑 杨　桦　张秀华
出版发行 哈尔滨工业大学出版社
社　　址 哈尔滨市南岗区复华四道街10号 邮编 150006
传　　真 0451-86414749
网　　址 http://hitpress.hit.edu.cn
印　　刷 哈尔滨市石桥印务有限公司
开　　本 787mm×960mm 1/16 印张 15.25
版　　次 2013年1月第1版 2013年1月第1次印刷
书　　号 ISBN 978-7-5603-3761-6
定　　价 48.00元

序 言

本书的编辑、作者、出版人都将庆祝这本卓著书籍的出版，这对于电子与光子材料领域的工作者也将是无法衡量的好消息。从以往编辑的系列手册看，我认为本书的出版是值得的，坚持出版这样一本书也是必要的。本书之所以显得特别重要，是因为它在这个领域，内容覆盖范围广泛，涉及的方法也是当今的最新研究进展。在这样一个迅速发展的领域，这是一个相当大的挑战，它已经赢得了人们的敬意。

早期的手册和百科全书也都注重阐述半导体材料的发展趋势，而且必须覆盖半导体材料广泛的研究范围和所涉及的现象。这是可以理解的，原因在于半导体材料在电子领域中的主导地位。但没有多少人有足够的勇气预测未来的发展趋势。1992年，Mahajan和Kimerling在其《简明半导体材料百科全书和相关技术》一书的引言中做了尝试，并且预测未来的挑战将是纳米电子领域、低位错密度的III–V族衬底技术、半绝缘III–V族衬底技术、III–V族图形外延技术、替换电介质和硅接触技术、离子注入和扩散技术的发展。这些预测或多或少地成为了现实，但是这也同样说明做出这样的预测是多么的困难。

十年前没有多少人会想到III族氮化物在这本书中将成为重要的部分。与制备相关的问题是，作为高熔点材料，在受欢迎的能在光谱蓝端作光发射器的材料中它们的熔点并不高。这是一个很有意思的话题，至少与解决早期光谱红端的固体激光器工作寿命短的问题一样有趣。总地说来，光电子学和光子学在前十年中已经呈现出一些令人瞩目的研究进展，这些在本书中得到了体现，范围从可见光发光器件材料到红外线材料。书中Part D的内容范围很宽，包括III–V族和II–VI族光电子材料和能带隙工程，以及光子玻璃、液态晶体、有机光电导体和光子晶体的新领域。整个部分反映了材料的光产生、工艺、光传输和光探测，包括所有用光取代电子的必要内容。

在电子材料这一章（Part C）探讨了硅的进展。毋庸置疑地是，硅是占据了电子功能和电子电路整个范围主导地位的材料，包括新电介质和其他关于缩减电路和器件的几何尺寸以实现更高密度的封装方面的内容，以及其他书很少涉及的领域，薄膜、高温电子材料、非晶和微晶材料。增加硅使用寿命的新技术成果（包括硅/锗合金）在书中也有介绍，并且又一次提出了同样问题，即，预测硅过时时间是否过于超前！铁电体——一类与硅非常有效结合的材料同样也出现在书中。

Part E章节中（新型材料和选择性的应用）使用了一些极好的新方法开辟了新领地。我们大都知道且频繁使用信息记录器件，但是很少知道，涉及器件使用的材料或原理，比如说CD、视频、DVD等。本书介绍了磁信息存储材料，同样介绍了相变光记录材料，使我们充分与当前发展步伐保持同步更新。该章也同样介绍了太阳能电池、传感器、光导体和碳纳米管的应用，这样大量的工作也体现出编写内容汲取到了世界范围的广度。本章各节中的分子电子和封装材料从研究到应用都得到了呈现。

本书的突出优点在于它的内容覆盖了从基础科学（Part A）到材料的制备、特性（Part B）再到材料的应用（Part C~E）。实际上，书中介绍了涉及的所有材料的广泛应用，这就是本书为什么将会实用的原因之一。就像我之前提及的那样，我们之中没有多少人能够成功地预测未来的发展方向和趋势，在未来十年占领这个领域的主导地位。但是，本书教给我们关于材料的基本性能，可用它们去满足将来的需要。我热切地把这本书推荐给你们。

Prof. Arthur Willoughby
Materials Research Group,
University of Southampton,
UK

前　言

不同学科各种各样的手册，例如电子工程、电子学、生物医学工程、材料科学等手册被广大学生、教师、专业人员很好地使用着，大部分的图书馆也都藏有这些手册。这类手册一般包含许多章（至少50章）内容，在已确定的学科内覆盖广泛的课题；学科选材和论述水平吸引着本科生、研究生、研究员，乃至专业工程人员；最新课题提供广泛的信息，这对该领域所有初学者和研究人员是非常有帮助的；每隔几年，就会有增加新内容的新版本更新之前的版本。

电子和光子材料领域没有类似手册的出版，我们出版这本《电子与光子材料手册》的想法是源自于对手册的需求。它广泛覆盖当今材料领域内的课题，在工程学、材料科学、物理学和化学中都有需要。电子和光子材料真正是一门跨学科的学问，它包含了一些传统的学科，如材料科学、电子工程、化学工程、机械工程、物理学和化学。不难发现，机械工程人员对电子封装实施研究，而电子工程人员对半导体特性进行测量。只有很少的几所大学创建了电子材料或光子材料系。一般来说，电子材料作为一个“学科”是以研究组或跨学科的活动出现在“学院”中。有人可能会对此有异议，因为它事实上是一个跨学科领域，非常需要既包括基础学科又要有最新课题介绍的手册，这就是出版本手册的原因。

本手册是一部关于电子和光子材料的综合论述专著，每一章都是由该领域的专家编写的。本手册针对于大学四年级学生或研究生、研究人员和工作在电子、光电子、光子材料领域的专业人员。书中提供了必要的背景知识和内容广泛的更新知识。每一章都有对内容的一个介绍，并且有许多清晰的说明和大量参考文献。清晰的解释和说明使手册对所有层次的研究者有很大的帮助。所有的章节内容都尽可能独立。既有基础又有前沿的章节内容将吸引不同背景的读者。本手册特别重要的一个特点就是跨学科。例如，将会有这样一些读者，其背景（第一学历）是学化学工程的，工作在半导体工艺线上，而想要学习半导体物理的基础知识；第一学历是物理学的另外一些读者需要尽快更新材料科学的新概念，例如，液相外延等。只要可能，本手册尽量避免采用复杂的数学公式，论述将以半定量的形式给出。手册给出了名词术语表（Glossary of Defining Terms），可为读者提供术语定义的快速查找——这对跨学科工具书来说是必须的。

编者非常感激所有作者们卓越的贡献和相互合作，以及在不同阶段对撰写这本手册的奉献。真诚地感谢Springer Boston的Greg Franklin在文献整理以及手册出版的漫长的工作中给予的支持和帮助。Dr.Werner Skolaut在Springer Heidelberg非常熟练地处理了无数个出版问题，涉及审稿、绘图、书稿的编写和校样的修改，我们真诚地感谢他和他所做出的工作——使得手册能够吸引读者。他是我们见过的最有奉献精神和有效率的编者。

感谢Arthur Willoughby教授的诸多建设性意见使得本手册更加完善。他在材料科学杂志（Journal of Materials Science）积累了非常丰富的编辑经验：电子材料这一章在书中起着重要作用，不仅仅是选取章节，而且还要适应读者需要。

最后，编者感谢所有的成员（Marian, Samuel and Tomas; and Nicollette）在全部工作中的支持和付出的特别耐心。

Dr. Peter Capper
Materials Team Leader,
SELEX Sensors and Airborne Systems,
Southampton, UK

Prof. Safa Kasap
Professor and Canada Research Chair,
Electrical Engineering Department,
University of Saskatchewan,
Canada

Foreword

The Editors, Authors, and Publisher are to be congratulated on this distinguished volume, which will be an invaluable source of information to all workers in the area of electronic and photonic materials. Having made contributions to earlier handbooks, I am well aware of the considerable, and sustained work that is necessary to produce a volume of this kind. This particular handbook, however, is distinguished by its breadth of coverage in the field, and the way in which it discusses the very latest developments. In such a rapidly moving field, this is a considerable challenge, and it has been met admirably.

Previous handbooks and encyclopaedia have tended to concentrate on semiconducting materials, for the understandable reason of their dominance in the electronics field, and the wide range of semiconducting materials and phenomena that must be covered. Few have been courageous enough to predict future trends, but in 1992 Mahajan and Kimerling attempted this in the Introduction to their Concise Encyclopaedia of Semiconducting Materials and Related Technologies (Pergamon), and foresaw future challenges in the areas of nanoelectronics, low dislocation-density III-V substrates, semi-insulating III-V substrates, patterned epitaxy of III-Vs, alternative dielectrics and contacts for silicon technology, and developments in ion-implantation and diffusion. To a greater or lesser extent, all of these have been proved to be true, but it illustrates how difficult it is to make such a prediction.

Not many people would have thought, a decade ago, that the III-nitrides would occupy an important position in this book. As high melting point materials, with the associated growth problems, they were not high on the list of favourites for light emitters at the blue end of the spectrum! The story is a fascinating one – at least as interesting as the solution to the problem of the short working life of early solid-state lasers at the red end of the spectrum. Optoelectronics and photonics, in general, have seen one of the most spectacular advances over the last decade, and this is fully reflected in the book, ranging from visible light emitters, to infra-red materials. The book covers a wide range of work in Part D, including III-V and II-VI optoelectronic materials and band-gap engineering, as well as photonic glasses, liquid crystals, organic photoconductors, and the new area of photonic crystals. The whole Part reflects materials for light generation, processing, transmission and detection – all the essential elements for using light instead of electrons.

Prof. Arthur Willoughby
Materials Research Group,
University of Southampton,
UK

In the Materials for Electronics part (Part C) the book charts the progress in silicon – overwhelmingly the dominant material for a whole range of electronic functions and circuitry – including new dielectrics and other issues associated with shrinking geometry of circuits and devices to produce ever higher packing densities. It also includes areas rarely covered in other books – thick films, high-temperature electronic materials, amorphous and microcrystalline materials. The existing developments that extend the life of silicon technology, including silicon/germanium alloys, appear too, and raise the question again as to whether the predicted timetable for the demise of silicon has again been declared too early!! Ferroelectrics – a class of materials used so effectively in conjunction with silicon – certainly deserve to be here.

The chapters in Part E (Novel Materials and Selected Applications), break new ground in a number of admirable ways. Most of us are aware of, and frequently use, information recording devices such as CDs, videos, DVDs etc., but few are aware of the materials, or principles, involved. This book describes magnetic information storage materials, as well as phase-change optical recording, keeping us fully up-to-date with recent developments. The chapters also include applications such as solar cells, sensors, photoconductors, and carbon nanotubes, on which such a huge volume of work is presently being pursued worldwide. Both ends of the spectrum from research to applications are represented in chapters on molecular electronics and packaging materials.

A particular strength of this book is that it ranges from the fundamental science (Part A) through growth and characterisation of the materials (Part B) to applications (Parts C–E). Virtually all the materials covered here have a wide range of applications, which is one of the reasons why this book is going to be so useful. As I indicated before, few of us will be successful in predicting the future direction and trends, occupying the high-ground in this field in the coming decade, but this book teaches us the basic principles of materials, and leaves it to us to adapt these to the needs of tomorrow. I commend it to you most warmly.

Preface

Other handbooks in various disciplines such as electrical engineering, electronics, biomedical engineering, materials science, etc. are currently available and well used by numerous students, instructors and professionals. Most libraries have these handbook sets and each contains numerous (at least 50) chapters that cover a wide spectrum of topics within each well-defined discipline. The subject and the level of coverage appeal to both undergraduate and postgraduate students and researchers as well as to practicing professionals. The advanced topics follow introductory topics and provide ample information that is useful to all, beginners and researchers, in the field. Every few years, a new edition is brought out to update the coverage and include new topics.

There has been no similar handbook in electronic and photonic materials, and the present Springer Handbook of Electronic and Photonic Materials (SHEPM) idea grew out of a need for a handbook that covers a wide spectrum of topics in materials that today's engineers, material scientists, physicists, and chemists need. Electronic and photonic materials is a truly interdisciplinary subject that encompasses a number of traditional disciplines such as materials science, electrical engineering, chemical engineering, mechanical engineering, physics and chemistry. It is not unusual to find a mechanical engineering faculty carrying out research on electronic packaging and electrical engineers carrying out characterization measurements on semiconductors. There are only a few established university departments in electronic or photonic materials. In general, electronic materials as a "discipline" appears as a research group or as an interdisciplinary activity within a "college". One could argue that, because of the very fact that it is such an interdisciplinary field, there is a greater need to have a handbook that covers not only fundamental topics but also advanced topics; hence the present handbook.

This handbook is a comprehensive treatise on electronic and photonic materials with each chapter written by experts in the field. The handbook is aimed at senior undergraduate and graduate students, researchers and professionals working in the area of electronic, optoelectronic and photonic materials. The chapters provide the necessary background and up-to-date knowledge in a wide range of topics. Each chapter has an introduction to the topic, many clear illustrations and numerous references. Clear explanations and illustrations make the handbook useful to all levels of researchers. All chapters are as self-contained as possible. There are both fundamental and advanced chapters to appeal to readers with different backgrounds. This is particularly important for this handbook since the subject matter is highly interdisciplinary. For example, there will be readers with a background (first degree) in chemical engineering and working on semiconductor processing who need to learn the fundamentals of semiconductors physics. Someone with a first degree in physics would need to quickly update himself on materials science concepts such as liquid phase epitaxy and so on. Difficult mathematics has been avoided and, whenever possible, the explanations have been given semiquantitatively. There is a *"Glossary of Defining Terms"* at the end of the handbook, which can serve to quickly find the definition of a term – a very necessary feature in an interdisciplinary handbook.

Dr. Peter Capper
Materials Team Leader,
SELEX Sensors and Airborne Systems,
Southampton, UK

Prof. Safa Kasap
Professor and Canada Research Chair,
Electrical Engineering Department,
University of Saskatchewan,
Canada

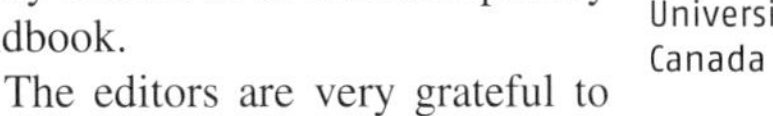

The editors are very grateful to all the authors for their excellent contributions and for their cooperation in delivering their manuscripts and in the various stages of production of this handbook. Sincere thanks go to Greg Franklin at Springer Boston for all his support and help throughout the long period of commissioning, acquiring the contributions and the production of the handbook. Dr. Werner Skolaut at Springer Heidelberg has very skillfully handled the myriad production issues involved in copy-editing, figure redrawing and proof preparation and correction and our sincere thanks go to him also for all his hard work in making the handbook attractive to read. He is the most dedicated and efficient editor we have come across.

It is a pleasure to thank Professor Arthur Willoughby for his many helpful suggestions that made this a better handbook. His wealth of experience as editor of the Journal of Materials Science: Materials in Electronics played an important role not only in selecting chapters but also in finding the right authors.

Finally, the editors wish to thank all the members of our families (Marian, Samuel and Thomas; and Nicollette) for their support and particularly their endurance during the entire project.

Peter Capper and Safa Kasap
Editors

List of Authors

Martin Abkowitz
1198 Gatestone Circle
Webster, NY 14580, USA
e-mail: *mabkowitz@mailaps.org*,
abkowitz@chem.chem.rochester.edu

Sadao Adachi
Gunma University
Department of Electronic Engineering,
Faculty of Engineering
Kiryu-shi 376-8515
Gunma, Japan
e-mail: *adachi@el.gunma-u.ac.jp*

Alfred Adams
University of Surrey
Advanced Technology Institute
Guildford, Surrey, GU2 7XH,
Surrey, UK
e-mail: *alf.adams@surrey.ac.uk*

Guy J. Adriaenssens
University of Leuven
Laboratorium voor Halfgeleiderfysica
Celestijnenlaan 200D
B-3001 Leuven, Belgium
e-mail: *guy.adri@fys.kuleuven.ac.be*

Wilfried von Ammon
Siltronic AG
Research and Development
Johannes Hess Strasse 24
84489 Burghausen, Germany
e-mail: *wilfried.ammon@siltronic.com*

Peter Ashburn
University of Southampton
School of Electronics and Computer Science
Southampton, SO17 1BJ, UK
e-mail: *pa@ecs.soton.ac.uk*

Mark Auslender
Ben-Gurion University of the Negev Beer Sheva
Department of Electrical
and Computer Engineering
P.O.Box 653
Beer Sheva 84105, Israel
e-mail: *marka@ee.bgu.ac.il*

Darren M. Bagnall
University of Southampton
School of Electronics and Computer Science
Southampton, SO17 1BJ, UK
e-mail: *dmb@ecs.soton.ac.uk*

Ian M. Baker
SELEX Sensors and Airborne Systems Infrared Ltd.
Southampton, Hampshire SO15 0EG, UK
e-mail: *ian.m.baker@selex-sas.com*

Sergei Baranovskii
Philipps University Marburg
Department of Physics
Renthof 5
35032 Marburg, Germany
e-mail: *baranovs@staff.uni-marburg.de*

Mark Baxendale
Queen Mary, University of London
Department of Physics
Mile End Road
London, E1 4NS, UK
e-mail: *m.baxendale@qmul.ac.uk*

Mohammed L. Benkhedir
University of Leuven
Laboratorium voor Halfgeleiderfysica
Celestijnenlaan 200D
B-3001 Leuven, Belgium
e-mail: *MohammedLoufti.Benkhedir*
@fys.kuleuven.ac.be

Monica Brinza
University of Leuven
Laboratorium voor Halfgeleiderfysica
Celestijnenlaan 200D
B-3001 Leuven, Belgium
e-mail: *monica.brinza@fys.kuleuven.ac.be*

Paul D. Brown
University of Nottingham
School of Mechanical, Materials and
Manufacturing Engineering
University Park
Nottingham, NG7 2RD, UK
e-mail: *paul.brown@nottingham.ac.uk*

Mike Brozel
University of Glasgow
Department of Physics and Astronomy
Kelvin Building
Glasgow, G12 8QQ, UK
e-mail: *mikebrozel@beeb.net*

Lukasz Brzozowski
University of Toronto
Sunnybrook and Women's Research Institute,
Imaging Research/
Department of Medical Biophysics
Research Building, 2075 Bayview Avenue
Toronto, ON, M4N 3M5, Canada
e-mail: *lukbroz@sten.sunnybrook.utoronto.ca*

Peter Capper
SELEX Sensors and Airborne Systems Infrared Ltd.
Materials Team Leader
Millbrook Industrial Estate, PO Box 217
Southampton, Hampshire SO15 0EG, UK
e-mail: *pete.capper@selex-sas.com*

Larry Comstock
San Jose State University
6574 Crystal Springs Drive
San Jose, CA 95120, USA
e-mail: *Comstock@email.sjsu.edu*

Ray DeCorby
University of Alberta
Department of Electrical
and Computer Engineering
7th Floor, 9107-116 Street N.W.
Edmonton, Alberta T6G 2V4, Canada
e-mail: *rdecorby@trlabs.ca*

M. Jamal Deen
McMaster University
Department of Electrical
and Computer Engineering (CRL 226)
1280 Main Street West
Hamilton, ON L8S 4K1, Canada
e-mail: *jamal@mcmaster.ca*

Leonard Dissado
The University of Leicester
Department of Engineering
University Road
Leicester, LE1 7RH, UK
e-mail: *lad4@le.ac.uk*

David Dunmur
University of Southampton
School of Chemistry
Southampton, SO17 1BJ, UK
e-mail: *d.a.dunmur@soton.ac.uk*

Lester F. Eastman
Cornell University
Department of Electrical
and Computer Engineering
425 Phillips Hall
Ithaca, NY 14853, USA
e-mail: *lfe2@cornell.edu*

Andy Edgar
Victoria University
School of Chemical and Physical Sciences SCPS
Kelburn Parade/PO Box 600
Wellington, New Zealand
e-mail: *Andy.Edgar@vuw.ac.nz*

Brian E. Foutz
Cadence Design Systems
1701 North Street, Bldg 257-3
Endicott, NY 13760, USA
e-mail: *foutz@cadence.com*

Mark Fox
University of Sheffield
Department of Physics and Astronomy
Hicks Building, Hounsefield Road
Sheffield, S3 7RH, UK
e-mail: *mark.fox@shef.ac.uk*

Darrel Frear
RF and Power Packaging Technology Development,
Freescale Semiconductor
2100 East Elliot Road
Tempe, AZ 85284, USA
e-mail: *darrel.frear@freescale.com*

Milan Friesel
Chalmers University of Technology
Department of Physics
Fysikgränd 3
41296 Göteborg, Sweden
e-mail: *friesel@chalmers.se*

Jacek Gieraltowski
Université de Bretagne Occidentale
6 Avenue Le Gorgeu, BP: 809
29285 Brest Cedex, France
e-mail: *Jacek.Gieraltowski@univ-brest.fr*

Yinyan Gong
Columbia University
Department of Applied Physics
and Applied Mathematics
500 W. 120th St.
New York, NY 10027, USA
e-mail: *yg2002@columbia.edu*

Robert D. Gould†
Keele University
Thin Films Laboratory, Department of Physics,
School of Chemistry and Physics
Keele, Staffordshire ST5 5BG, UK

Shlomo Hava
Ben-Gurion University of the Negev Beer Sheva
Department of Electrical
and Computer Engineering
P.O. Box 653
Beer Sheva 84105, Israel
e-mail: *hava@ee.bgu.ac.il*

Colin Humphreys
University of Cambridge
Department of Materials Science and Metallurgy
Pembroke Street
Cambridge, CB2 3!Z, UK
e-mail: *colin.Humphreys@msm.cam.ac.uk*

Stuart Irvine
University of Wales, Bangor
Department of Chemistry
Gwynedd, LL57 2UW, UK
e-mail: *sjc.irvine@bangor.ac.uk*

Minoru Isshiki
Tohoku University
Institute of Multidisciplinary Research
for Advanced Materials
1-1, Katahira, 2 chome, Aobaku
Sendai, 980-8577, Japan
e-mail: *isshiki@tagen.tohoku.ac.jp*

Robert Johanson
University of Saskatchewan
Department of Electrical Engineering
57 Campus Drive
Saskatoon, SK S7N 5A9, Canada
e-mail: *johanson@engr.usask.ca*

Tim Joyce
University of Liverpool
Functional Materials Research Centre,
Department of Engineering
Brownlow Hill
Liverpool, L69 3BX, UK
e-mail: *t.joyce@liv.ac.uk*

M. Zahangir Kabir
Concordia University
Department of Electrical and Computer Engineering
Montreal, Quebec S7N5A9, Canada
e-mail: *kabir@encs.concordia.ca*

Safa Kasap
University of Saskatchewan
Department of Electrical Engineering
57 Campus Drive
Saskatoon, SK S7N 5A9, Canada
e-mail: *safa.kasap@usask.ca*

Alexander Kolobov
National Institute of Advanced Industrial Science and Technology
Center for Applied Near-Field Optics Research
1-1-1 Higashi, Tsukuba
Ibaraki, 305-8562, Japan
e-mail: *a.kolobov@aist.go.jp*

Cyril Koughia
University of Saskatchewan
Department of Electrical Engineering
57 Campus Drive
Saskatoon, SK S7N 5A9, Canada
e-mail: *kik486@mail.usask.ca*

Igor L. Kuskovsky
Queens College, City University of New York (CUNY)
Department of Physics
65-30 Kissena Blvd.
Flushing, NY 11367, USA
e-mail: *igor_kuskovsky@qc.edu*

Geoffrey Luckhurst
University of Southampton
School of Chemistry
Southampton, SO17 1BJ, UK
e-mail: *g.r.luckhurst@soton.ac.uk*

Akihisa Matsuda
Tokyo University of Science
Research Institute for Science and Technology
2641 Yamazaki, Noda-shi
Chiba, 278-8510, Japan
e-mail: *amatsuda@rs.noda.tus.ac.jp*, *a.matsuda@aist.go.jp*

Naomi Matsuura
Sunnybrook Health Sciences Centre
Department of Medical Biophysics, Imaging Research
2075 Bayview Avenue
Toronto, ON M4N 3M5, Canada
e-mail: *matsuura@sri.utoronto.ca*

Kazuo Morigaki
University of Tokyo
C-305, Wakabadai 2-12, Inagi
Tokyo, 206-0824, Japan
e-mail: *k.morigaki@yacht.ocn.ne.jp*

Hadis Morkoç
Virginia Commonwealth University
Department of Electrical and Computer Engineering
601 W. Main St., Box 843072
Richmond, VA 23284-3068, USA
e-mail: *hmorkoc@vcu.edu*

Winfried Mönch
Universität Duisburg-Essen
Lotharstraße 1
47048 Duisburg, Germany
e-mail: *w.moench@uni-duisburg.de*

Arokia Nathan
University of Waterloo
Department of Electrical and Computer Engineering
200 University Avenue W.
Waterloo, Ontario N2L 3G1, Canada
e-mail: *anathan@uwaterloo.ca*

Gertrude F. Neumark
Columbia University
Department of Applied Physics
and Applied Mathematics
500W 120th St., MC 4701
New York, NY 10027, USA
e-mail: *gfn1@columbia.edu*

Stephen K. O'Leary
University of Regina
Faculty of Engineering
3737 Wascana Parkway
Regina, SK S4S 0A2, Canada
e-mail: *stephen.oleary@uregina.ca*

Chisato Ogihara
Yamaguchi University
Department of Applied Science
2-16-1 Tokiwadai
Ube, 755-8611, Japan
e-mail: *ogihara@yamaguchi-u.ac.jp*

Fabien Pascal
Université Montpellier 2/CEM2-cc084
Centre d'Electronique
et de Microoptoélectronique de Montpellier
Place E. Bataillon
34095 Montpellier, France
e-mail: *pascal@cem2.univ-montp2.fr*

Michael Petty
University of Durham
Department School of Engineering
South Road
Durham, DH1 3LE, UK
e-mail: *m.c.petty@durham.ac.uk*

Asim Kumar Ray
Queen Mary, University of London
Department of Materials
Mile End Road
London, E1 4NS, UK
e-mail: *a.k.ray@qmul.ac.uk*

John Rowlands
University of Toronto
Department of Medical Biophysics
Sunnybrook and Women's College
Health Sciences Centre
S656-2075 Bayview Avenue
Toronto, ON M4N 3M5, Canada
e-mail: *john.rowlands@sri.utoronto.ca*

Oleg Rubel
Philipps University Marburg
Department of Physics
and Material Sciences Center
Renthof 5
35032 Marburg, Germany
e-mail: *oleg.rubel@physik.uni-marburg.de*

Harry Ruda
University of Toronto
Materials Science and Engineering,
Electrical and Computer Engineering
170 College Street
Toronto, M5S 3E4, Canada
e-mail: *ruda@ecf.utoronto.ca*

Edward Sargent
University of Toronto
Department of Electrical
and Computer Engineering
ECE, 10 King's College Road
Toronto, M5S 3G4, Canada
e-mail: *ted.sargent@utoronto.ca*

Peyman Servati
Ignis Innovation Inc.
55 Culpepper Dr.
Waterloo, Ontario N2L 5K8, Canada
e-mail: *pservati@uwaterloo.ca*

Derek Shaw
Hull University
Hull, HU6 7RX, UK
e-mail: *DerekShaw1@compuserve.com*

Fumio Shimura
Shizuoka Institute of Science and Technology
Department of Materials and Life Science
2200-2 Toyosawa
Fukuroi, Shizuoka 437-8555, Japan
e-mail: *shimura@ms.sist.ac.jp*

Michael Shur
Rensselaer Polytechnic Institute
Department of Electrical, Computer,
and Systems Engineering
CII 9017, RPI, 110 8th Street
Troy, NY 12180, USA
e-mail: *shurm@rpi.edu*

Jai Singh
Charles Darwin University
School of Engineering and Logistics,
Faculty of Technology, B-41
Ellengowan Drive
Darwin, NT 0909, Australia
e-mail: *jai.singh@cdu.edu.au*

Tim Smeeton
Sharp Laboratories of Europe
Edmund Halley Road, Oxford Science Park
Oxford, OX4 4GB, UK
e-mail: *tim.smeeton@sharp.co.uk*

Boris Straumal
Russian Academy of Sciences
Institute of Sold State Physics
Institutskii prospect 15
Chernogolovka, 142432, Russia
e-mail: *straumal@issp.ac.ru*

Stephen Sweeney
University of Surrey
Advanced Technology Institute
Guildford, Surrey GU2 7XH, UK
e-mail: *s.sweeney@surrey.ac.uk*

David Sykes
Loughborough Surface Analysis Ltd.
PO Box 5016, Unit FC, Holywell Park, Ashby Road
Loughborough, LE11 3WS, UK
e-mail: *d.e.sykes@lsaltd.co.uk*

Keiji Tanaka
Hokkaido University
Department of Applied Physics,
Graduate School of Engineering
Kita-ku, N13 W8
Sapporo, 060-8628, Japan
e-mail: *keiji@eng.hokudai.ac.jp*

Charbel Tannous
Université de Bretagne Occidentale
LMB, CNRS FRE 2697
6 Avenue Le Gorgeu, BP: 809
29285 Brest Cedex, France
e-mail: *tannous@univ-brest.fr*

Ali Teke
Balikesir University
Department of Physics, Faculty of Art and Science
Balikesir, 10100, Turkey
e-mail: *ateke@balikesir.edu.tr*

Junji Tominaga
National Institute of Advanced Industrial
Science and Technology, AIST
Center for Applied Near-Field Optics Research,
CAN-FOR
Tsukuba Central 4 1-1-1 Higashi
Tsukuba, 3.5-8562, Japan
e-mail: *j-tomonaga@aist.go.jp*

Dan Tonchev
University of Saskatchewan
Department of Electrical Engineering
57 Campus Drive
Saskatoon, SK S7N 5A9, Canada
e-mail: *dan.tonchev@usask.ca*

Harry L. Tuller
Massachusetts Institute of Technology
Department of Materials Science and Engineering,
Crystal Physics and Electroceramics Laboratory
77 Massachusetts Avenue
Cambridge, MA 02139, USA
e-mail: *tuller@mit.edu*

Qamar-ul Wahab
Linköping University
Department of Physics,
Chemistry, and Biology (IFM)
SE-581 83 Linköping, Sweden
e-mail: *quw@ifm.liu.se*

Robert M. Wallace
University of Texas at Dallas
Department of Electrical Engineering
M.S. EC 33, P.O.Box 830688
Richardson, TX 75083, USA
e-mail: *rmwallace@utdallas.edu*

Jifeng Wang
Tohoku University
Institute of Multidisciplinary Research
for Advanced Materials
1-1, Katahira, 2 Chome, Aobaku
Sendai, 980-8577, Japan
e-mail: *wang@tagen.tohoku.ac.jp*

David S. Weiss
NexPress Solutions, Inc.
2600 Manitou Road
Rochester, NY 14653-4180, USA
e-mail: *David_Weiss@Nexpress.com*

Rainer Wesche
Swiss Federal Institute of Technology
Centre de Recherches en Physique des Plasmas
CRPP (c/o Paul Scherrer Institute), WMHA/C31,
Villigen PS
Lausanne, CH-5232, Switzerland
e-mail: *rainer.wesche@psi.ch*

Roger Whatmore
Tyndall National Institute
Lee Maltings, Cork , Ireland
e-mail: *roger.whatmore@tyndall.ie*

Neil White
University of Southampton
School of Electronics and Computer Science
Mountbatten Building
Highfield, Southampton SO17 1BJ, UK
e-mail: *nmw@ecs.soton.ac.uk*

Magnus Willander
University of Gothenburg
Department of Physics
SE-412 96 Göteborg, Sweden
e-mail: *mwi@fy.chalmers.se*

Jan Willekens
University of Leuven
Laboratorium voor Halfgeleiderfysica
Celestijnenlaan 200D
B-3001 Leuven, Belgium
e-mail: *jan.willekens@kc.kuleuven.ac.be*

Acknowledgements

B.17 Structural Characterization

by Paul D. Brown

As ever, there are many people one wishes to acknowledge for their involvement in the growth, processing and underpinning characterisation research programmes drawn from to illustrate this chapter. University of Nottingham: with thanks to Tom Foxon, T.S. Cheng, Sergei Novikov and Chris Statton for the provision of MBE GaN samples and supporting XRD analysis; and to Mike Fay for GaAs CBED patterns. University of Cambridge: with thanks to Colin Humphreys for provision of instrumentation; Chris Boothroyd for EDX and HAADF data on the SiGe samples; Michael Natusch for GaN EELS data; Robin Taylor for RHEED stage development; David Tricker for the Si-doped GaN micrograph; and Yan Xin for the GaN images used for dislocation analysis. University of Warwick: with thanks to Richard Kubiak and E.H.C. Parker for supplying SiGe/Si samples. Polish academy of Sciences, Warsaw: with thanks to Jan Weyher for homoepitaxial GaN samples. With thanks also to the EPSRC for funding support.

B.19 Thermal Properties and Thermal Analysis: Fundamentals, Experimental Techniques and Applications

by S.O. Kasap

The authors thank NSERC for financial support.

B.20 Electrical Characterization of Semiconductor Materials and Devices

by M. Jamal Deen, Fabien Pascal

The authors are very grateful to Drs. O. Marinov and D. Landheer for their careful review of the manuscript and their assistance. They are also grateful to several previous students and researchers whose collaborative research is discussed here. Finally, they are grateful to NSREC of Canada, the Canada Research Chair program and the CNRS of France for supporting this research.

目 录

缩略语

Part B 制备和特性

Contents

Part B Growth and Characterization

List of Abbreviations

2DEG	two-dimensional electron gas

A

AC	alternating current
ACCUFET	accumulation-mode MOSFET
ACRT	accelerated crucible rotation technique
AEM	analytical electron microscopes
AES	Auger electron spectroscopy
AFM	atomic force microscopy
ALD	atomic-layer deposition
ALE	atomic-layer epitaxy
AMA	active matrix array
AMFPI	active matrix flat-panel imaging
AMOLED	amorphous organic light-emitting diode
APD	avalanche photodiode

B

b.c.c.	body-centered cubic
BEEM	ballistic-electron-emission microscopy
BEP	beam effective pressure
BH	buried-heterostructure
BH	Brooks–Herring
BJT	bipolar junction transistor
BTEX	m-xylene
BZ	Brillouin zone

C

CAIBE	chemically assisted ion beam etching
CB	conduction band
CBE	chemical beam epitaxy
CBED	convergent beam electron diffraction
CC	constant current
CCD	charge-coupled device
CCZ	continuous-charging Czochralski
CFLPE	container-free liquid phase epitaxy
CKR	cross Kelvin resistor
CL	cathodoluminescence
CMOS	complementary metal-oxide-semiconductor
CNR	carrier-to-noise ratio
COP	crystal-originated particle
CP	charge pumping
CPM	constant-photocurrent method
CR	computed radiography
CR-DLTS	computed radiography deep level transient spectroscopy
CRA	cast recrystallize anneal
CTE	coefficient of thermal expansion
CTO	chromium(III) trioxalate
CuPc	copper phthalocyanine
CuTTBPc	tetra-tert-butyl phthalocyanine
CV	chemical vapor
CVD	chemical vapor deposition
CVT	chemical vapor transport
CZ	Czochralski
CZT	cadmium zinc telluride

D

DA	Drude approximation
DAG	direct alloy growth
DBP	dual-beam photoconductivity
DC	direct current
DCPBH	double-channel planar buried heterostructure
DET	diethyl telluride
DFB	distributed feedback
DH	double heterostructure
DIL	dual-in-line
DIPTe	diisopropyltellurium
DLC	diamond-like carbon
DLHJ	double-layer heterojunction
DLTS	deep level transient spectroscopy
DMCd	dimethyl cadmium
DMF	dimethylformamide
DMOSFET	double-diffused MOSFET
DMS	dilute magnetic semiconductors
DMSO	dimethylsulfoxide
DMZn	dimethylzinc
DOS	density of states
DQE	detective quantum efficiency
DSIMS	dynamic secondary ion mass spectrometry
DTBSe	ditertiarybutylselenide
DUT	device under test
DVD	digital versatile disk
DWDM	dense wavelength-division multiplexing
DXD	double-crystal X-ray diffraction

E

EBIC	electron beam induced conductivity
ED	electrodeposition
EDFA	erbium-doped fiber amplifier
EELS	electron energy loss spectroscopy
EFG	film-fed growth
EHP	electron–hole pairs
ELO	epitaxial lateral overgrowth
ELOG	epitaxial layer overgrowth
EM	electromagnetic
EMA	effective media approximation

ENDOR electron–nuclear double resonance
EPD etch pit density
EPR electron paramagnetic resonance
ESR electron spin resonance spectroscopy
EXAFS extended X-ray absorption fine structure

F

FCA free-carrier absorption
f.c.c. face-centered cubic
FET field effect transistor
FIB focused ion beam
FM Frank–van der Merwe
FPA focal plane arrays
FPD flow pattern defect
FTIR Fourier transform infrared
FWHM full-width at half-maximum
FZ floating zone

G

GDA generalized Drude approximation
GDMS glow discharge mass spectrometry
GDOES glow discharge optical emission spectroscopy
GF gradient freeze
GMR giant magnetoresistance
GOI gate oxide integrity
GRIN graded refractive index
GSMBE gas-source molecular beam epitaxy
GTO gate turn-off

H

HAADF high-angle annular dark field
HB horizontal Bridgman
HBT hetero-junction bipolar transistor
HDC horizontal directional solidification crystallization
HEMT high electron mobility transistor
HF high-frequency
HOD highly oriented diamond
HOLZ high-order Laue zone
HPc phthalocyanine
HPHT high-pressure high-temperature
HRXRD high-resolution X-ray diffraction
HTCVD high-temperature CVD
HVDC high-voltage DC
HWE hot-wall epitaxy

I

IC integrated circuit
ICTS isothermal capacitance transient spectroscopy
IDE interdigitated electrodes
IFIGS interface-induced gap states
IFTOF interrupted field time-of-flight
IGBT insulated gate bipolar transistor
IMP interdiffused multilayer process
IPEYS internal photoemission yield spectroscopy
IR infrared
ITO indium-tin-oxide

J

JBS junction barrier Schottky
JFET junction field-effect transistors
JO Judd–Ofelt

K

KCR Kelvin contact resistance
KKR Kramers–Kronig relation
KLN $K_3Li_2Nb_5O_{12}$
KTPO $KTiOPO_4$

L

LB Langmuir–Blodgett
LD laser diodes
LD lucky drift
LDD lightly doped drain
LEC liquid-encapsulated Czochralski
LED light-emitting diodes
LEIS low-energy ion scattering
LEL lower explosive limit
LF low-frequency
LLS laser light scattering
LMA law of mass action
LO longitudinal optical
LPE liquid phase epitaxy
LSTD laser light scattering tomography defect
LVM localized vibrational mode

M

MBE molecular beam epitaxy
MCCZ magnetic field applied continuous Czochralski
MCT mercury cadmium telluride
MCZ magnetic field applied Czochralski
MD molecular dynamics
MEED medium-energy electron diffraction
MEM micro-electromechanical systems
MESFET metal-semiconductor field-effect transistor
MFC mass flow controllers
MIGS metal-induced gap states
ML monolayer
MLHJ multilayer heterojunction
MOCVD metal-organic chemical vapor deposition
MODFET modulation-doped field effect transistor

MOMBE metalorganic molecular beam epitaxy
MOS metal/oxide/semiconductor
MOSFET metal/oxide/semiconductor field effect transistor
MOVPE metalorganic vapor phase epitaxy
MPc metallophthalocyanine
MPC modulated photoconductivity
MPCVD microwave plasma chemical deposition
MQW multiple quantum well
MR magnetoresistivity
MS metal–semiconductor
MSRD mean-square relative displacement
MTF modulation transfer function
MWIR medium-wavelength infrared

N

NDR negative differential resistance
NEA negative electron affinity
NeXT nonthermal energy exploration telescope
NMOS n-type-channel metal–oxide–semiconductor
NMP N-methylpyrrolidone
NMR nuclear magnetic resonance
NNH nearest-neighbor hopping
NSA naphthalene-1,5-disulfonic acid
NTC negative temperature coefficient
NTD neutron transmutation doping

O

OLED organic light-emitting diode
OSF oxidation-induced stacking fault
OSL optically stimulated luminescence
OZM overlap zone melting

P

PAE power added efficiency
PAni polyaniline
pBN pyrolytic boron nitride
Pc phthalocyanine
PC photoconductive
PCA principal component analysis
PCB printed circuit board
PDMA poly(methylmethacrylate)/poly(decyl methacrylate)
PDP plasma display panels
PDS photothermal deflection spectroscopy
PE polysilicon emitter
PE BJT polysilicon emitter bipolar junction transistor
PECVD plasma-enhanced chemical vapor deposition
PEN polyethylene naphthalate
PES photoemission spectroscopy
PET positron emission tomography
pHEMT pseudomorphic HEMT
PL photoluminescence
PM particulate matter
PMMA poly(methyl-methacrylate)
POT poly(*n*-octyl)thiophene
ppb parts per billion
ppm parts per million
PPS polyphenylsulfide
PPY polypyrrole
PQT-12 poly[5,5'-bis(3-alkyl-2-thienyl)-2,2'-bithiophene]
PRT platinum resistance thermometers
PSt polystyrene
PTC positive temperature coefficient
PTIS photothermal ionisation spectroscopy
PTS 1,1-dioxo-2-(4-methylphenyl)-6-phenyl-4-(dicyanomethylidene)thiopyran
PTV polythienylene vinylene
PV photovoltaic
PVD physical vapor transport
PVDF polyvinylidene fluoride
PVK polyvinylcarbazole
PVT physical vapor transport
PZT lead zirconate titanate

Q

QA quench anneal
QCL quantum cascade laser
QCSE quantum-confined Stark effect
QD quantum dot
QHE quantum Hall effect
QW quantum well

R

RAIRS reflection adsorption infrared spectroscopy
RBS Rutherford backscattering
RCLED resonant-cavity light-emitting diode
RDF radial distribution function
RDS reflection difference spectroscopy
RE rare earth
RENS resolution near-field structure
RF radio frequency
RG recombination–generation
RH relative humidity
RHEED reflection high-energy electron diffraction
RIE reactive-ion etching
RIU refractive index units
RTA rapid thermal annealing
RTD resistance temperature devices
RTS random telegraph signal

S

SA self-assembly
SAM self-assembled monolayers

SAW surface acoustic wave
SAXS small-angle X-ray scattering
SCH separate confinement heterojunction
SCVT seeded chemical vapor transport
SE spontaneous emission
SEM scanning electron microscope
SIMS secondary ion mass spectrometry
SIPBH semi-insulating planar buried heterostructure
SIT static induction transistors
SK Stranski–Krastanov
SNR signal-to-noise ratio
SO small outline
SOA semiconductor optical amplifier
SOC system-on-a-chip
SOFC solid oxide fuel cells
SOI silicon-on-insulator
SP screen printing
SPECT single-photon emission computed tomography
SPR surface plasmon resonance
SPVT seeded physical vapor transport
SQW single quantum wells
SSIMS static secondary ion mass spectrometry
SSPC steady-state photoconductivity
SSR solid-state recrystallisation
SSRM scanning spreading resistance microscopy
STHM sublimation traveling heater method
SVP saturated vapor pressure
SWIR short-wavelength infrared

T

TAB tab automated bonding
TBA tertiarybutylarsine
TBP tertiarybutylphosphine
TCE thermal coefficient of expansion
TCNQ tetracyanoquinodimethane
TCR temperature coefficient of resistance
TCRI temperature coefficient of refractive index
TDCM time-domain charge measurement
TE transverse electric
TED transient enhanced diffusion
TED transmission electron diffraction
TEGa triethylgallium
TEM transmission electron microscope
TEN triethylamine
TFT thin-film transistors
THM traveling heater method
TL thermoluminescence
TLHJ triple-layer graded heterojunction
TLM transmission line measurement
TM transverse magnetic
TMA trimethyl-aluminum
TMG trimethyl-gallium
TMI trimethyl-indium
TMSb trimethylantimony
TO transverse optical
TOF time of flight
ToFSIMS time of flight SIMS
TPC transient photoconductivity
TPV thermophotovoltaic
TSC thermally stimulated current
TSL thermally stimulated luminescence

U

ULSI ultra-large-scale integration
UMOSFET U-shaped-trench MOSFET
UPS uninterrupted power systems
UV ultraviolet

V

VAP valence-alternation pairs
VB valence band
VCSEL vertical-cavity surface-emitting laser
VCZ vapor-pressure-controlled Czochralski
VD vapor deposition
VFE vector flow epitaxy
VFET vacuum field-effect transistor
VGF vertical gradient freeze
VIS visible
VOC volatile organic compounds
VPE vapor phase epitaxy
VRH variable-range hopping
VUVG vertical unseeded vapor growth
VW Volmer–Weber

W

WDX wavelength dispersive X-ray
WXI wide-band X-ray imager

X

XAFS X-ray absorption fine-structure
XANES X-ray absorption near-edge structure
XEBIT X-ray-sensitive electron-beam image tube
XPS X-ray photon spectroscopy
XRD X-ray diffraction
XRSP X-ray storage phosphor

Y

YSZ yttrium-stabilized zirconia

Part B

Growth a

Part B Growth and Characterization

12. Bulk Crystal Growth – Methods and Materials

This chapter covers the field of bulk single crystals of materials used in electronics and optoelectronics. These crystals are used in both active and passive modes (to produce devices directly in/on bulk-grown slices of material, or as substrates in epitaxial growth, respectively). Single-crystal material usually provides superior properties to polycrystalline or amorphous equivalents. The various bulk growth techniques are outlined, together with specific critical features, and examples are given of the types of materials (and their current typical sizes) grown by these techniques. Materials covered range from Group IV (Si, Ge, SiGe, diamond, SiC), Group III–V (such as GaAs, InP, nitrides) Group II–IV (including CdTe, ZnSe, MCT) through to a wide range of oxide/halide/phosphate/borate materials. This chapter is to be treated as a snapshot only; the interested reader is referred to the remainder of the chapters in this Handbook for more specific growth and characterization details on the various materials outlined in this chapter. This chapter also does not cover the more fundamental aspects of the growth of the particular materials covered; for these, the reader is again referred to relevant chapters within the Handbook, or to other sources of information in the general literature.

Despite the widespread progress in several epitaxial growth techniques for producing electronic and optoelectronic device-quality material, various bulk growth methods are still used to produce tens of thousands of tons of such materials each year. These crystals are used in both active and passive modes; in other words, to produce devices directly in/on bulk-grown slices of material, or used as substrates in epitaxial growth, respectively.

This chapter covers the field of bulk single crystals of materials used in electronics and optoelectronics. Single-crystal material usually provides superior properties to polycrystalline or amorphous equivalents. The various bulk growth techniques are outlined, together with specific critical features, and examples are given of the types of materials grown by these techniques, as well as their current typical sizes. We cover materials ranging from Group IV (Si, Ge, SiGe, diamond, SiC), Group III–V (GaAs, InP, nitrides, among others) and Group II–IV (such as CdTe, ZnSe, MCT) materials through to various oxides/halides/phosphates/borates. However, this chapter should only be treated as a brief foray into the field. The reader interested in more detail is referred to the chapters following this one (or to other sources of information in the general literature) for more on the growth and characterization of various materials outlined in this chapter. Many of the crystals grown at small sizes, mainly for R&D purposes, particularly in universities, are not discussed; neither, in general, are the organic materials that are being studied in this area.

12.1 History

Several very useful studies on the history of crystal growth in general can be found in the literature [12.1–7]. Many significant contributions were made to the fundamentals of crystal growth during the eighteenth and nineteenth centuries, including the development of thermodynamics, undercooling and supersaturation [12.6, 7]. In terms of crystal growth techniques, it is accepted that the first method applied to produce usable crystals on a large scale was that of flame fusion by *Verneuil* [12.8].

Before World War II, synthetic crystals (other than ruby) were mainly used in scientific instruments. However, between 1900 and 1940 there were enormous advances in both the theoretical area of this field and in producing samples for scientific study. The diffusion boundary layer was applied by *Nernst* [12.9], while ideas on the growth of perfect crystals were proposed by *Volmer* [12.10], *Kossel* [12.11] and *Stranski* [12.12].

Table 12.1 Estimated worldwide annual production rates of crystals (as at 1986, [12.1]). Note that some materials are mainly used for non-(electronic/optoelectronic) purposes; for example cubic zirconia and much of the ruby are used in jewelry and watches (items in brackets are small in volume but high in value)

Crystal	Rate (t/yr)	Growth methods
Silicon	4000	Czochralski, float-zone, (VPE)
Metals	4000	Bridgman, Strain anneal
Quartz	800	Hydrothermal
III–V compounds	600	Czochralski, Bridgman (VPE, LPE)
Alkali halides	500	Bridgman, Kyropoulos
Ruby	500	Verneuil
Germanium	400	Czochralski, Bridgman
Garnets	200	Czochralski
Lithium niobate	100	Czochralski
Phosphates	50	Low-temperature solution
Lithium tantalate	20	Czochralski
Cubic zirconia	15	Skull melting
TGS	10	Low-temperature solution
Diamond	10	High-temperature solution
II–VI compounds	5	Vapor, Bridgman

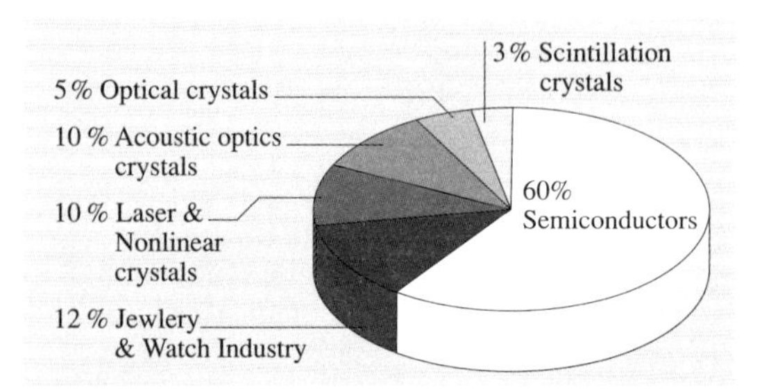

Fig. 12.1 Estimated share of the world production of 20 000 t of bulk crystals (1999). After [12.2], copyright Elsevier Science (used with permission)

Many of the growth techniques now used were also initially developed during this period. The flame fusion of *Verneuil* [12.8] was followed by hydrothermal growth [12.13, 14], crystal pulling [12.15], *Kyropoulos* [12.16], *Bridgman* growth [12.17–20], and vertical gradient freeze [12.21].

During World War II synthetic crystals were used as piezo-electric transducers (in sonar) as resonant devices (in radar) or as infra-red detectors. However, it was the invention of the transistor (announced in 1948) that heralded the modern era of crystal growth for practical purposes rather than for pure scientific interest. *Teal* and *Little* developed crystal pulling of Ge in 1950 [12.22] and *Dash* [12.23] improved this with the 'necking' technique to produce dislocation-free material. A key issue at this stage was measuring levels of impurities. The semiconductors used in solid-state devices such as transistors work because they have selected dopants in them. Techniques were developed at this time to improve the purity of materials, such as zone refining [12.24] and float-zone refining [12.25,26]. Theoretical aspects developed in this period include the role of screw dislocations in growth [12.27] and the generalized theory of *Burton* et al. [12.28]. Reducing melt inclusions in crystals was discussed by *Ivantson* [12.29, 30] in terms of diffusional undercooling, and by *Tiller* et al. [12.31] in terms of constitutional supercooling. Similar work was carried out in solution growth by *Carlson* [12.32], who studied flow effects across crystal faces, and *Scheel* and *Elwell* [12.33], who derived the maximum stable growth rate and optimized supersaturation to produce inclusion-free crystals. Segregation effects, which are related to mass and heat transfer, were studied by *Burton* et al. [12.34] for melt growth and by *van Erk* [12.35] for solution growth, while the experimental conditions needed to produce striation-free material were established by *Rytz* and *Scheel* [12.36]. Forced convection in diffusion-limited growth was recognized as being beneficial for open systems with stirrers [12.37–40], while stirring in sealed containers was accomplished using the accelerated crucible rotation technique [12.41].

In 1986, *Brice* [12.1] estimated the annual production rates of crystals, and Table 12.1 is a reproduction of that data. Semiconductor materials clearly dominated at that stage, in particular silicon and III–V compounds. Brice also tabulated the uses of the various crystals. Later, *Scheel* [12.6] gave estimates of 5000 t in 1979, 11 000 t in 1986 (from [12.1]) and approximately 20 000 t in 1999, see Fig. 12.1. By 1999 the balance had shifted somewhat from the earlier estimates, but semiconductors continued to dominate, at $\approx 60\%$. There were roughly equal percentages of scintillator, optical and acousto-optical crystals, at around 10–12%. The remainder were made up of laser and nonlinear optical crystals and crystals for jewelry and the watch industries.

A recent book edited by *Capper* [12.42] gives a comprehensive update on the bulk growth of many of the materials used in the electronic, optical and optoelectronic fields.

12.2 Techniques

12.2.1 Verneuil

This is the fastest growth method and was the first found to be capable of controlling nucleation and thus producing large crystals of high melting point oxide crystals, such as sapphire and ruby. Currently, a large number of high melting point materials have been grown by this technique, including ZrO_2 (2700 °C), SrO (2420 °C) and Y_2O_3 (2420 °C), but the largest use is still Al_2O_3 (often doped to produce ruby, sapphire, and so on). Figure 12.2 shows a schematic of the equipment used. Many different heat sources have been used, such as solar furnaces, glow discharges, plasma torches, arc images and radio-frequency heating, but the original gas flame technique is still the most popular. An oxy-hydrogen flame heats the seed crystal, which is placed on a ceramic pedestal. Powder from a hopper is shaken through the flame and melts, forming a melt surface on the seed. During growth, the seed is lowered, controlling the linear growth rate, while the volume growth rate is governed by the powder feed rate. The balance of these two rates controls the crystal diameter and the crystal is normally rotated slowly. Ruby crystals up to 200 mm in diameter can be grown. Drawbacks of the technique include high dislocation densities and concentration variations.

12.2.2 Czochralski

Czochralski is a fast growth method widely used for both semiconductors and oxide/fluoride materials intended for optical applications. It usually also produces the most homogeneous crystals and those with the fewest flaws. However, it is only really applicable to those materials that melt congruently or nearly congruently, so that the solid and the melt compositions are the same at equilibrium. A crucible material is normally needed that is compatible with the melt, but the crucible material most commonly used for silicon – silica – dissolves slowly in the melt and this raises certain process issues about

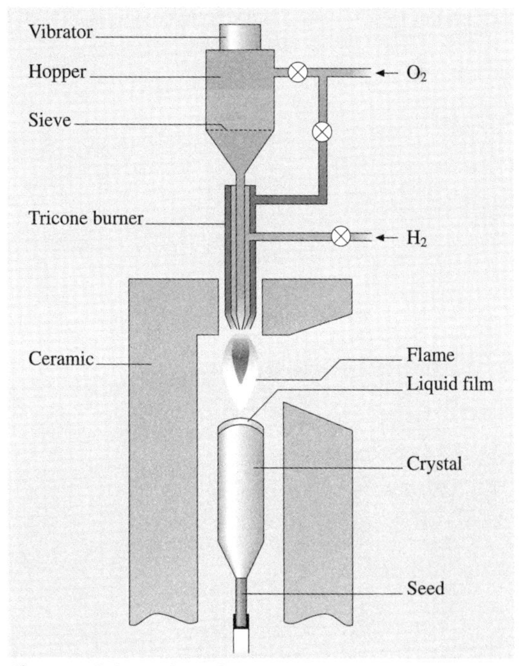

Fig. 12.2 Schematic of Verneuil growth equipment. (After [12.1])

the growth of this material. Capital costs for Czochralski pulling are higher than most other techniques, but it is used when the greatest perfection is required.

The largest use is clearly for silicon, but ≈ 100 materials are grown commercially using this method, while many more are at the research stage. As the diameter of the crystal increases, for example from 200 mm to 300 mm for silicon, controlling dislocations becomes more difficult due to increased radial temperature gradients. Figure 12.3 shows a schematic of Czochralski growth equipment. The basic method is relatively simple: the solid charge is placed in the crucible and heated to a temperature several degrees above the melting point. The seed crystal, rotating slowly, is lowered to contact the melt and the seed then slowly melts. After a short delay (a few minutes), pulling is commenced and new material (with slightly reduced diameter) begins to grow. A long narrow 'neck' region is grown to reduce dislocations, and then the melt temperature is reduced to increase the diameter. When the crystal attains full diameter, growth is maintained until the desired length of crystal has grown, and growth is terminated by sharply increasing the pull rate or increasing the melt temperature so that the diameter reduces to zero.

Constant diameter is maintained by adjusting the power input to the melt. This is done automatically, either by directly monitoring the diameter optically (by observing the bright ring around the crystal periphery), or by indirectly measuring the diameter via weighing methods. Crucibles are normally round-based for semiconductors but should be relatively flat for oxide/halide growth (for ionic materials). Often the depth of the initial melt is the same as or slightly less than the diameter of the crystal.

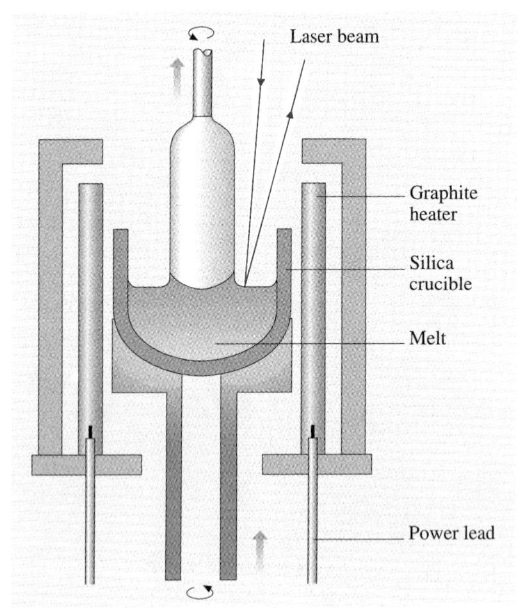

Fig. 12.3 Schematic of Czochralski growth equipment. (After [12.1])

Hurle [12.43] discusses the effects, in detail, of convection, flows in melts, heat transport, mass transport, solute segregation, use of magnetic fields, systems dynamics and automatic diameter control, morphological stability and defect control. In general, as melt sizes have increased natural convective flows have also increased, leading to turbulence, which causes growth rate fluctuations leading to dopant concentration variations (so-called "striations").

For the growth of III–V compounds, liquid-encapsulated Czochralski (LEC) was pioneered by *Mullin* and coworkers [12.44] for GaAs and GaP. Suppressing the volatility of As and P was crucial to the successful growth of these types of compounds. This technique involves the use of an inert layer of a transparent liquid, usually B_2O_3, which floats on the melt surface and acts as a liquid seal. Most importantly, the encapsulant should wet both the crucible and crystal so that a thin film adheres to the crystal as growth proceeds. The latter prevents dissociation of the hot crystal. PBN crucibles are often used these days, and either resistance or RF heating is employed. Semi-insulating GaAs is produced in a high-pressure puller (100–200 atm). The main advantage here is that elemental Ga and As can be used as the starting materials. Diameter is often controlled via crystal-weight measurements, rather than by meniscus observation. Pull rates are typically < 1 cm/h. Axial or transverse magnetic fields can again be used to control melt turbulence and dopant segregation.

Precious metal (Pt, Ir) crucibles are used for oxide crystal growth, as high temperatures are used. To prevent the reduction of oxide melts, a partial pressure of oxygen is needed. Normally, the crucible is not rotated or translated and the growth rate is greater than the pulling rate as the melt height decreases. Stable growth rates are obtained for crystal diameters of less than half the crucible diameter. Growth rates are a few mm/h at best, so growth times are normally > 1 week. Large temperature gradients can lead to melt turbulence, which in turn produces a banded structure of solute concentration and varying stoichiometry in the crystals. Controlling convection in these systems is therefore of critical importance.

12.2.3 Kyropoulos

This simple technique is used where a large diameter is more important than the length, as in the cases of windows, prisms, lenses and other optical components, and for scintillator materials. The set-up and method of growth are similar to those of Czochralski (Fig. 12.4), but after the seed is brought into contact with the melt it is not raised much during growth. As in Czochralski, a short necked region is still grown but then the seed removal is stopped and growth proceeds by reducing the input power to the melt. The resulting crystals normally have diameters of $\approx 80-90\%$ of the crucible diameter. Although the control systems used for this process are relatively simple, the temperature distribution over the crucible is critical. For alkali halides the crystal density is greater than the melt density, so the melt level decreases with growth, and the desired temperature distribution is one of increasing temperature as the base of the crucible is approached. Average linear growth rates are a few mm/h, corresponding to cooling rates of $< 1\,°C/h$. The only other process parameter of concern is the seed rotation rate, which is normally low or zero.

While the method appears to be very attractive economically, there are technical deficiencies. The isothermal surfaces are curved, resulting in high dislocation densities, and the growth interface is composed of different crystal faces, with consequent inhomogeneities in impurities and vacancy concentrations. High levels of impurities result from the majority of the melt (which is not stirred well) being consumed. Despite these drawbacks, many tons of alkali halides are grown for optical applications each year by this technique.

12.2.4 Stepanov

In this technique, a crystal is pulled from a crucible containing a crystal-shaped aperture (Fig. 12.5). Crystals can be pulled vertically upwards, downwards, or even horizontally. Growth rates are below those of the normal Czochralski technique, but dislocation densities can be reasonably low.

12.2.5 Edge-Defined Film Growth

A die with a central capillary is placed on the surface of the melt (Fig. 12.6). Surface tension forces cause the melt to wet the die and be drawn up the capillary. A pointed seed crystal is lowered to contact this melt and then pulled upwards. The melt is cooled slightly to increase the crystal diameter until it reaches the size of

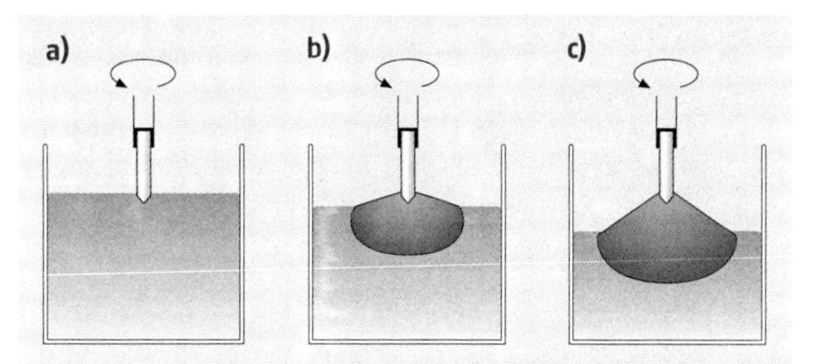

Fig. 12.4a–c Schematic of Kyropoulos growth equipment. (**a**) The seed crystal contacts the melt, a small amount melts and then cooling is commenced to produce (**b**) and (**c**). (After [12.1])

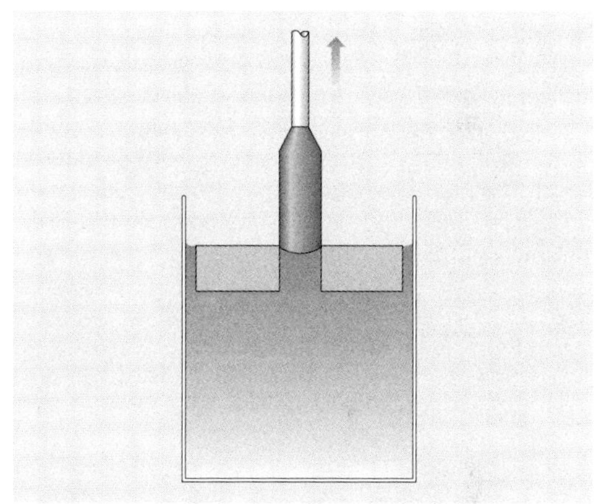

Fig. 12.5 Schematic of Stepanov growth equipment, in which a crystal is pulled through an aperture that defines its shape. (After [12.1])

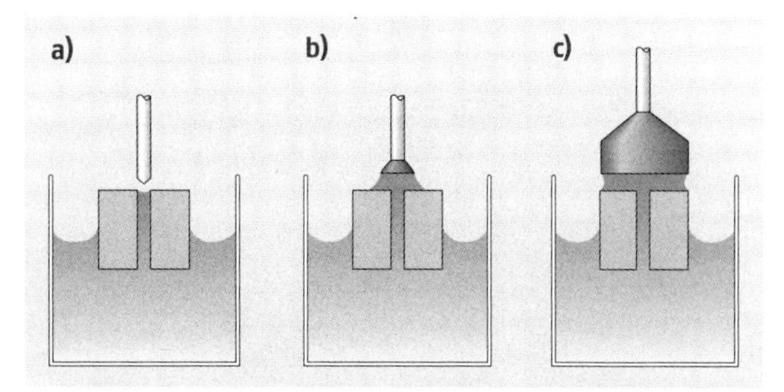

Fig. 12.6a–c Schematic of edge-defined film growth equipment. (**a**) The melt wets the die and is drawn up the capillary; (**b**) the seed is contacted to the melt and pulled up and cooling starts to increase the crystal size; (**c**) the crystal reaches the size of the die. (After [12.1])

the die. Die can be designed to produce various shapes of crystals, including tubes and sheets. Rapid growth is possible, but crystal quality normally suffers. One product is alumina tape (1 mm thick by several centimeters wide) used as a substrate for the production of high-frequency circuits. In this technique, however, purity can be limited.

12.2.6 Bridgman

In essence, this is a method of producing a crystal from a melt by progressively freezing it from one end to the other (Fig. 12.7). Crystals can be obtained with good dimensional control and the method uses relatively simple technology requiring little supervision. However, as the diameter increases, controlling heat flow becomes progressively more difficult. A wide range of materials have been produced by this technique, including sapphire at a melting point of $\approx 2370\,°\mathrm{C}$. One major requirement is that neither the melt nor its vapor must attack the crucible material significantly. Dislocation densities can also be limited to $> 10^4\,\mathrm{cm}^{-2}$, and many materials contain low-angle grain boundaries.

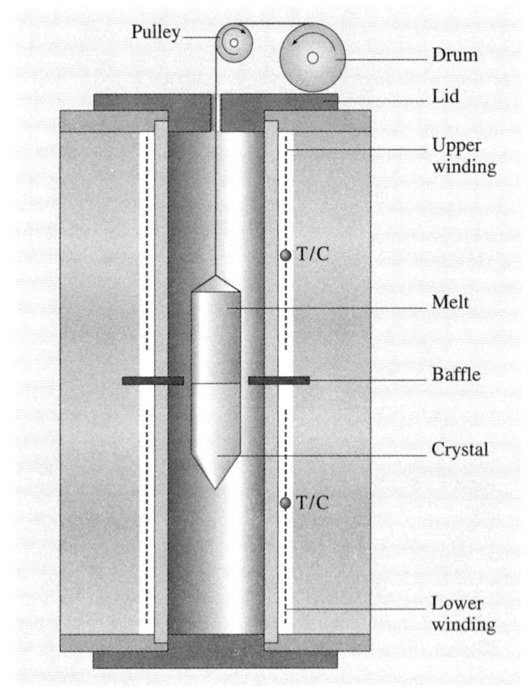

Fig. 12.7 Schematic of Bridgman growth equipment. (After [12.1])

Growth rates can be in the range 0.1–30 mm/h. Either the crucible or the furnace can be moved to achieve movement of the freezing isotherm, and both vertical and horizontal orientations are used. If vibration is an issue in a particular growth system, then it is usually preferable to move the furnace rather than the crystal. Crucible materials include silica (for covalently bonded crystals), graphite and some metals (such as Mo for sapphire growth). The crucibles traditionally have tapered tips in order to try to restrict nucleation to one crystal, although seed crystals can also be used. If spurious nucleation occurs it usually forms at the crucible walls, but these can be suppressed by making the growth face concave into the melt. Baffles are often used to separate upper and lower parts of the furnace to ensure thermal isolation. A low radial temperature gradient is needed to reduce dislocation densities. Calculating the temperature distribution is possible [12.45]. The most obvious requirements are a large temperature gradient at the growth interface and low temperature gradients in the radial direction.

Impurity distributions in crystals are, in general, governed by the type of mixing in the melt. In stirred melts – so-called normal freezing – the relevant equation is

$$C_s = k_{\mathrm{eff}} C_0 (1 - x) k_{\mathrm{eff}} - 1\,,$$

where C_s = concentration in the solid, C_0 = initial concentration, k_{eff} = effective distribution coefficient, and x is the fraction solidified. In unstirred melts, the interface segregation coefficient k^* governs

$$C_s = k^* C_0 (1 - [(1 - k^*)/k] \times \{1 - \exp[-(1 - k^*) k^* f_z / D_L]\})\,,$$

where f is the growth rate, z the axial distance, and D_L is the diffusion coefficient. For a molten zone of length L, the equation is

$$C_s = C_0 [1 + (k_{\mathrm{eff}} - 1) \exp(-k_{\mathrm{eff}} x)]\,.$$

For materials with a volatile component there are several possible refinements of the basic process, such as sealed ampoules, overpressure and liquid encapsulation (for example using B_2O_3). In sealed-ampoule growth there is normally a lack of control over stirring. The accelerated crucible rotation technique (ACRT) can be used for vertical systems, and this author has developed this refinement for the growth of cadmium mercury telluride up to 20 mm in diameter [12.46]. Faster stable growth rates are possible with ACRT [12.46], as are larger diameter crystals, improved uniformity and better crystallinity.

It is worth noting that the majority of melt-grown crystals are produced by the Bridgman and Czochralski techniques (and their variants). In general, if no suitable crucible can be found, then other methods are required, such as the Verneuil method.

12.2.7 Vertical Gradient Freeze

The vertical gradient freeze (VGF) technique involves the progressive freezing of the lower end of a melt upwards. This freezing process can be controlled by moving the furnace past the melt or, preferable, by moving the temperature gradient in a furnace with several independently controlled zones. Low-temperature gradients are normally obtained, leading to reduced dislocation densities, and the crystal is of a defined shape and size. Difficulties include furnace design, the choice of boat material, and the issue of seeding.

12.2.8 Float Zone

In this technique a molten zone is maintained between two solid rods (Fig. 12.8). By moving the zone relative to the rods, one of them grows, and a single crystal can be grown if a seed is used. Silicon is the only material grown on a large scale by this technique. The only other use for it is in the small-scale growth of very pure crystals, as no crucible contact is involved. For silicon, RF heating is used, with frequencies of 2–3 MHz for diameters > 70 mm. There is a steep temperature gradient that induces flows in the molten zone, and if both the seed and feed rods are rotated then the shape of the solid/liquid interface can be controlled. The molten zone normally moves upwards.

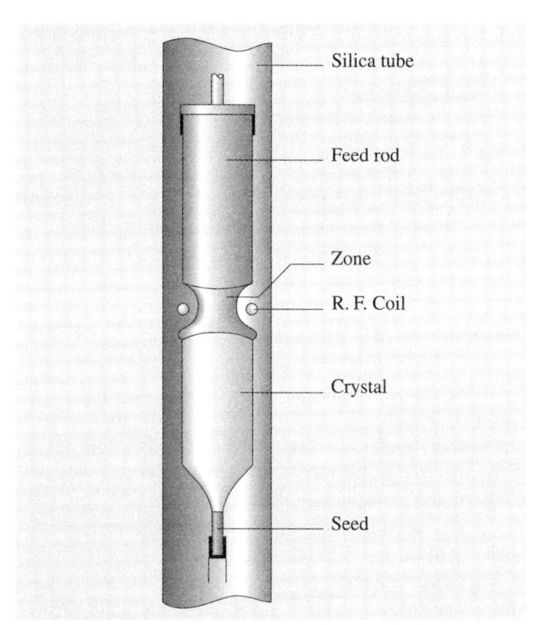

Fig. 12.8 Schematic of float-zone growth equipment. (After [12.1])

12.2.9 Travelling Heater Method (THM)

This technique was developed for II–VI alloy growth (Te-based) by *Triboulet* [12.47]. In the technique, a molten zone is made to migrate through homogeneous solid source material. This is normally accomplished by slowly moving the ampoule relative to the heater (Fig. 12.9). The key requirement here is to obtain the appropriate temperature profile, also shown in Fig. 12.9. Matter transport is by convection and diffusion across the solvent zone under the influence of the temperature gradient resulting from the movement. For alloy growth, a steady state can be reached where the solvent dissolves a solid of composition C_0 at the upper growth interface and deposits, at near-equilibrium conditions, a material of the same composition at the lower growth interface. Growth occurs at a constant temperature below the solidus temperature and hence shows all the advantages of low-temperature growth. When tellurium was used as the molten zone there

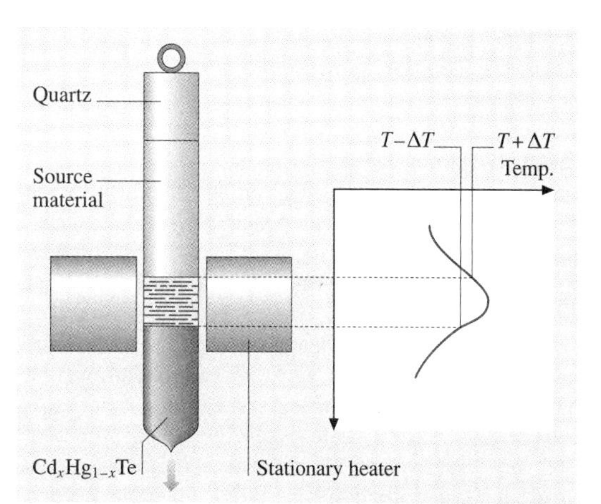

Fig. 12.9 Schematic of travelling heater method growth equipment. (After [12.1])

was a marked purification effect. Seeding was also found to be possible with the technique, as was growth to a larger diameter than achieved with the other common melt techniques used for these compounds.

Problems with THM include the availability of suitable feed material with the required composition and dimensions, although several different routes have been used to overcome this problem. Natural convection is the dominant mechanism of material transport, which led to the addition of ACRT to the basic THM method for some of the ternary alloys.

Graded composition alloys were also made by producing two bevelled cylinders of binary compounds. These were then used to assess the effects of composition on various electrical and optical properties in the given system. Other modifications to the basic process included the 'cold THM' process, in which the relevant metallic elements were used as the source material. This produced a process that accomplished synthesis, growth and purification at low temperature all in a single run. To avoid the problems resulting from solvent excess in the crystal, 'sublimation THM' was developed, in which an empty space of the same dimension as the molten zone is used. This method was successfully applied to the growth of ZnSe. Repeated runs on the same material were used to improve the purification effect. A drawback of this is repeated handling, which in turn was solved by using the 'multipass THM' technique, which can be thought of as a variety of the classical zone-melting method.

12.2.10 Low-Temperature Solution Growth

Most of the crystals grown by this technique are water soluble. This limits their use to applications in which moisture can be excluded. Growth rates are low (0.1–10 mm/d), as the growth faces are unstable. This is due to the concentration gradient near the growth face, in addition to which the crystal is normally totally immersed in the solution so that latent heat evolved increases the adverse supersaturation gradient (Fig. 12.10).

Water (both light and heavy) is used in $\approx 95\%$ of the cases, and must be highly pure. All equipment must be carefully cleaned and protected from dust and the solutions must be stirred vigorously while being prepared. Both slow cooling and solvent evaporation techniques are used. Large volumes of solution (> 100 l) are needed for very large crystals of, for example, KH_2PO_4 (KDP). These large volumes imply very long thermal time

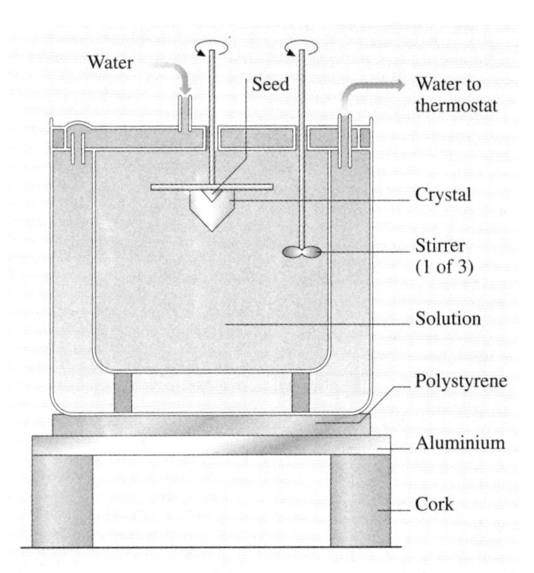

Fig. 12.10 Schematic of low-temperature solution growth equipment. (After [12.1])

constants, which can be increased by adding thermal insulation, up to a limit. Large time constants make it easy to obtain good temperature control. The growth equipment is normally held in temperature-controlled rooms. Although temperature stability is key, other parameters, such as fluctuations in stirring, can lead to large changes in growth rate too.

Solvent evaporation simplifies the temperature control system and makes it more reliable. However, it is difficult to ensure a constant acceptable rate of loss of solvent. This can be approached by using a water-cooled condensation region. One drawback of this technique is that concentrations of impurities increase as the growth proceeds.

12.2.11 High-Temperature Solution Growth (Flux)

This method is used for those materials that melt incongruently; the solvent (flux) reduces the freezing point below the relevant temperature to produce the desired phase of the compound. Both liquid metals (such as Ga, In, Sn) used in semiconductors and oxides/halides (such as PbO and PbF_2) used for ionic materials are employed as solvents. Often, an excess of one of the components will be used (Ga for GaAs and GaP for instance) or a common ionic material (such as K_2CO_3 for $KTa_xNb_{1-x}O_3$). Alternatively, the solvents con-

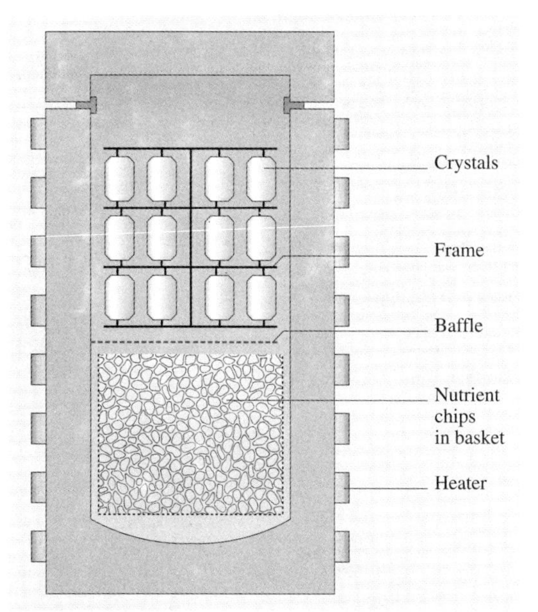

Fig. 12.11 Schematic of hydrothermal growth equipment. (After [12.1])

tain large atoms, such as Pb and Bi, which are too large to enter the lattice of the desired crystal. Mixed solvents, such as alkali metal ions, break the chains that exist in B_2O_3, SiO_2 and Bi_2O_3, lowering viscosities and surface tensions. The book by *Elwell* and *Scheel* [12.5] contains much useful information about this topic.

Slow cooling of high-temperature solutions was used between 1950 and 1970 to produce hundreds of different materials. Small crystals were normally obtained but these were sufficient to obtain useful measurements of magnetic, optical and dielectric properties. Later, seeded growth and stirring using ACRT led to much larger crystals. A Czochralski-type pulling technique can also be used in high-temperature solution growth to produce larger crystals.

Diamond is produced at high temperature ($>$ 1500 K) and high pressure ($>$ 50 kbar). Solvents used include Pt, Pd, Mn, Cr and Ta, but better results were obtained with Ni, Co and Fe and their alloys.

12.2.12 Hydrothermal

This is growth from aqueous solution at high temperature and pressure. Most materials grown have low solubilities in pure water, so other materials (called

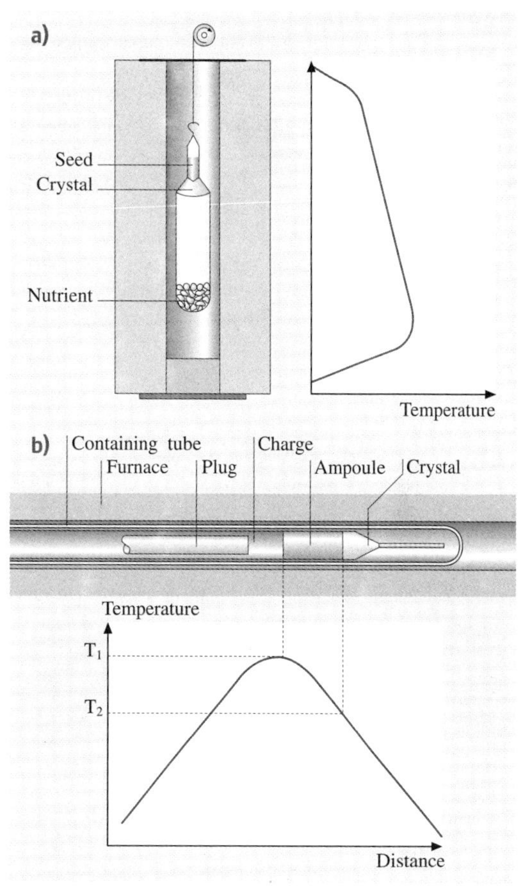

Fig. 12.12a,b Schematic of vapor growth equipment. (**a**) Seeded growth; (**b**) Piper–Polich set-up. (After [12.1])

mineralizers, such as NaOH and Na_2CO_3) are added to increase the solubility. Quartz is the only material grown on a significant scale by this method, although ZnO is also produced commercially by this technique.

Natural quartz crystals are used as seeds, as they have lower dislocation densities. Growth proceeds in the temperature gradient transfer mode. Nutrient is held in the lower part of the liquid in an autoclave. This region is held at between 5 and 50 °C above the upper portion of the autoclave. Convection carries the hot supersaturated liquid to the cooler upper regions, where deposition occurs on the seed crystals (Fig. 12.11). These autoclaves can be operated at up to 400 °C and 2000 atm pressure for approximately 10 years without degradation. Autoclaves can be up to 1 m in diameter and 2.5 m high for low-pressure growth from Na_2CO_3 solution, while 40 cm-diameter and 8 m-high autoclaves are typical of growth from NaOH.

Resistive heating elements are strapped to the outside of the autoclave and connected to two control systems, allowing independent control of the temperatures in the solution and growth zones. *Brice* [12.1] presents a table that shows some typical hydrothermal growth conditions for a wide range of materials.

The essential problem in the growth of quartz is to produce useful devices with reasonable yields with OH concentration (measured by IR absorption) of < 100 ppma. Several parameters restrict the maximum growth rate to ≈ 0.5 mm/d.

12.2.13 Growth from the Vapor

Vapor growth has some applications, particularly in high melting point materials and in II–VI compounds such as ZnSe and ZnS (Fig. 12.12). High-temperature vapor transport, or sublimation, of SiC has progressed recently and is now able to produce 1–2 inch crystals. Growth proceeds by sublimation of a SiC source at 2000–2500 °C in near-vacuum conditions onto a seed held at ≈ 1000 °C lower than the source. 6H-polytype material oriented at ⟨0001⟩ is produced. Growth rates are ≈ 2 mm/h. Remaining problems include impurities ($\approx 10^{16}$ cm^{-3}), voids, high dislocation densities and mixed polymorphism.

12.3 Materials Grown

This section attempts to summarize the current position regarding the wide range of materials used in electronics and optoelectronics. Again the reader is referred to the following relevant chapters for more in-depth coverage of a particular material. The section is subdivided roughly into Group IV, Group III–V, Group II–VI, oxides, halides, and finally phosphates/borates.

12.3.1 Group IV

Silicon and Germanium

The growth of silicon still dominates the electronics industry in terms of size of activity. *Hurle* [12.43] describes in detail the growth of silicon by the Czochralski method. In essence a polysilicon charge is placed in a high-purity quartz crucible mounted in prebaked graphite supports. Chains or cables are used in the pulling system, rather than rods, to limit the height of modern pullers. Both the crystal and the crucible can be rotated, although the details of this aspect vary from manufacturer to manufacturer and are proprietary. As the growth proceeds, the crucible is raised to maintain the position of the melt surface in the heater.

Silicon has two major advantages from a crystal growth perspective: a high thermal conductivity (which permits the latent heat of solidification to be removed) and a high critical resolved shear stress (which allows high thermal gradients without dislocation generation). These two factors enable high growth rates, leading to economic benefits. Heat is provided by 'picket fence'-type graphite heaters. Most commercial pullers are highly automated with full computer control. In most cases, diameter control is achieved via monitoring of the 'bright ring' around the meniscus of the growing crystal. Oxygen is a particular problem and is introduced via erosion of the silica crucible by the melt. Convection moves the oxygen within the melt towards the crystal, where it is incorporated. Slightly reduced pressure over the melt can help to remove oxygen from the melt as silicon monoxide. A similar situation arises with carbon monoxide loss from the melt. Problems with convection have increased as the diameters of silicon crystals have increased over the past several decades. This convection controls the concentrations and distributions of oxygen, carbon and other unwanted impurities.

Although current production of large-diameter material is at the 300 mm level, there have been recent reports of 400 mm-diameter material [12.48]. In this case it was necessary to grow a secondary neck below the Dash dislocation-reducing one, as the thin Dash neck could not hold the weight of the crystal. Crystals of up to 438 kg and 1100 mm in length were reported. A recent report [12.49] shows how codoping Si with B and Ge can obviate the need for this narrow neck region, while maintaining dislocation-free growth.

The highest grade of silicon (the highest purity) is probably still produced by the float-zone method, rather than the Czochralski technique [12.50, 51]. The higher purity (particularly regarding carbon and oxygen) and better microdefect control result in higher solar cell efficiencies in FZ material. Faster growth rates and heating/cooling times together with the absence of crucibles and hot-zone consumables also give FZ material an economic advantage over Czochralski material.

Table 12.2 summarizes the present position on sizes of various crystals, including silicon and germanium.

Table 12.2 Typical sizes (diameters) of a selection of bulk-grown crystals (in mm and/or weight in kg) currently in production and used in R&D

Crystal	Commercial	R&D
Silicon (Cz)	300 (250 kg)	400 (438 kg)
Silicon (FZ)	100	
Germanium	75	200–300
Silicon/Germanium		35 (120 mm long)
SiC	50	100
Diamond	10	10 × 10
GaAs	150 (50 kg)	200
InP	100	150
GaSb	50	75
InSb	75	100
InAs	50	
InGaAs		10
GaN (sublimation)	50	
MCT	20	40
CdZnTe	140 (10 kg)	140 (12 kg)
Bi_2Te_3	12	28
ZnSe (vapor, LEC)	60	
ZnTe	80	
ZnO	50	
HgMnTe	30	
HgZnTe	30	
$LiNbO_3$		40
$PbZnNbTiO_3$		75
YAG	20	75
$LiTaO_3$	100	125
Ruby	200	
BGO		130
$La_3Ga_5SiO_{14}$		100
Quartz	15 – 30 × 150 – 250, (0.5–18 kg)	
CaF_2	385 (100 kg)	
Alkali halides	500 (550 kg)	1000?
Phosphates	45 × 45 × 86	
TGS	40	

The author does not claim that this list is complete nor is it authoritative, it merely serves to show the range of crystals and sizes currently being grown by bulk growth techniques. The information is taken from a number of recently published conference proceedings [12.52–57] and other sources [12.42, 58–60].

Silicon Carbide

The growth of SiC for semiconductor devices, as opposed to abrasive applications, became successful when substrates could be reproducibly grown up to large sizes [12.61–63]. Currently, substrates up to 50 mm in diameter are commercially available and 100 mm-diameter substrates have been reported. The current method is that of seeded sublimation growth. A water-cooled quartz reactor enclosure surrounds a graphite crucible, which is heated by means of RF. Crucible sizes are slightly larger than the required crystal size and SiC powder (particle size 20–200 μm) and/or sintered polysilicon is used as the source. A distance of 1–20 mm between the seed and the source is typically used. At the high process temperatures of 1800–2400 °C, volatile species of Si_2C, SiC_2 and Si evaporate from the source and deposit on the seed. The source temperature determines the rate of evaporation, and the temperature difference between the source and the seed determines the diffusion transport rate; together these two temperatures govern the growth rate. The Si:C ratio is a key control parameter for proper growth of the desired material. Sublimation is normally performed in vacuum or inert gas, such as argon. The pressure of this gas also controls the growth rate to an extent. Growth rates can reach up to 0.5–1.0 mm/h, and crystal lengths of up to 20–40 mm are achieved. Doping can be accomplished via nitrogen and aluminium addition for n- and p-type material, respectively.

Polytype control is obtained by carefully choosing the seed orientation, with growth on the (0001)Si face giving 6H material, while 4H material grows on the $(000\bar{1})$ face. Both growth temperature and pressure and intentional/unintentional impurities can also affect the polytype obtained. One remaining problem is that of micropipes within the material. These are hollow core defects of 1–10 μm diameter, although their density has been reduced from 1000 to $1\,cm^{-2}$ in the best recent samples. Maintaining good control over the nucleation conditions can reduce micropipe density.

Diamond

High-pressure diamond is produced under conditions where it is thermodynamically more stable than graphite [12.64]. However, very high temperatures and pressures are required for growth, unless catalysts are also employed. Clearly a detailed knowledge of the diamond–graphite pressure–temperature phase diagram is a prerequisite for successful growth. For example, in static compression, 8–20 GPa of pressure and temperatures of 1000–3000 °C are needed. Carbon sources include graphite, amorphous carbon, glassy carbon and C_{60}. Dynamic compression techniques employ pressures of 7–150 GPa. A wide variety of catalysts have been used, ranging from conventional ones, such as transition metals, to carbide-forming elements (including Ti and Zr), to Mg, to oxygen-containing materials (carbonates and hydroxides for instance), to inert elements (such as P, Cu, Zn) to hydrides.

Crystals above 1 mm in size are grown by the temperature gradient method. Growth temperature, temperature gradient, growing time, type of catalyst and impurities all affect the growth. Crystals up to 10 mm in size are commercially available. Growth rates can reach 10 mg/h and are governed by the growth temperature and the temperature gradient. If a high-quality diamond seed is used, crystals free of major defects (such as inclusions, stacking faults and dislocations) can be obtained. Impurities, mainly from the catalysts used, are still an issue, as is their nonuniform distribution.

12.3.2 Groups III–V

Gallium Arsenide (GaAs)

Mullin [12.44] provides a comprehensive discussion on all aspects of the growth of III–V and II–VI compounds. He starts his treatment with the early work done on purification in Ge and Si, which highlighted the need for both high purity and for single-crystal growth. He notes that *Pfann* [12.65] initiated zone melting, where, in the simplest case, a horizontal molten charge is progressively frozen from one end. Both zone leveling (in which a liquid zone within a solid bar is moved through the bar in one direction and then in the reverse direction, thus producing a uniform distribution of a dopant) and zone melting (in which liquid zones are repeatedly passed through a solid bar to force impurities to segregate to the ends of the bar, and then these are subsequently removed before the bar is used in growth) are used. Horizontal techniques were also used to produce single-crystal material by slight back-melting of a seed crystal, followed by progressive freezing. These various horizontal techniques were subsequently applied to the purification of both elements and compounds in materials from Groups III–V and II–VI.

Table 12.3, taken from [12.44], summarizes the growth techniques that have been used for the various compounds in Groups III–V and II–VI. There is much current debate about the relative advantages and disadvantages of these techniques as applied to compound crystal growth.

GaAs is the most important Group III–V compound. Horizontal growth can be used to produce low temperature gradients at the solid/liquid interface, which reduces stress-induced slip and hence dislocations. This is particularly beneficial for the use of GaAs in laser diodes, where dislocation densities are required to be very low. However, there is a potential problem in using these low temperature gradients – constitutional supercooling – which may result in unstable growth and second-phase inclusions. In addition to this, there is the issue of impurity uptake from the crucible and nucleation of twins, grain boundaries and even polycrystalline growth from the crucible walls. These problems can be minimized by taking care over experimental details, but they still make horizontal growth unattractive where large-area uniformity is required, particularly as scaling up to ever-larger diameters is difficult.

Table 12.3 General applicability of growth techniques to compound semiconductors. The more asterisks, the more appropriate the technique. *P*: potentially applicable; *C*: conventional VGF; *L*: LEC VGF. (From [12.44])

Technique/ Compound	Zone melting, horizontal Bridgman	VGF, vertical Bridgman	Conventional vertical pulling	Liquid encapsulation pulling	Vapor growth
InSb	***	P	***		P
GaSb	***	P	***		P
InAs	***	P		***	P
GaAs	***	C***: L***		***	P
InP	*	C*: L**		***	P
GaP	*	C*: L**		***	P
HgSe		**			P
HgTe		***			P
CdSe	*	**			***
CdTe	***	***			***
ZnSe		**			***
ZnTe		**			***
HgS					***
ZnS					***
CdS					***

Early efforts to produce this material by the vertical Czochralski method failed due to dissociation of the melt. This problem was solved by the LEC technique [12.44]. A layer of boric oxide floats on the melt and the seed crystal is dipped through this to contact the melt (Fig. 12.13). PBN crucibles are normally used and heating can be by resistance or RF means, but commercial systems use graphite heaters. Most semi-insulating GaAs, which is required for integrated circuits, is produced at pressures of 100–200 atm. The advantage of this modification is that elemental As and Ga can be used as the source. Diameter control is normally via crystal weighing. Currently, LEC produces material 150 mm in diameter [12.66] from 28–40 kg melts using 400 mm-diameter PBN crucibles. Also reported were modifications to the basic Czochralski technique to reduce stresses, and hence dislocations, in crystals up to 100–150 mm in diameter. These are vapor pressure controlled Czochralski, fully encapsulated Czochralski and hot-wall Czochralski. The reader is referred to the reference for more details of these techniques.

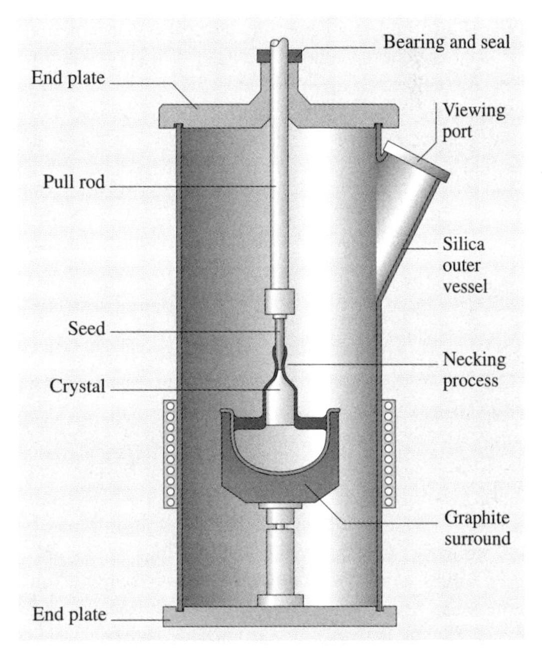

Fig. 12.13 Schematic of liquid-encapsulated Czochralski growth equipment. A silica outer vessel with viewing port is held between end-plates. The induction heating coils couple into the graphite surround. The seed is fixed in the chuck on the pull rod that rotates and moves through the bearing and seal. The crystal grows from the seed through a necking process, and on withdrawal pulls out a layer of B_2O_3 over its surface. (After [12.44]; copyright Wiley-VCH (1998), used with permission)

An alternative to Czochralski is that of vertical gradient freeze. For reproducible growth of large-diameter crystals by this technique, it was found necessary to use B_2O_3 in a PBN crucible. Freezing is accomplished by moving the temperature gradient via furnace controller changes, rather than movement of the furnace itself. This naturally produces low temperature gradients, which give low dislocation densities, and it produces a crystal of the right size and shape for subsequent slicing and processing. *Rudolph* [12.66] reports that in 1999 LEC growth of GaAs accounted for $\approx 90\%$ of all SI GaAs, with the remaining 10% produced via vertical Bridgman or VGF growth. By 2000, he puts the figures at $\approx 50\%$ LEC and $\approx 50\%$ VGF. *Rudolph* also notes that VGF has been reported to grow material up to 150 mm in diameter.

Indium and Gallium Phosphides

As for GaAs, high-pressure LEC growth is normally used for both InP and GaP [12.44]. The problems involved are analogous to those for GaAs growth, with the major addition of an increased tendency for twinning. Loss of phosphorus (P_4) is still a problem and leads to a deterioration in crystal quality. Dislocation densities are still seen to be high. *Asahi* et al. [12.67] report that some of these problems have been alleviated by modifications to the basic LEC process, such as thermal-baffled LEC, phosphorus vapor pressure controlled LEC, and vapor pressure controlled Czochralski (VCZ). Dislocations tend to increase with crystal diameter, although crystals up to 100 mm in size with low dislocation densities have been grown recently by VCZ. Lower dislocation densities are produced by the VGF method at a given crystal diameter. High-pressure VGF growth of InP is difficult and temperature fluctuations cause twinning. By improving the temperature control (to within $\pm 0.03\,°C$) *Asahi* et al. have shown that twin-free $\langle 100 \rangle$ InP single crystals 100 mm in diameter and 80 mm in length can be obtained. Temperature gradients of $< 10\,°C/cm$ reduced the etch pit densities to below those of comparable LEC material.

Indium and Gallium Antimonides

Mullin [12.44] has detailed the early work on InSb; much less work has been done on GaSb. The low melting point combined with the negligible vapor pressure of Sb over the melt make zone-refining methods attractive. However, certain impurities render this route troublesome, although ways of overcoming this issue were found. Crystal-pulling methods were preferred for their versatility, although twinning was seen as a major issue. Growth on the (111)Sb face produces this twinning and it is recommended that growth should occur on the [211]Sb or [311]Sb faces instead.

Micklethwaite and *Johnson* [12.68] have summarized the current status of the production of InSb for IR detector applications. InSb expands by $\approx 13\%$ on freezing, leading to a net flow away from the growth interface. This 'density effusion' must be accounted for when modeling this system. Hydrogen is used as the atmosphere, as it reduces floating InO_x. The basic Czochralski method is used for commercial production and diameters of 75 mm are routine. Defect densities can be as low as $10\,cm^{-2}$. Many of the details of the growth process are proprietary. Growth proceeds without a liquid encapsulant, and as such resembles Si growth more closely than its arsenide and phosphide cousins. However, there are significant differences from the silicon case in terms of reduced radiative heat transfer and increased convective flow in the hydrogen atmosphere. Fluid mechanics in the melt are complicated by both 'density effusion' and the facet effect [12.44].

Bridgman growth is not used for InSb growth any longer, but float-zone growth has been achieved in microgravity conditions using an oxide 'skin' for containment.

Group III Nitrides

There has been a rapid increase in the interest in these compounds following the successful development of blue laser diodes. Ideally, epitaxial thin films of these nitrides would be grown on bulk substrates of similar compounds, to minimize lattice mismatch. However, growth of these compounds by the more normal Czochralski and Bridgman techniques is not possible due to the high melting temperatures (2200–3500 K) and high decomposition pressures (0.2–60 kbar) [12.69,70]. One growth method that can be used is high-pressure solution growth. In this technique, liquid Ga (plus 0.2–0.5 at.% Mg, Ca, Be or Zn) is held in a high-pressure chamber with a N_2 atmosphere. The maximum pressure is ≈ 20 kbar and the maximum temperature is 2000 K. Both pressure and temperature are controlled within tight limits and in situ annealing in vacuum is also allowed for. A temperature gradient of $2–20\,°C/cm$ is maintained along the crucible axis and the N_2 dissolves in the hotter end and GaN crystallizes in the cooler end. No seeding was used and after 120–150 hours of growth hexagonal needles or platelets are produced of size 10×10 mm. Growth by MOVPE and MBE of GaN and InGaN MQWs on these bulk GaN samples was also reported by *Grzegory* et al. [12.70]. LEDs and laser diodes based on these epitaxial layers were also described.

This technique only produces very small crystals of both AlN and InN.

Nishino and *Sakai* [12.71] describe the sublimation growth of both GaN and AlN. Initially, the source powder is synthesized by heating Ga metal in NH_3. This powder is then annealed in NH_3 before loading into the growth chamber. Temperatures are 1000–1100 °C but atmospheric pressure can now be employed. Both NH_3 and N_2 are introduced into the growth chamber. Either small (3 mm × several hundred μm) 'bulk' crystals or free-standing films (10–30 μm thick) on sapphire substrates are produced. Thicker samples (up to 500 μm thick) can be produced by the sublimation sandwich method in which the distance from the source to substrate is reduced to 2–5 mm. A hot-pressed polycrystalline source of AlN and a SiC-coated graphite crucible was used to grow 0.3–1.0 mm-thick films on the 6H-SiC seed (10 × 10 mm in size) at temperatures of 1950–2250 °C.

12.3.3 Groups II–VI

Several recent reviews of the growth of a wide range of binary and ternary II–VI compounds have been published [12.44, 58, 72, 73]. One of the key concepts in the growth of these compounds is the ease with which phase transitions occur. This imposes limits on certain growth techniques for particular compounds.

There are two main types of bulk growth technique, namely growth from the liquid and growth from the vapor. Most narrow-gap II–VI compounds are characterized by high melting points and/or high component partial pressures, and early work concentrated on various forms of vapor growth, particularly for S- and Se-containing compounds. Problems with low growth rates and/or small crystal sizes and purities led to a switch to melt growth techniques, although there have been some recent significant developments in the growth of larger crystals from the vapor.

Many techniques have been used to grow narrow-gap II–VI compounds. Solid state recrystallization (SSR) is used for ternary systems where there is a wide separation between solidus and liquidus, such as MCT. Other names that have been used for this process are quench anneal (QA) and cast recrystallize anneal (CRA). The term anneal is used in the first case to define a high-temperature grain growth process, while in CRA it is a low-temperature process used to adjust stoichiometry. Strictly speaking, SSR is crystal growth from the solid phase at temperatures close to the melting point, but it is included here for convenience. In the basic technique, pure elements are loaded into a cleaned silica ampoule and the charge is melted and rocked to ensure complete mixing. Charges are then normally quenched rapidly, into air or oil, to room temperature in the vertical orientation. This produces a dendritic structure that is reduced/removed by the recrystallization step, which proceeds at temperatures just below the melting point for many days. Grain growth occurs and micro-inhomogeneities in composition are removed. Care must be taken in the quenching stage to avoid pipes/voids that cannot be removed by the recrystallization step.

Bridgman growth is the most widely used of the bulk growth techniques applied to narrow-gap II–VI compounds. It is the only technique that has produced crystals of all the binary and ternary compounds studied. Numerous modifications have been applied to the basic process, but the three principal means of achieving growth are to move an ampoule through a temperature gradient, to move the furnace past a stationary ampoule and to move a temperature profile from high to low temperatures with both the furnace and ampoule stationary, the so-called vertical gradient freeze (VGF) method.

Silica ampoules are usually used but various coatings have been applied, and the use of other materials such as PBN has been reported [12.74]. Charges are prepared either using the appropriate elements or, for ternaries and quaternaries, preformed compounds of the binaries. Both vertical and horizontal configurations have been used and large crystal sizes coupled with a relatively high growth rate (1–2 mm/h) make the basic technique relatively cheap and versatile. Marked segregation of both matrix elements in ternaries and quaternaries and impurities can occur. Difficulties with controlling component vapor pressures can be overcome by employing a reservoir of one of the components at a temperature lower than that of the growing crystal. High partial pressures in Hg-containing compounds can cause ampoule failure and ways have been developed to cope with this situation [12.75].

The traveling heater method (THM) has mainly been used for Te-based binary and ternary compounds. Growth by THM combines the advantages of low-temperature solution growth with steady-state conditions, as in zone melting. A homogeneous alloy can be used as the starting ingot, or segments of the binaries can be employed. If the Te zone height is made equal to the ring-heater size, then a flat interface is obtained for particular geometries of crystal and furnace and growth parameters [12.76]. Very low growth rates (0.1 mm/h) are typical, but diameters of up to 40 mm have been accomplished. As in Bridgman, many variations on the basic THM method have been attempted. These include multipass THM [12.77] and 'cold' THM [12.78].

Hydrothermal growth has mainly been applied to S- and Se-based binary compounds and has been reviewed by *Kuznetsov* [12.79]. The main advantages are that it reduces growth temperature (to 200–250 °C) and that it deals with the high partial pressures. Growth rates are very low (0.05 mm/d), and crystals are small (few mm) and contain subgrains and inclusions, but can still provide useful fundamental data.

Growth from the vapor is usually carried out at much lower temperatures than growth from the liquid, below unwanted phase transition temperatures, and can lead to less defective crystals. However, growth rates are generally much lower and uncontrolled nucleation and twinning are more prevalent. Techniques can be split essentially into unseeded and seeded methods. The simplest technique is that of chemical vapor transport (CVT), usually using iodine as the transport agent. The compound to be grown is reacted at a high temperature with the iodine, transported to a cooler region and deposited. The reverse reaction produces iodine, which diffuses back to the high-temperature region and the process repeats itself. The method uses a two-zone furnace but has the disadvantages of iodine incorporation and small crystal sizes. *Mimila* and *Triboulet* [12.80] have experimented with sublimation and CVT using water as a transporting agent for ZnSe, and Triboulet (private communication, 1996) has succeeded in growing CdTe using this method.

Most other vapor techniques are in some sense a derivative of the Piper–Polich method (Fig. 12.12b). Although this was developed originally for the growth of CdS [12.81], the technique may be applied to the vapor growth of any compound that sublimes readily below the melting point temperature. The crystal is grown in a closed crucible that is moved through a steep temperature gradient, such that the source material is always hotter than the growing surface and mass transport occurs from source to crystal.

Vapor growth under the controlled partial pressure of one of the constituents has been obtained by including small quantities of one of the elements in an extended part of the capsule located in a cooler part of the furnace [12.82]. The technique, termed 'Durham', is a vertical unseeded growth procedure, where the source is sublimed from bottom to top in an evacuated silica capsule over several days. Constituent partial pressures are independently controlled by placing one of the elements in a separately heated reservoir connected to the main capsule via a small orifice.

The vertical unseeded vapor growth (VUVG) method differs from the Durham technique in that the partial pressures are controlled by the initial deviation from stoichiometry of the charge material. It was initially developed to grow CdTe. The crystal is grown in an evacuated cylindrical silica capsule in a vertical furnace. A long quartz tube, which extends outside the furnace, is attached to the top of the capsule, providing a heat pipe to create a cooler area for nucleation, which is closely monitored to ensure that the nucleus is single-grained. If it is not, then the capsule is returned to the starting position and the process repeated.

The sublimation traveling heater method (STHM) was developed from THM by *Triboulet* and *Marfaing* [12.77]. The Te-rich molten zone in conventional THM is replaced by an empty space or vapor zone. A problem arises in that the constituent partial pressures are not only functions of the temperature. However, the vapor can be maintained near the stoichiometric $P_{\min}$ condition if a small capillary, one end of which is at room temperature, is connected to the vapor chamber [12.83]. Excess species are sublimed down the capillary, preserving the relative constituent partial pressures and hence the growth rate.

In one version of the seeded technique, an oriented seed is placed on top of a long sapphire rod inside a sealed quartz capsule. The source material is placed in a basket above the seed and sublimation is top-down. The sapphire rod is carefully centered within the growth capsule to ensure a narrow gap between it and the walls of the capsule. Crystals are grown in vacuum or in about 200 Torr of Ar, H_2 or NH_4I. The crystals grow clear of the container walls and generally consist of two or three large grains, some of which maintain the seed orientation. This 'Markov' technique has been adapted for commercial production by the ELMA Research and Development Association of Moscow (cited in *Durose* et al. [12.84]). The seed is mounted on an optically heated sapphire rod heat pipe, in a continuously pumped chamber, which can be back-filled with He to control growth rate. Large crystals of CdTe, 50 mm in diameter and up to 10 mm in length have been grown.

It is clear that HgS is grown by vapor means or by hydrothermal techniques, both of which are at low temperatures and hence low S partial pressures. HgSe has been produced by vapor growth and also by the Bridgman technique. Thick-walled silica ampoules are required in the growth of the Hg-containing compounds to contain the high Hg partial pressures. More details on the growth of Hg-based binaries can be found in [12.78] for vapor and Bridgman growth, while *Kuznetsov* has reviewed the hydrothermal growth of HgS [12.79].

Early work on the growth of the Cd-based binaries was summarized by *Lorenz* [12.85]. Vapor growth was favored for CdS and CdSe, although hydrothermal growth was also discussed. The latter technique, as applied to CdS, was discussed in [12.79]. CdTe was grown by various melt growth techniques, including seeded zone melting and Bridgman, and by vapor growth methods. *Zanio* [12.86] has given a review of the growth, properties and applications of CdTe, including radiation detectors. By that time, vapor growth had been essentially supplanted by both Bridgman and THM growth. A more recent review of CdTe can be found in *Capper* and *Brinkman* [12.87]. The main uses of CdTe (and CdZnTe/CdTeSe) are as a substrate material for the epitaxial growth of MCT and, more recently, as radiation detectors [12.88]. *Triboulet* et al. [12.89,90] applied SSR to the growth of both CdSe and CdTe. The prime reasons for the use of this technique were the high melting point of CdSe and the solid-state phase transitions in the case of CdTe.

A wide variety of growth techniques from the liquid have been used to produce crystals of CdTe. These include solvent evaporation [12.91], liquid encapsulated Czochralski [12.92], zone refining [12.93], VGF [12.94,95], solution growth [12.96], heat exchanger method [12.97], float-zoning [12.98], Bridgman [12.87] and THM [12.47], and *Triboulet* et al. [12.89] have reviewed these materials as substrates for MCT epitaxial growth and compared them to the alternatives based on GaAs, sapphire and Si. Problems with the growth of CdTe include low thermal conductivity, difficulty in seeding due to the need for superheated melts, ease of twin formation, tilts and rotations in the lattice, stoichiometry control and impurities. However, *Triboulet* et al. [12.90] concluded that despite the progress made in the alternative substrate materials, lattice-matched substrates based on CdTe produced MCT epitaxial layers that gave the best device performance.

Within the Bridgman growth technique, many alternatives have been attempted. These include holding excess Cd at a lower temperature in a separate reservoir [12.99], vibration-free growth [12.100], vibrational stirring [12.101], horizontal growth [12.102], low temperature gradients [12.103], high-pressure (100 atm) growth [12.104] and the addition of the accelerated crucible rotation technique (ACRT) [12.105]. Work in the US [12.102] aimed at improving both the horizontal and vertical techniques and scaling-up to large melts, up to 8 kg (horizontal). Recent work by *Szeles* et al. [12.106] has shown that crystals up to 140 mm in diameter and 10 kg in weight can be grown by careful control of the temperature fields.

In the Bridgman process, elemental Cd and Te are loaded into a clean carbon-coated silica ampoule, homogenized by melting/rocking, heated to temperatures in excess of 1100 °C, and then frozen at rates of a few mm/h in a vertical or horizontal system. Single crystals or, more normally, large-grained ingots of size ≈ 75 mm in diameter and 10–15 cm in length are produced. Seeding on the (111) is often utilized but it is not clear that it results in better growth in every case.

Growth by variations on the basic THM process have included the addition of ACRT [12.107,108], focused radiant heating [12.109], sublimation THM and multipass THM [12.77] and 'cold' THM at 780 °C [12.78]. Crystals are grown either from the pre-compounded binaries or from the elements (as in cold THM). Growth rates tend to be low, a few mm/d, but the crystals are very pure due to segregation effects and high-resistivity material results (10^7 Ω cm in [12.77]). This makes THM material ideally suited for gamma and X-ray detectors.

The vast majority of work on bulk growth of MCT has been from the melt. Although rapid progress has taken place in epitaxial growth techniques for MCT, material grown by several bulk methods is still in use for infra-red detection, particularly for photoconductive detectors (this is the case in this author's lab and various others worldwide, such as AIM in Germany). Several historical reviews of the development of bulk MCT have been published [12.110–112]. Many techniques were tried in the early years but three prime techniques survived: SSR, Bridgman and THM.

Two fundamentally different approaches have been followed to improve the basic Bridgman process. These are based on controlling melt mixing and heat flows, respectively. The former has been studied by this author and coworkers [12.113] while the latter includes the work of *Szofran* and *Lehoczky* [12.114], among others. In the Bridgman process, elemental Cd, Hg and Te are loaded into a clean silica ampoule, homogenized by melting/rocking and then frozen slowly from one end in a vertical system to produce a single crystal or, more normally, a largegrained ingot. Marked segregation of CdTe with respect to HgTe occurs in the axial direction, but this leads to an advantage of the Bridgman process over other techniques: material in both ranges of interest (3–5 and 8–12 μm, for $x = 0.3$ and 0.2, respectively) is produced in a single run.

A means of stirring melts contained in sealed, pressurized ampoules was needed and the ACRT of *Scheel* and *Schulz-Dubois* [12.41], in which the melt is subjected to periodic acceleration/deceleration at rotation rates of up to 60 rpm, was chosen. The first report of ACRT in MCT Bridgman growth was given in a patent [12.115]. These effects were developed and discussed in more detail in later papers, which are reviewed in [12.113]. Crystals were produced up to 20 mm in diameter and with x values of up to 0.6 in the tip regions of some crystals (this has recently been increased to ≈ 0.7 at 20 mm diameter in the author's laboratory [12.42, 116]).

Triboulet [12.47] developed this technique for MCT, where diameters of up to 40 mm have been accomplished as well as x values of up to 0.7 [12.117] for optical communication devices. *Durand* et al. [12.118] employed seeds to produce large, oriented crystals, an advantage of this technique over other bulk methods. *Gille* et al. [12.119] adopted a slightly different approach. A pre-THM step is carried out to quench a Te-rich (53–60%) MCT melt. A first THM run at 2 mm/d is then used to provide the source ingot, after removal of the tip and tail sections, for the final THM run. The entire growth procedure takes several months but gives uniform material. This group has also used rotation in the horizontal growth by THM with some success [12.120]. Attempts to use 'cold' THM (starting from the elements) led to little success for MCT nor did use of a Hg reservoir to control the process. Two groups have applied ACRT to THM and obtained enhanced material properties. *Royer* et al. [12.121] used a saw-tooth ACRT sequence and obtained improved radial and axial compositional uniformity. *Bloedner* and *Gille* [12.122] used ACRT although no significant dependence of crystallinity on rotation sequence was seen. They achieved an increase in growth rate from 1.5 to 8.5 mm/d.

Bridgman is the main melt growth technique for the Hg- and Cd-based ternary and quaternary compounds, while work on the vapor growth of these compounds has been limited. The various Hg-based ternaries (such as HgZnTe and HgMnTe) have been studied as potential alternatives to MCT for infrared detectors (*Rogalski* [12.123] has more details including phase equilibria plots). This work was initiated by a theoretical prediction that the Hg–Te bond is stabilized by the addition of Zn, in particular. Rogalski concluded that THM produced the best quality HgZnTe while Bridgman growth of HgMnTe was described as being similar to MCT growth but with reduced segregation. Crystals of HgMnTe up to 40 mm in diameter were produced. *Triboulet* [12.124] has discussed alternatives to MCT. He concluded that HgMgTe, HgCdSe and HgZnSe are not suitable, for a variety of reasons, but that HgMnTe and HgZnTe are potentially suitable.

The main Cd-based ternaries are CdZnTe and CdTeSe, which are used as substrates for epitaxial MCT growth. A vast literature exists on these compounds and it is still a very active area of research. Vertical and horizontal Bridgman are the main techniques used [12.87], although VGF has been used [12.87], as has THM [12.47]. High-quality crystals up to 100 mm (and even 125 mm) in diameter can be produced by VGF [12.67, 95]. Even larger crystals are being grown by horizontal Bridgman [12.106, 125]. Two main problems remain in these compounds. These are the uniformity of Zn or Se, and impurities.

Recent reports by *Pelliciari* et al. [12.126] have shown how solvent evaporation from a Te-rich solution of CdTe in an open tube system can produce large-grained and even single crystals of CdTe up to 300 mm in diameter. This large-area material is aimed at the fabrication of X-ray and γ-ray detectors.

Dilute magnetic semiconductors (DMSs) are a class of semiconductor where the semiconducting properties are changed through the addition of a magnetic ion (such as Mn^{2+} or Fe^{2+}). These materials therefore display normal electrical and magnetic properties, and novel ones such as large Faraday rotations and giant magnetoresistance. *Pajaczkowska* [12.127] gave an early review of the phase diagrams, lattice parameters and growth of many of the Mn- and Fe-based DMS compounds. Vapor growth and hydrothermal growth were mainly used, although the Te-based compounds were being grown by the Bridgman technique. Only small additions of the magnetic ions are required, so essentially the same techniques can be used as those applied to the parent binaries. The most comprehensive review of DMSs is given in [12.128]. In that book, *Giriat* and *Furdyna* review the crystal structures of, and growth methods used for, the Mn-based compounds. The majority of the compounds, particularly the Te-based ones, were grown using Bridgman or unspecified modifications of it. CdMnS was grown in graphite crucibles under high inert gas pressure, or by chemical vapor transport (using iodine as the transport agent). HgMnS was grown hydrothermally, yielding 2 mm-diameter crystals. Similar preparation methods to those used for the Mn-based compounds were applied for Fe-based compounds, but less Fe can be incorporated in the lattice.

Bismuth telluride (strictly speaking a V–VI compound) is used in thermoelectric applications close to room temperature [12.129]. This material is also grown by the Bridgman technique. Both n-type and p-type materials are required to make the thermoelectric devices, and these can be achieved either by stoichiometric deviations or deliberate doping. Typical growth rates are 0.5–2.9 mm/h for 8 mm-diameter ingots of up to 85 mm length. In this author's lab, bismuth telluride is grown at similar rates for 12 mm-diameter ingots up to 100 mm length using antimony and iodine as acceptor and donor dopants, respectively.

12.3.4 Oxides/Halides/Phosphates/Borates

Oxides

Hurle [12.43] summarized issues concerning the growth of oxides by the Czochralski technique. The high melting points of most oxides of interest necessitate the use of precious metal crucibles, such as iridium or platinum, although molybdenum and tungsten ones can be used for lower melting point materials. Extensive and efficient thermal insulation is required, using alumina, magnesia, zirconia and thoria. Some materials require a partial pressure of oxygen in the growth chamber to prevent reduction of the melt. However, this can lead to oxidation of the crucible, which in turn enters the melt. Careful management of the oxygen pressure and gas flow is needed, together with physical baffles. Heating is normally by means of RF. Mounting of the seeds is problematic and long seeds are necessary to avoid thermal degradation of the seed holder and pull rod. Rotation or translation of the crucible is not normally used so the rate of growth has to be above the pulling rate to account for the fall in melt height. Crystals are normally limited to less than half the crucible diameter for stable growth. Growth rates are low, a few mm/h or less, to avoid unstable growth. Afterheaters are also employed to reduce the built-in stress in the crystal during the later stages of growth and subsequent cool-down. Growth runs extend over several days, so automatic diameter control, in this instance using weighing techniques, is mandatory. The high temperatures lead to high temperature gradients and thus strong buoyancy-driven convection that is turbulent by nature. This turbulence gives rise to nonuniformities in the crystals. In addition, catastrophic failure can result due to the coupling between this turbulent flow and the flow due to the crystal rotation. Thus, an understanding of the convective flows present during oxide growth is critical if large-diameter highly perfect crystals are to be produced.

More recently, *Fukuda* et al. [12.130, 131] have presented detailed reviews of the growth of a wide range of oxide materials of current interest for optical and optoelectronic applications. They firstly note that only $LiNbO_3$ (LN), $LiTaO_3$ (LT) and $Y_3Al_5O_{12}$ (YAG) have as yet reached production status. Many of the other materials exhibit superior properties but are not available in large quantities at sufficient perfection to warrant them being used in devices. Recently, crystals of new nonlinear optical materials, piezoelectric materials and new laser materials have attracted growing interest. In general, the growth methods are based on liquid–solid transitions, either from the melt or from high-temperature solutions (flux), for congruently and incongruently melting materials, respectively. Czochralski growth is still the mainstay of oxide growth, although problems with shape (diameter) control of some oxides (including rare earth vanadates and rutile) necessitate the use of edge-defined film-fed growth (EFG). Growth from fluxes is used for a range of oxide materials, although some incongruently melting materials can be produced from melts of the same composition as the crystal by a micro-pulling down technique.

In the micro-pulling down technique, the melt is held in a Pt crucible and a micronozzle (die) is arranged at the bottom of the melt. An after heater is placed below this arrangement. This combination produces a steep temperature gradient (300 °C/mm), which ensures stable growth at high rates. Single crystals of 0.05–1.0 mm diameter are produced onto seeds. LN crystals, for example, are grown at 12–90 mm/h from stoichiometric melts up to 100 mm in length, assumed to be limited only by the melt volume used. Dislocations are introduced if the diameter of the crystal exceeds 0.8 mm. Growth of $K_3Li_2Nb_5O_{12}$ (KLN), which has outstanding electro-optic and nonlinear optical properties, is problematic by either Czochralski or Kyropoulos methods due to segregation effects at the solid/liquid interface and cracking. This material was grown by the micro-pulling down technique at rates of 20–80 mm/h with diameters of 0.15–0.5 mm. The crystal composition was found to be near to the melt composition and crystals were single-domain.

Crystals in the $KTiOPO_4$ (KTPO) family were grown by the flux method using slow-cooling from 1100 to 800 °C. Crystals a few millimeters in size were obtained for second harmonic-generation measurements.

Crystals from the langasite family ($La_3Ga_3SiO_{12}$) are used for lasers. These were grown by the Czochralski method to 50 mm in diameter and 130 mm in length. High-purity oxide powders were used as starting materials and an Ar/O_2 mixture was used to suppress evaporation of the gallium suboxide. A pulling rate of

1.0 mm/h was used together with a crystal rotation rate of 10 rpm. The resulting crystals were single-phase and crack-free. These crystals show superior properties, as filters, to quartz. Crystals of up to 100 mm in diameter have been grown recently [12.132].

Both rare earth vanadates and some garnet crystals show a tendency toward spiral growth when the Czochralski method is used. This tendency can be circumvented by adjusting the temperature gradient in the melt.

Another way of overcoming problems with the Czochralski growth of these materials is to use the EFG method. Rutile (TiO_2) is used as a polarizer in optical isolators and prisms for optical communication systems. Crystals of this material can be grown by Verneuil and float-zone methods, but growth by Czochralski is difficult. Iridium is used for the die material and crystals up to 8×1 cm and 0.5 mm thick were grown. A modified die was used to grow a rod of 15 mm diameter and ≈ 40 mm in length. A double-die system was shown that was used to produce a core-doped crystal of LN, using Nd^{3+} and Cr^{3+} as the dopants. A 5 mm-diameter crystal some 60 mm in length was produced. Such crystals have potential in optical applications where pumping energy is only absorbed in the core region, reducing heat-induced energy losses.

Quartz is used in both piezoelectric and in various optical applications [12.133]. *Balitsky* [12.134] noted that some 2500 t are produced per year. Hydrothermal growth is the main growth technique used in order to keep the growth temperature below the β- (non-piezoelectric) to α-quartz (piezoelectric) phase transition. Growth at temperatures of 250–500 °C and pressures of 50–200 MPa are used in both alkaline and fluoride solutions. Growth rates of 0.5–1.0 mm/d are achieved on (0001) or (1120) seeds and crystals up to $200 \times 70 \times 60$ mm in size weighing from 1–18 kg.

Various oxide materials are used widely in scintillator applications. The most common are $Bi_4Ge_2O_{12}$ (BGO) and $PbWO_4$ (PWO) [12.135, 136]. These materials are grown by normal Czochralski and Bridgman techniques and the requirement is for thousands of each for large high-energy/astrophysics projects in several centers throughout the world. For example, some 85 000 PWO crystals are required for one application at the Large Hadron Collider at CERN. High-purity starting materials are important, as are certain growth parameters and post-growth annealing, for example in oxygen, to reduce stress in the material.

Sapphire is grown by several techniques for a wide variety of optical, microwave and microelectronics applications [12.137]. Verneuil [12.138], Czochralski and EFG [12.43] have been used extensively to produce large single crystals, particularly for laser rods, and shaped pieces using various forms of dies. In addition, the horizontal directional solidification crystallization (HDC) technique [12.137] has been used as it is a simple and inexpensive technique to set up and use. A molybdenum boat filled with the charge is simply moved through the temperature gradient. Low temperature gradients lead to crystals with low stress and low dislocation densities (compared with the Verneuil and pulling techniques. The growth rate is high at 8 mm/h, and crystals of up to 10 kg and up to $30 \times 30 \times 2.5$ cm are produced. Even larger crystals, up to 20 kg with a diameter of 200 mm, can be produced by a modified Kyropolous method [12.139].

Another technique used is the heat exchanger method, which is a solidification technique from the melt [12.140]. Large crystals can be grown onto seeds by independently controlling the heat input and the heat output. The seed is prevented from melting by a helium gas flow at the bottom of the melt. By progressively increasing the helium flow, following partial melting of the seed, and/or decreasing the furnace temperature, growth is progressed. After complete solidification, an annealing step is performed prior to cooling to room temperature. The submerged solid/liquid interface damps out any thermal and mechanical variations and also produces a low temperature gradient at the interface. There is impurity segregation into the last-to-freeze parts of the crystal that form near the crucible walls. These areas are removed before machining takes place. Crystals up to 340 mm in diameter (65 kg) have been produced in this way. Slabs cut from these crystals are used as optical windows, between 5–25 mm thick.

Halides

Several halides form another class of scintillator materials. These are used in a range of applications, from high-energy physics and nuclear medicine to environmental monitoring and security systems [12.141]. Both large diameters (comparable with the human body) and large numbers are required for high-energy physics applications. The simple Bridgman growth process is still used for many of these materials, although problems do arise due to the contact between crystal and crucible. Insufficient melt mixing also leads to inclusions and striations in crystals. The alternative technique for growth is that of continuously fed Czochralski/Kyropolous growth. A melt-level sensor is normally used to control the diameter over the tens/hundreds of hours of growth needed for large halide crystals. Replenishment of the melt by molten material from a closed feeder enables the starting materials to be purified, and a conical crucible that is not too deep is used. Crystals of CsI up to 500 mm in diameter and 750 mm in height weighing ≈ 550 kg have been produced [12.142]. The ultimate aim is to produce crystals > 1 m in diameter so that entire crystals can be used to monitor the human body.

Phosphates and Borates

Nonlinear optical materials are very important for laser frequency conversion applications. One of the most important of the phosphates is potassium dihydrogen phosphate (KDP), which is used for higher harmonic generation in large laser systems for fusion experiments [12.94, 143]. Growth takes place at room temperature to 60 °C, and growth rates can be as high as 10–20 mm/day, with sizes of ≈ 40 cm × 40 cm × 85 cm [12.143] or 45 cm × 45 cm × 70 cm [12.94], the latter quoted as taking over a year to grow!

Another important phosphate is potassium titanyl phosphate (KTP), used to obtain green light by frequency doubling a Nd:YAG laser. Growth in this case is from high-temperature solution at about 950 °C [12.94]. Sizes of up to 32 mm × 42 mm × 87 mm (weight 173 g) can be grown in 40 days.

Borates, including barium borate, lithium borate, cesium borate and coborates such as cesium lithium borate are used in UV-generation applications. Crystals are again grown by the high-temperature solution method up to 14 cm × 11 cm × 11 cm in size, weighing 1.8 kg, in 3 weeks [12.94].

12.4 Conclusions

This chapter has summarized the current status of the bulk growth of crystals for optoelectronic and electronic applications. It is not intended to be a completely comprehensive view of the field, merely serving to introduce the reader to the wide range of materials produced and the numerous crystal growth techniques that have been developed to grow single crystals. An historical perspective has been attempted to give the reader a feel for the scale of some of the activities. The sections on specific materials try to summarize the particular growth techniques employed, and those that cannot in some cases, and outline the typical sizes currently produced in the commercial and R&D sectors. For more details on current developments, the reader should refer to the books given in references [12.42, 60].

References

12.1 J. C. Brice: *Crystal Growth Processes* (Blackie, London 1986)
12.2 H. J. Scheel: J. Cryst. Growth **211**, 1 (2000)
12.3 H. E. Buckley: *Crystal Growth* (Wiley, New York 1951)
12.4 J. G. Burke: *Origins of the Science of Crystals* (Univ. California Press, Berkeley 1966)
12.5 D. Elwell, H. J. Scheel: *Crystal Growth from High-Temperature Solutions* (Academic, New York 1975)
12.6 H. J. Scheel: *The Technology of Crystal Growth and Epitaxy*, ed. by H. J. Scheel, T. Fukuda (Wiley, Chichester 2003)
12.7 A. A. Chernov: J. Mater. Sci. Mater. El. **12**, 437 (2001)
12.8 A. V. L. Verneuil: Compt. Rend. (Paris) **135**, 791 (1902)
12.9 W. Nernst: Z. Phys. Chem. **47**, 52 (1904)
12.10 M. Volmer: Z. Phys. Chem. **102**, 267 (1927)
12.11 W. Kossel: Nachr. Gesellsch. Wiss. Göttingen Math.-Phys. Kl, 135 (1927)
12.12 I. N. Stranski: Z. Phys. Chem. **136**, 259 (1928)
12.13 G. Spezia: Acad. Sci. Torino Atti **30**, 254 (1905)
12.14 G. Spezia: Acad. Sci. Torino Atti **44**, 95 (1908)
12.15 J. Czochralski: Z. Phys. Chem. **92**, 219 (1918)
12.16 S. Kyropoulos: Z. Anorg. Chem. **154**, 308 (1926)
12.17 P. W. Bridgman: Proc. Am. Acad. Arts Sci. **58**, 165 (1923)
12.18 P. W. Bridgman: Proc. Am. Acad. Arts Sci. **60**, 303 (1925)
12.19 F. Stöber: Z. Kristallogr. **61**, 299 (1925)
12.20 D. C. Stockbarger: Rev. Sci. Instrum. **7**, 133 (1936)
12.21 H. C. Ramsberger, E. H. Malvin: J. Opt. Soc. Am. **15**, 359 (1927)
12.22 G. K. Teal, J. B. Little: Phys. Rev. **78**, 647 (1950)
12.23 W. C. Dash: J. Appl. Phys. **30**, 459 (1959)
12.24 W. G. Pfann: Trans. AIME **194**, 747 (1952)
12.25 H.C. Theurer: US Patent, 3 060 123 (1952)
12.26 P. H. Keck, M. J. E. Golay: Phys. Rev. **89**, 1297 (1953)
12.27 F. C. Frank: Discuss. Farad. Soc. **5**, 48 (1949)

12.28 W. K. Burton, N. Cabrera, F. C. Frank: Philos. Trans. A **243**, 299 (1951)
12.29 G. P. Ivantsov: Dokl. Akad. Nauk SSSR **81**, 179 (1952)
12.30 G. P. Ivantsov: Dokl. Akad. Nauk SSSR **83**, 573 (1953)
12.31 W. A. Tiller, K. A. Jackson, J. W. Rutter, B. Chalmers: Acta Metall. Mater. **1**, 428 (1953)
12.32 A.E. Carlson: PhD Thesis, Univ. of Utah (1958)
12.33 H. J. Scheel, D. Elwell: J. Cryst. Growth **12**, 153 (1972)
12.34 J. A. Burton, R. C. Prim, W. P. Slichter: J. Chem. Phys. **21**, 1987 (1953)
12.35 W. van Erk: J. Cryst. Growth **57**, 71 (1982)
12.36 D. Rytz, H. J. Scheel: J. Cryst. Growth **59**, 468 (1982)
12.37 L. Wulff: Z. Krystallogr. (Leipzig) **11**, 120 (1886)
12.38 L. Wulff: Z. Krystallogr. (Leipzig) **100**, 51 (1886)
12.39 F. Krüger, W. Finke: Kristallwachstumsvorrichtung, Deutsches Reichspatent DRP 228 246 (5.11.1910) (1910)
12.40 A. Johnsen: *Wachstum und Auflösung der Kristalle* (Wilhelm Engelmann, Leipzig 1910)
12.41 H. J. Scheel, E. O. Schulz-Dubois: J. Cryst. Growth **8**, 304 (1971)
12.42 P. Capper (Ed.): *Bulk Crystal Growth of Electronic, Optical and Optoelectronic Materials* (Wiley, Chichester 2005)
12.43 D. T. J. Hurle: *Crystal Pulling from the Melt* (Springer, Berlin, Heidelberg 1993)
12.44 J. B. Mullin: *Compound Semiconductor Devices: Structures and Processing*, ed. by K. A. Jackson (Wiley, Weinheim 1998)
12.45 C. J. Jones, P. Capper, J. J. Gosney, I. Kenworthy: J. Cryst. Growth **69**, 281 (1984)
12.46 P. Capper: Prog. Cryst. Growth Ch. **28**, 1 (1994)
12.47 R. Triboulet: Prog. Cryst. Growth Ch. **28**, 85 (1994)
12.48 M. Shiraishi, K. Takano, J. Matsubara, N. Iida, N. Machida, M. Kuramoto, H. Yamagishi: J. Cryst. Growth **229**, 17 (2001)
12.49 K. Hoshikawa, Huang Xinming, T. Taishi: J. Cryst. Growth **275**, 276 (2004)
12.50 L. Jensen: Paper given at 1st International School on Crystal Growth and Technology, Beatenberg, Switzerland (1998)
12.51 T. Ciszek: *The Technology of Crystal Growth and Epitaxy*, ed. by H. J. Scheel, T. Fukuda (Wiley, Chichester 2003)
12.52 J. B. Mullin, D. Gazit, Y. Nemirovsky (Eds.): J. Cryst. Growth **189** (1999)
12.53 J. B. Mullin, D. Gazit, Y. Nemirovsky (Eds.): J. Cryst. Growth **199** (1999)
12.54 T. Hibiya, J. B. Mullin, W. Uwaha (Eds.): J. Cryst. Growth **237** (2002)
12.55 T. Hibiya, J. B. Mullin, W. Uwaha (Eds.): J. Cryst. Growth **239** (2002)
12.56 K. Nakajima, P. Capper, S. D. Durbin, S. Hiyamizu (Eds.): J. Cryst. Growth **229** (2001)
12.57 T. Duffar, M. Heuken, J. Villain (Eds.): J. Cryst. Growth **275** (2005)
12.58 P. Capper (Ed.): *Narrow-Gap II–VI Compounds for Optoelectronic and Electromagnetic Applications* (Chapman Hall, London 1997)
12.59 P. Capper, C. T. Elliott (Eds.): *Infrared Detectors and Emitters: Materials and Devices* (Kluwer, Boston 2001)
12.60 H. J. Scheel, T. Fukuda (Eds.): *The Technology of Crystal Growth and Epitaxy* (Wiley, Chichester 2003)
12.61 S. Nishino: *Properties of Silicon Carbide*, EMIS Datarev. Ser. **13**, ed. by G. L. Harris (IEE, London 1995)
12.62 A. O. Konstantinov: *Properties of Silicon Carbide*, EMIS Datarev. Ser. **13**, ed. by G. L. Harris (IEE, London 1995)
12.63 N. Nordell: *Process Technology for Silicon Carbide Devices*, ed. by C. M. Zetterling (IEE, London 2002)
12.64 H. Kanda, T. Sekine: *Properties, Growth and Applications of Diamond*, EMIS Datarev. Ser. **26**, ed. by M. H. Nazare, A. J. Neves (IEE, London 2001)
12.65 W. G. Pfann: *Zone Melting*, 2nd edn. (Wiley, New York 1966)
12.66 P. Rudolph: *The Technology of Crystal Growth and Epitaxy*, ed. by H. J. Scheel, T. Fukuda (Wiley, Chichester 2003)
12.67 T. Asahi, K. Kainosho, K. Kohiro, A. Noda, K. Sato, O. Oda: *The Technology of Crystal Growth and Epitaxy*, ed. by H. J. Scheel, T. Fukuda (Wiley, Chichester 2003)
12.68 W. F. J. Micklethwaite, A. J. Johnson: *Infrared Detectors and Emitters: Materials and Devices*, ed. by P. Capper, C. T. Elliott (Kluwer, Boston 2001)
12.69 I. Grzegory, S. Porowski: *Properties, Processing and Applications of Gallium Nitride and Related Semiconductors*, EMIS Datarev. Ser. **23**, ed. by J. H. Edgar, S. Strite, I. Akasaki, H. Amano, C. Wetzel (IEE, London 1999)
12.70 I. Grzegory, S. Krukowski, M. Leszczynski, P. Perlin, T. Suski, S. Porowski: *Nitride Semiconductors: Handbook on Materials and Devices*, ed. by P. Ruterana, M. Albrecht, J. Neugebauer (Wiley, Weinheim 2003)
12.71 K. Nishino, S. Sakai: *Properties, Processing and Applications of Gallium Nitride and Related Semiconductors*, EMIS Datarev. Ser. **23**, ed. by J. H. Edgar, S. Strite, I. Akasaki, H. Amano, C. Wetzel (IEE, London 1999)
12.72 P. Rudolph: *Recent Developments of Bulk Crystal Growth 1998*, ed. by M. Isshiki (Research Signpost, Trivandrum, India 1998) p. 127
12.73 H. Hartmann, K. Bottcher, D. Siche: *Recent Developments of Bulk Crystal Growth 1998*, ed. by M. Isshiki (Research Signpost, Trivandrum, India 1998) p. 165
12.74 B. J. Fitzpatrick, P. M. Harnack, S. Cherin: Philips J. Res. **41**, 452 (1986)
12.75 P. Capper, J. E. Harris, D. Nicholson, D. Cole: J. Cryst. Growth **46**, 575 (1979)

12.76 R. Triboulet, T. Nguyen Duy, A. Durand: J. Vac. Sci. Technol. A **3**, 95 (1985)
12.77 R. Triboulet, Y. Marfaing: J. Cryst. Growth **51**, 89 (1981)
12.78 R. Triboulet, K. Pham Van, G. Didier: J. Cryst. Growt **101**, 216 (1990)
12.79 V.A. Kuznetsov: Prog. Cryst. Growth Ch. **21**, 163 (1990)
12.80 J. Mimila, R. Triboulet: Mater. Lett. **24**, 221 (1995)
12.81 W.W. Piper, S.J. Polich: J. Appl. Phys. **32**, 1278 (1961)
12.82 G.J. Russell, J. Woods: J. Cryst. Growth **46**, 323 (1979)
12.83 P. Blanconnier, P. Henoc: J. Cryst Growth **17**, 218 (1972)
12.84 K. Durose, A. Turnbull, P. D. Brown: Mater. Sci. Eng. B **16**, 96 (1993)
12.85 M.R. Lorenz: *Physics and Chemistry of II–VI Compounds*, ed. by M. Aven, J.S. Prener (North-Holland, Amsterdam 1967) Chap. 2
12.86 K. Zanio: Semicond. Semimet. **13** (1978)
12.87 P. Capper, A. Brinkman: *Properties of Narrow Gap Cadmium-Based Compounds*, EMIS Datarev. Ser. **10**, ed. by P. Capper (IEE, London 1994) p. 369
12.88 A.W. Brinkman: *Narrow-Gap II–VI Compounds for Optoelectronic and Electromagnetic Applications*, ed. by P. Capper (Chapman & Hall, London 1997)
12.89 R. Triboulet, J.O. Ndap A. El Mokri et al.: J. Phys. IV **5**, C3–141 (1995)
12.90 R. Triboulet, A. Tromson-Carli, D. Lorans, T. Nguyen Duy: J. Electron. Mater. **22**, 827 (1993)
12.91 J.B. Mullin, C.A. Jones, B.W. Straughan, A. Royle: J. Cryst. Growth **59**, 135 (1982)
12.92 H.M. Hobgood, B.W. Swanson, R.N. Thomas: J. Cryst. Growth **85**, 510 (1987)
12.93 R. Triboulet, Y. Marfaing: J. Electrochem. Soc. **120**, 1260 (1973)
12.94 T. Sasaki, Y. Mori, M. Yoshimura: *The Technology of Crystal Growth and Epitaxy*, ed. by H.J. Scheel, T. Fukuda (Wiley, Chichester 2003)
12.95 R. Hirano, H. Kurita: *Bulk Crystal Growth of Electronic, Optical and Optoelectronic Materials*, ed. by P. Capper (Wiley, Chichester 2005)
12.96 K. Zanio: J. Electron. Mater. **3**, 327 (1974)
12.97 C.P. Khattak, F. Schmid: Proc. SPIE **1106**, 47 (1989)
12.98 W. M. Chang, W. R. Wilcox, L. Regel: Mater. Sci. Eng. B **16**, 23 (1993)
12.99 N. R. Kyle: J. Electrochem. Soc. **118**, 1790 (1971)
12.100 J.C. Tranchart, B. Latorre, C. Foucher, Y. LeGouce: J. Cryst. Growth **72**, 468 (1985)
12.101 Y.-C. Lu, J.-J. Shiau, R.S. Fiegelson, R.K. Route: J. Cryst. Growth **102**, 807 (1990)
12.102 J.P. Tower, S.B. Tobin, M. Kestigian: J. Electron. Mater. **24**, 497 (1995)
12.103 S. Sen, S.M. Johnson, J.A. Kiele: Mater. Res. Soc. Symp. Proc. **161**, 3 (1990)
12.104 J. F. Butler, F. P. Doty, B. Apotovsky: Mater. Sci. Eng. B **16**, 291 (1993)
12.105 P. Capper, J.E. Harris, E. O'Keefe, C.L. Jones, C.K. Ard, P. Mackett, D.T. Dutton: Mater. Sci. Eng. B **16**, 29 (1993)
12.106 C. Szeles, S. E. Cameron, S.A. Soldner, J.-O. Ndap, M. D. Reed: J. Electron. Mater. **33/6**, 742 (2004)
12.107 A. El Mokri, R. Triboulet, A. Lusson: J. Cryst. Growth **138**, 168 (1995)
12.108 R. U. Bloedner, M. Presia, P. Gille: Adv. Mater. Opt. Electron. **3**, 233 (1994)
12.109 R. Schoenholz, R. Dian, R. Nitsche: J. Cryst. Growth **72**, 72 (1985)
12.110 W. F. H. Micklethwaite: Semicond. Semimet. **18**, 3 (1981)
12.111 P.W. Kruse: Semicond. Semimet. **18**, 1 (1981)
12.112 H. Maier: *N.A.T.O. Advanced Research Workshop on the Future of Small-Gap II–VI Semiconductors* (Liege, Belgium 1988)
12.113 P. Capper: Prog. Cryst. Growth Ch. **19**, 259 (1989)
12.114 F. R. Szofran, S. L. Lehoczky: J. Cryst. Growth **70**, 349 (1984)
12.115 P. Capper, J.J.G. Gosney: U.K. Patent 8115911 (1981)
12.116 P. Capper, C. Maxey, C. Butler, M. Grist, J. Price: J. Mater. Sci. Mater. El. **15**, 721 (2004)
12.117 Y. Nguyen Duy, A. Durand, J.L. Lyot: Mater. Res. Soc. Symp. Proc. **90**, 81 (1987)
12.118 A. Durand, J. L. Dessus, T. Nguyen Duy, J. F. Barbot: Proc. SPIE **659**, 131 (1986)
12.119 P. Gille, F. M. Kiessling, M. Burkert: J. Cryst. Growth **114**, 77 (1991)
12.120 P. Gille, M. Pesia, R.U. Bloedner, N. Puhlman: J. Cryst. Growth **130**, 188 (1993)
12.121 M. Royer, B.R. Jean, A.R. Durand, R. Triboulet: French Patent 8804370 (1988)
12.122 R.U. Bloedner, P. Gille: J. Cryst. Growth **130**, 181 (1993)
12.123 A. Rogalski: *New Ternary Alloy Systems for Infrared Detectors* (SPIE, Bellingham 1994)
12.124 R. Triboulet: Semicond. Sci. Technol. **5**, 1073 (1990)
12.125 R. Korenstein, R.J. Olson Jr., D. Lee: J. Electron. Mater. **24**, 511 (1995)
12.126 B. Pelliciari, F. Dierre, D. Brellier, B. Schaub: J. Cryst. Growth **275**, 99 (2005)
12.127 A. Pajaczkowska: Prog. Cryst. Growth Ch. **1**, 289 (1978)
12.128 W. Giriat, J.K. Furdyna: Semicond. Semimet. **25**, 1 (1988)
12.129 M.C.C. Custodio, A.C. Hernandes: J. Cryst. Growth **205**, 523 (1999)
12.130 T. Fukuda, V.I. Chani, K. Shimamura: *Recent Developments of Bulk Crystal Growth 1998*, ed. by M. Isshiki (Research Signpost, Trivandrum, India 1998) p. 191
12.131 T. Fukuda, V.I. Chani, K. Shimamura: *The Technology of Crystal Growth and Epitaxy*, ed. by H.J. Scheel, T. Fukuda (Wiley, Chichester 2003)

12.132 S. Uda, S.Q. Wang, N. Konishi, H. Inaba, J. Harada: J. Cryst. Growth **237/239**, 707 (2002)
12.133 F. Iwasaki, H. Iwasaki: J. Cryst. Growth **237/239**, 820 (2002)
12.134 V. S. Balitsky: Paper given at 1st International School on Crystal Growth and Technology, Beatenberg, Switzerland (1998)
12.135 M. Korzhik: Paper given at 1st International School on Crystal Growth and Technology, Beatenberg, Switzerland (1998)
12.136 P. J. Li, Z. W Yin, D. S. Yan: Paper given at 1st International School on Crystal Growth and Technology, Beatenberg, Switzerland (1998)
12.137 Kh. S. Bagdasarov, E. V. Zharikov: Paper given at 1st International School on Crystal Growth and Technology, Beatenberg, Switzerland (1998)
12.138 L. Lytvynov: Paper given at the 2nd International School on Crystal Growth and Technology, Zao, Japan (2000)
12.139 M. I. Moussatov, E. V. Zharikov: Paper given at 1st International School on Crystal Growth and Technology, Beatenberg, Switzerland (1998)
12.140 F. Schmid, Ch. P. Khattak: Paper given at 1st International School on Crystal Growth and Technology, Beatenberg, Switzerland (1998)
12.141 A. V. Gektin, B. G. Zaslavsky: Paper given at 1st International School on Crystal Growth and Technology, Beatenberg, Switzerland (1998)
12.142 A. V. Gektin: Paper given at the 2nd International School on Crystal Growth and Technology, Zao, Japan (2000)
12.143 N. Zaitseva, L. Carman, I. Smolsky: J. Cryst. Growth **241**, 363 (2002)

13. Single-Crystal Silicon: Growth and Properties

It is clear that silicon, which has been the dominant material in the semiconductor industry for some time, will carry us into the coming ultra-large-scale integration (ULSI) and system-on-a-chip (SOC) eras, even though silicon is not the optimum choice for every electronic device. Semiconductor devices and circuits are fabricated through many mechanical, chemical, physical, and thermal processes. The preparation of silicon single-crystal substrates with mechanically and chemically polished surfaces is the first step in the long and complex device fabrication process. In this chapter, the approaches currently used to prepare silicon materials (from raw materials to single-crystalline silicon) are discussed.

Silicon, which has been and will continue to be the dominant material in the semiconductor industry for some time to come [13.1], will carry us into the ultra-large-scale integration (ULSI) era and the system-on-a-chip (SOC) era.

As electronic devices have become more advanced, device performance has become more sensitive to the quality and the properties of the materials used to construct them.

Germanium (Ge) was originally utilized as a semiconductor material for solid state electronic devices. However, the narrow bandgap (0.66 eV) of Ge limits the operation of germanium-based devices to temperatures of approximately 90 °C because of the considerable leakage currents observed at higher temperatures. The wider bandgap of silicon (1.12 eV), on the other hand, results in electronic devices that are capable of operating at up to $\approx 200\,°C$. However, there is a more serious problem than the narrow bandgap: germanium does not readily provide a stable passivation layer on the surface. For example, germanium dioxide (GeO_2) is water-soluble and dissociates at approximately 800 °C. Silicon, in contrast to germanium, readily accommodates surface passivation by forming silicon dioxide (SiO_2), which provides a high degree of protection to the underlying device. This stable SiO_2 layer results in a decisive advantage for silicon over germanium as the basic semiconductor material used for electronic device fabrication. This advantage has lead to a number of new technologies, including processes for diffusion doping and defining intricate patterns. Other advantages of silicon are that it is completely nontoxic, and that silica (SiO_2), the raw material from which silicon is obtained, comprises approximately 60% of the mineral content of the Earth's crust. This implies that the raw material from which silicon is obtained is available in plentiful supply to the IC industry. Moreover, electronic-grade silicon can be obtained at less than one-tenth the cost of germanium. All of these advantages have caused silicon to almost completely replace germanium in the semiconductor industry.

Although silicon is not the optimum choice for every electronic device, its advantages mean that it will almost certainly dominate the semiconductor industry for some time yet.

13.1 Overview

Very fruitful interactions have occurred between the users and manufacturers of semiconductor material since the invention of the point-contact transistor in 1947, when the necessity for "perfect and pure" crystals was recognized. The competition was often such that the crystal quality demanded by new devices could only be met by controlling crystal growth using electronic equipment built with these new devices. Since dislocation-free silicon crystals were grown as early as the 1960s using the *Dash technique* [13.3], semiconductor material research and developmental efforts have concentrated on material purity, production yields, and problems related to device manufacture.

Semiconductor devices and circuits are fabricated using a wide variety of mechanical, chemical, physical and thermal processes. A flow diagram for typical semiconductor silicon preparation processes is shown in Fig. 13.1. The preparation of silicon single-crystal substrates with mechanically and chemically polished surfaces is the first step in the long and complex process of device fabrication.

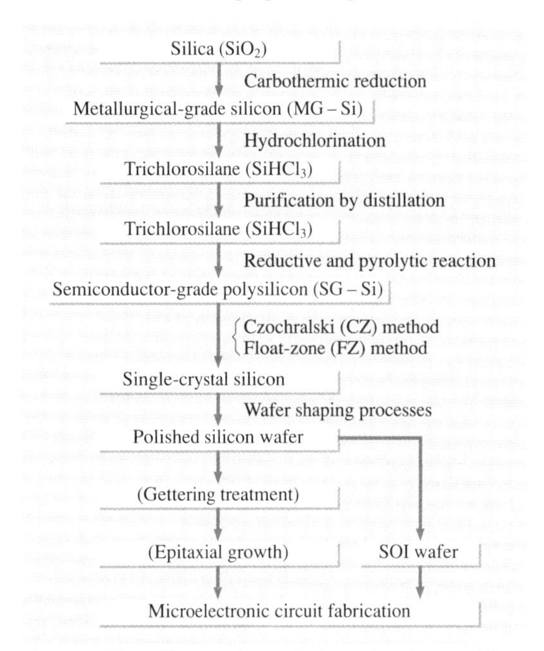

Fig. 13.1 Flow diagram for typical semiconductor silicon preparation processes. (After *Shimura* [13.1])

As noted above, silicon is the second most abundant element on Earth; more than 90% of the Earth's crust is composed of silica and silicates. Given this boundless supply of raw material, the problem is then to transform silicon into the usable state required by the semiconductor technology. The first and main requirement is that the silicon used for electronic devices must be extremely pure, since very small amounts of some impurities have a strong influence on the electronic characteristics of silicon, and therefore the performance of the electronic device. The second requirement is for large-diameter crystals, since the chip yield per wafer increases substantially with larger diameters, as shown in Fig. 13.2 for the case of DRAM [13.2], one of the most common electronic devices. Besides the purity and the diameter, the cost of production and the specifications of the material, including the grown-in defect density and the resistive homogeneity, must meet current industrial demands.

In this chapter, current approaches to the preparation of silicon – converting the raw material into single-crystalline silicon (see Fig. 13.1) – are discussed.

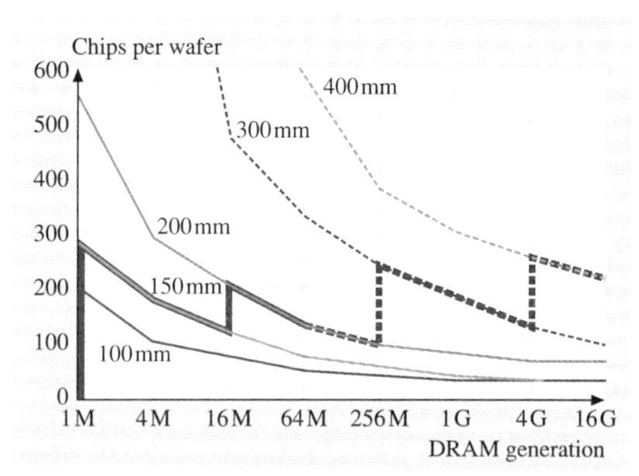

Fig. 13.2 Chips per wafer as a function of DRAM generation. (After *Takada* et al. [13.2])

13.2 Starting Materials

13.2.1 Metallurgical-Grade Silicon

The starting material for high-purity silicon single crystals is silica (SiO_2). The first step in silicon manufacture is the melting and reduction of silica. This is accomplished by mixing silica and carbon in the form of coal, coke or wood chips and heating the mixture to high temperatures in a submerged electrode arc furnace. This carbothermic reduction of silica produces fused silicon:

$$SiO_2 + 2C \rightarrow Si + 2CO\,. \qquad (13.1)$$

A complex series of reactions actually occur in the furnace at temperatures ranging from 1500 to 2000 °C. The lumps of silicon obtained from this process are called metallurgical-grade silicon (MG-Si), and its purity is about 98–99%.

13.2.2 Polycrystalline Silicon

Intermediate Chemical Compounds

The next step is to purify MG-Si to the level of semiconductor-grade silicon (SG-Si), which is used as the starting material for single-crystalline silicon. The basic concept is that powdered MG-Si is reacted with anhydrous HCl to form various chlorosilane compounds in a fluidized-bed reactor. Then the silanes are purified by distillation and chemical vapor deposition (CVD) to form SG-polysilicon.

A number of intermediate chemical compounds have been considered, such as monosilane (SiH_4), silicon tetrachloride ($SiCl_4$), trichlorosilane ($SiHCl_3$) and dichlorosilane (SiH_2Cl_2). Among these, trichlorosilane is most commonly used for subsequent polysilicon deposition for the following reasons: (1) it can be easily formed by the reaction of anhydrous hydrogen chloride with MG-Si at reasonably low temperatures (200–400 °C); (2) it is liquid at room temperature, so purification can be accomplished using standard distillation techniques; (3) it is easy to handle and can be stored in carbon steel tanks when dry; (4) liquid trichlorosilane is easily vaporized and, when mixed with hydrogen, it can be transported in steel lines; (5) it can be reduced at atmospheric pressure in the presence of hydrogen; (6) its deposition can take place on heated silicon, eliminating the need for contact with any foreign surfaces that may contaminate the resulting silicon; and (7) it reacts at lower temperatures (1000–1200 °C) and at faster rates than silicon tetrachloride.

Hydrochlorination of Silicon

Trichlorosilane is synthesized by heating powdered MG-Si at around 300 °C in a fluidized-bed reactor. That is, MG-Si is converted into $SiHCl_3$ according to the following reaction:

$$Si + 3HCl \rightarrow SiHCl_3 + H_2\,. \qquad (13.2)$$

The reaction is highly exothermic and so heat must be removed to maximize the yield of trichlorosilane. While converting MG-Si into $SiHCl_3$, various impurities such as Fe, Al, and B are removed by converting them into their halides ($FeCl_3$, $AlCl_3$, and BCl_3, respectively), and byproducts such as $SiCl_4$ and H_2 are also produced.

Distillation and Decomposition of Trichlorosilane

Distillation has been widely used to purify trichlorosilane. The trichlorosilane, which has a low boiling point (31.8 °C), is fractionally distilled from the impure halides, resulting in greatly increased purity, with an electrically active impurity concentration of less than 1 ppba. The high-purity trichlorosilane is then vaporized, diluted with high-purity hydrogen, and introduced into the deposition reactor. In the reactor, thin silicon rods called slim rods supported by graphite electrodes are available for surface deposition of silicon according to the reaction

$$SiHCl_3 + H_2 \rightarrow Si + 3HCl\,. \qquad (13.3)$$

In addition this reaction, the following reaction also occurs during polysilicon deposition, resulting in the formation of silicon tetrachloride (the major byproduct of the process):

$$HCl + SiHCl_3 \rightarrow SiCl_4 + H_2\,. \qquad (13.4)$$

This silicon tetrachloride is used to produce high-purity quartz, for example.

Needless to say, the purity of the slim rods must be comparable to that of the deposited silicon. The slim rods are preheated to approximately 400 °C at the start of the silicon CVD process. This preheating is required in order to increase the conductivity of high-purity (high-resistance) slim rods sufficiently to allow for resistive heating. Depositing for 200–300 h at around 1100 °C results in high-purity polysilicon rods of 150–200 mm in diameter. The polysilicon rods are shaped into various forms for subsequent crystal growth processes, such as chunks for Czochralski melt growth and long cylindrical rods for float-zone growth. The process for reducing trichlorosilane on a heated silicon rod using hydrogen was described in the late 1950s and early 1960s in a number of process patents assigned to Siemens; therefore, this process is often called the "Siemens method" [13.4].

The major disadvantages of the Siemens method are its poor silicon and chlorine conversion efficiencies, relatively small batch size, and high power consumption. The poor conversion efficiencies of silicon and chlorine are associated with the large volume of silicon tetrachloride produced as the byproduct in the CVD process. Only about 30% of the silicon provided in the CVD reaction is converted into high-purity polysilicon. Also, the cost of producing high-purity polysilicon may depend on the usefulness of the byproduct, $SiCl_4$.

Monosilane Process

A polysilicon production technology based on the production and pyrolysis of monosilane was established in the late 1960s. Monosilane potentially saves energy because it deposits polysilicon at a lower temperature and produces purer polysilicon than the trichlorosilane process; however, it has hardly been used due to the lack of an economical route to monosilane and due to processing problems in the deposition step [13.5]. However, with the recent development of economical routes to high-purity silane and the successful operation of a large-scale plant, this technology has attracted the attention of the semiconductor industry, which requires higher purity silicon.

In current industrial monosilane processes, magnesium and MG-Si powder are heated to 500 °C under a hydrogen atmosphere in order to synthesize magenesium silicide (Mg_2Si), which is then made to react with ammonium chloride (NH_4Cl) in liquid ammonia (NH_3) below 0 °C to form monosilane (SiH_4). High-purity polysilicon is then produced via the pyrolysis of the monosilane on resistively heated polysilicon filaments at 700–800 °C. In the monosilane generation process, most of the boron impurities are removed from silane via chemical reaction with NH_3. A boron content of 0.01–0.02 ppba in polysilicon has been achieved using a monosilane process. This concentration is very low compared to that observed in polysilicon prepared from trichlorosilane. Moreover, the resulting polysilicon is less contaminated with metals picked up through chemical transport processes because monosilane decomposition does not cause any corrosion problems.

Granular Polysilicon Deposition

A significantly different process, which uses the decomposition of monosilane in a fluidized-bed deposition reactor to produce free-flowing granular polysilicon, has been developed [13.5]. Tiny silicon seed particles are fluidized in a monosilane/hydrogen mix, and polysilicon is deposited to form free-flowing spherical particles that are an average of 700 μm in diameter with a size distribution of 100 to 1500 μm. The fluidized-bed seeds were originally made by grinding SG-Si in a ball or hammer mill and leaching the product with acid, hydrogen peroxide and water. This process was time-consuming and costly, and tended to introduce undesirable impurities into the system through the metal grinders. However, in a new method, large SG-Si particles are fired at each other by a high-speed stream of gas causing them to break into particles of a suitable size for the fluidized bed. This process introduces no foreign materials and requires no leaching.

Because of the greater surface area of granular polysilicon, fluidized-bed reactors are much more efficient than traditional Siemens-type rod reactors. The quality of fluidized-bed polysilicon has been shown to be equivalent to polysilicon produced by the more conventional Siemens method. Moreover, granular polysilicon of a free-flowing form and high bulk density enables crystal growers to obtain the most from of each production run. That is, in the Czochralski crystal growth process (see the following section), crucibles can be quickly and easily filled to uniform loadings which typically exceed those of randomly stacked polysilicon chunks produced by the Siemens method. If we also consider the potential of the technique to move from batch operation to continuous pulling (discussed later), we can see that free-flowing polysilicon granules could provide the advantageous route of a uniform feed into a steady-state melt. This product appears to be a revolutionary starting material of great promise for silicon crystal growth.

13.3 Single-Crystal Growth

Although various techniques have been utilized to convert polysilicon into single crystals of silicon, two techniques have dominated the production of them for electronics because they meet the requirements of the microelectronics device industry. One is a zone-melting method commonly called the *floating-zone (FZ) method*,

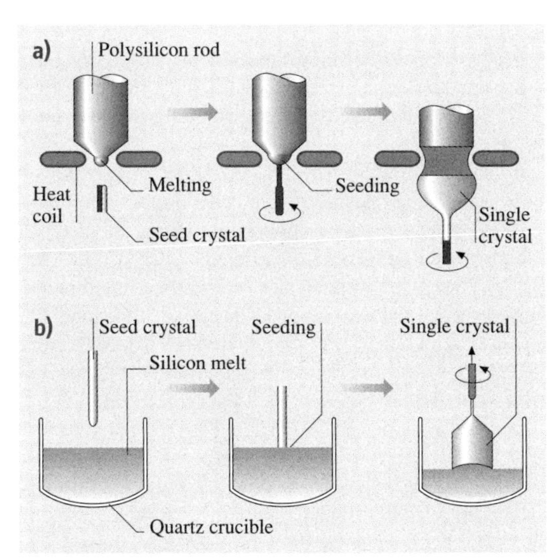

Fig. 13.3a,b Principles of single-crystal growth by (**a**) floating-zone method and (**b**) Czochralski method. (After *Shimura* [13.1])

and the other is a pulling method traditionally called the *Czochralski (CZ) method*, although it should actually be called the *Teal-Little method*. The principles behind these two crystal growth methods are depicted in Fig. 13.3. In the FZ method, a molten zone is passed through a polysilicon rod to convert it into a single-crystal ingot; in the CZ method, a single crystal is grown by pulling from a melt contained in a quartz crucible. In both cases, the *seed crystal* plays a very important role in obtaining a single crystal with a desired crystallographic orientation.

It is estimated that about 95% of all single-crystal silicon is produced by the CZ method and the rest mainly by the FZ method. The silicon semiconductor industry requires high purity and minimum defect concentrations in their silicon crystals to optimize device manufacturing yield and operational performance. These requirements are becoming increasingly stringent as the technology changes from LSI to VLSI/ULSI and then SOC. Besides the quality or perfection of silicon crystals, crystal diameter has also been steadily increasing in order to meet the demands of device manufacturers. Since microelectronic chips are produced via a "batch system," the diameters of the silicon wafers used for device fabrication significantly affect the productivity (as shown in Fig. 13.2), and in turn the production cost.

In the following sections, we first discuss the FZ method and then move on to the CZ method. The latter will be discussed in more detail due to its extreme importance to the microelectronics industry.

13.3.1 Floating-Zone Method

General Remarks

The FZ method originated from zone melting, which was used to refine binary alloys [13.6] and was invented by *Theuerer* [13.7]. The reactivity of liquid silicon with the material used for the crucible led to the development of the FZ method [13.8], which permits the crystallization of silicon without the need for any contact with the crucible material, which is needed to be able to grow crystals of the required semiconductor purity.

Outline of the Process

In the FZ process, a polysilicon rod is converted into a single-crystal ingot by passing a molten zone heated by a needle-eye coil from one end of the rod to the other, as shown in Fig. 13.3a. First, the tip of the polysilicon rod is contacted and fused with a seed crystal with the desired crystal orientation. This process is called *seeding*. The seeded molten zone is passed through the polysilicon rod by simultaneously moving the single crystal seed down the rod. When the molten zone of silicon solidifies, polysilicon is converted into single-crystalline silicon with the help of the seed crystal. As the zone travels along the polysilicon rod, single-crystal silicon freezes at its end and grows as an extension of the seed crystal.

After seeding, a thin neck about 2 or 3 mm in diameter and 10–20 mm long is formed. This process is called *necking*. Growing a neck eliminates dislocations that can be introduced into newly grown single-crystal silicon during the seeding operation due to thermal shock. This necking process, called the *Dash technique* [13.3], is therefore fundamental to growing dislocation-free crystals and is used universally in both the FZ and the CZ methods. An X-ray topograph of the seed, neck and conical part of a silicon single-crystal grown by the FZ method is shown in Fig. 13.4. It is apparent that dislocations generated at the melt contact are completely eliminated by necking. After the conical part is formed, the main body with the full target diameter is grown. During the entire FZ growth process, the shape of the molten zone and the ingot diameter are determined by adjusting the power to the coil and the travel rate, both of which are under computer control. The technique most commonly used to automatically control the diameter in both the FZ and CZ methods employs an infrared sen-

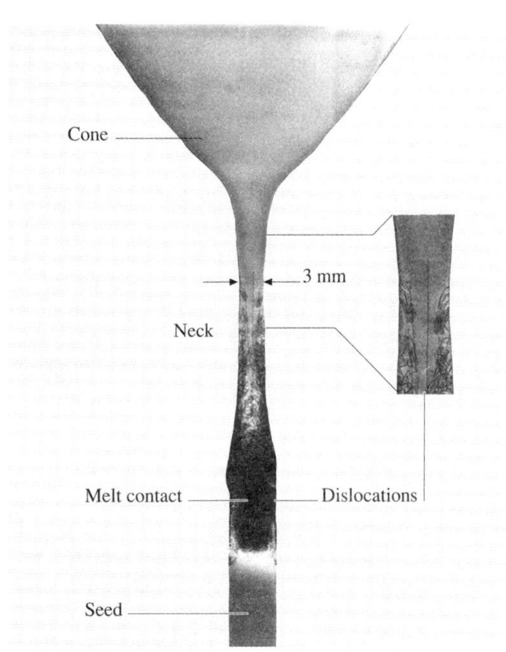

Fig. 13.4 X-ray topography of seed, neck and conical part of floating-zone silicon. (Courtesy of Dr. T. Abe)

sor focused on the meniscus. The shape of the meniscus on a growing crystal depends on its angle of contact at the three-phase boundary, the crystal diameter, and the magnitude of the surface tension. A change in meniscus angle (and therefore crystal diameter) is sensed, and the information is fed back in order to automatically adjust the growth conditions.

In contrast with CZ crystal growth, in which the seed crystal is dipped into the silicon melt and the growing crystal is pulled upward, in the FZ method the thin seed crystal sustains the growing crystal, as does the polysilicon rod from the bottom (Fig. 13.3). As a result, the rod is balanced precariously on the thin seed and neck during the entire growth process. The seed and neck can support a crystal of up to a 20 kg so long as the center of gravity of the growing crystal remains at the center of the growth system. If the center of gravity moves away from the center line, the seed will easily fracture. Hence, it was necessary to invent a crystal stabilizing and supporting technique before long and heavy FZ silicon crystals could be grown. For large crystals, it is necessary to support the growing crystal in the way shown in Fig. 13.5 [13.9], particularly in the case of recent FZ crystals with large diameters (150–200 mm), since their weights easily exceed 20 kg.

Doping

In order to obtain n- or p-type silicon single-crystals of the required resistivity, either the polysilicon or the growing crystal must be doped with the appropriate donor or acceptor impurities, respectively. For FZ silicon growth, although several doping techniques have been tried, the crystals are typically doped by blowing a dopant gas such as phosphine (PH_3) for n-type silicon or diborane (B_2H_6) for p-type silicon onto the molten zone. The dopant gas is usually diluted with a carrier gas, such as argon. The great advantage of this method is that the silicon crystal manufacturer does not need to store polysilicon sources with different resistivities.

Since the segregation (discussed in the next subsection) of elemental dopants for n-type silicon is much less than unity, FZ crystals doped by the traditional method have radial dopant gradients. Moreover, since the crystallization rate varies in the radial direction on the microscopic scale, the dopant concentrations distribute cyclically and give rise to so-called *dopant striations*, resulting in radial resistivity inhomogeneities. In order to obtain more homogeneously doped n-type silicon, neu-

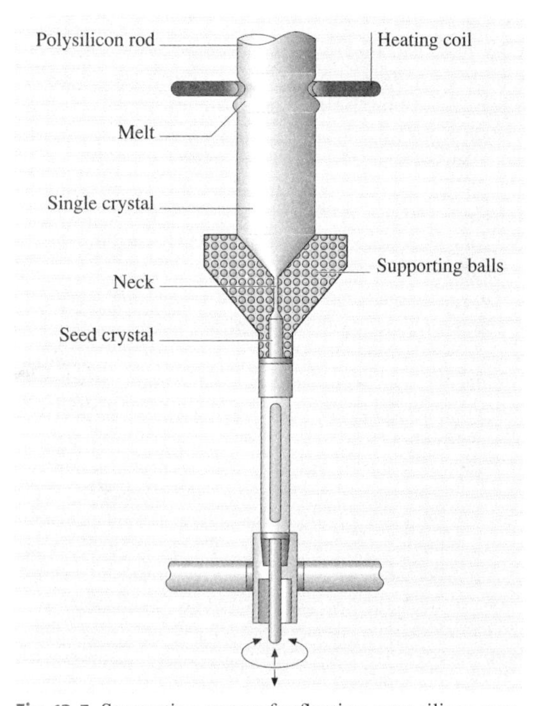

Fig. 13.5 Supporting system for floating-zone silicon crystal. (After *Keller* and *Mühlbauer* [13.9])

tron transmutation doping (NTD) has been applied to FZ silicon crystals [13.10]. This procedure involves the nuclear transmutation of silicon to phosphorus by bombarding the crystal with thermal neutrons according to the reaction

$$^{30}\mathrm{Si}(\mathrm{n},\gamma) \rightarrow {}^{31}\mathrm{Si} \xrightarrow{2.6\,\mathrm{h}} {}^{31}\mathrm{P}+\beta\,. \quad (13.5)$$

The radioactive isotope ^{31}Si is formed when ^{30}Si captures a neutron and then decays into the stable isotope ^{31}P (donor atoms), whose distribution is not dependent on crystal growth parameters. Immediately after irradiation the crystals exhibit high resistivity, which is attributed to the large number of lattice defects arising from radiation damage. The irradiated crystal, therefore, must be annealed in an inert ambient at temperatures of around 700 °C in order to annihilate the defects and to restore the resistivity to that derived from the phosphorus doping. Under the NTD scheme, crystals are grown without doping and are then irradiated in a nuclear reactor with a large ratio of thermal to fast neutrons in order to enhance neutron capture and to minimize damage to the crystal lattice.

The application of NTD has been almost exclusively limited to FZ crystals because of their higher purity compared to CZ crystals. When the NTD technique was applied to CZ silicon crystals, it was found that oxygen donor formation during the annealing process after irradiation changed the resistivity from that expected, even though phosphorus donor homogeneity was achieved [13.11]. NTD has the additional shortcoming that no process is available for p-type dopants and that an excessively long period of irradiation is required for low resistivities (in the range of 1–10 Ω cm).

Properties of FZ-Silicon Crystal

During FZ crystal growth, the molten silicon does not come into contact with any substance other than the ambient gas in the growth chamber. Therefore, an FZ silicon crystal is inherently distinguished by its higher purity compared to a CZ crystal which is grown from the melt – involving contact with a quartz crucible. This contact gives rise to high oxygen impurity concentrations of around 10^{18} atoms/cm^3 in CZ crystals, while FZ silicon contains less than 10^{16} atoms/cm^3. This higher purity allows FZ silicon to achieve high resistivities not obtainable using CZ silicon. Most of the FZ silicon consumed has a resistivity of between 10 and 200 Ω cm, while CZ silicon is usually prepared to resistivities of 50 Ω cm or less due to the contamination from the quartz crucible. FZ silicon is therefore mainly used to fabricate semiconductor power devices that support reverse voltages in excess of 750–1000 V. The high-purity crystal growth and the precision doping characteristics of NTD FZ-Si have also led to its use in infrared detectors [13.12], for example.

However, if we consider mechanical strength, it has been recognized for many years that FZ silicon, which contains fewer oxygen impurities than CZ silicon, is mechanically weaker and more vulnerable to thermal stress during device fabrication [13.13, 14]. High-temperature processing of silicon wafers during electronic device manufacturing often produces enough thermal stress to generate slip dislocations and warpage. These effects bring about yield loss due to leaky junctions, dielectric defects, and reduced lifetime, as well as reduced photolithographic yields due to the degradation of wafer flatness. Loss of geometrical planarity due to warpage can be so severe that the wafers are not processed any further. Because of this, CZ silicon wafers have been used much more widely in IC device fabrication than FZ wafers have. This difference in mechanical stability against thermal stresses is the dominant reason why CZ silicon crystals are exclusively used for the fabrication of ICs that require a large number of thermal process steps.

In order to overcome these shortcomings of FZ silicon, the growth of FZ silicon crystals with doping impurities such as oxygen [13.15] and nitrogen [13.16] has been attempted. It was found that doping FZ silicon crystals with oxygen or nitrogen at concentrations of $1-1.5\times10^{17}$ atoms/cm^3 or 1.5×10^{15} atoms/cm^3, respectively, results in a remarkable increase in mechanical strength.

13.3.2 Czochralski Method

General Remarks

This method was named after J. Czochralski, who established a technique for determining the crystallization velocities of metals [13.17]. However, the actual pulling method that has been widely applied to single-crystal growth was developed by *Teal* and *Little* [13.18], who modified Czochralski's basic principle. They were the first to successfully grow single-crystals of germanium, 8 inches in length and 0.75 inches in diameter, in 1950. They subsequently designed another apparatus for the growth of silicon at higher temperatures. Although the basic production process for single-crystal silicon has changed little since it was pioneered by Teal and coworkers, large-diameter (up to 400 mm) silicon single-crystals with a high degree of perfection that meet state-of-the-art device demands have been grown by incorporating the Dash technique and successive technological innovations into the apparatus.

Today's research and development efforts concerning silicon crystals are directed toward achieving microscopic uniformity of crystal properties such as the resistivity and the concentrations of impurities and microdefects, as well as microscopic control of them, which will be discussed elsewhere in this Handbook.

Outline of the Process

The three most important steps in CZ crystal growth are shown schematically in Fig. 13.3b. In principle, the process of CZ growth is similar to that of FZ growth: (1) melting polysilicon, (2) seeding and (3) growing. The CZ pulling procedure, however, is more complicated than that of FZ growth and is distinguished from it by the use of a quartz crucible to contain the molten silicon. Figure 13.6 shows a schematic view of typical modern CZ crystal growth equipment. Important steps in the actual or standard CZ silicon crystal growth sequence are as follows:

1. Polysilicon chunks or grains are placed in a quartz crucible and melted at temperatures higher than the melting point of silicon (1420 °C) in an inert ambient gas.
2. The melt is kept at a high temperature for a while in order to ensure complete melting and ejection of tiny bubbles, which may cause voids or negative crystal defects, from the melt.
3. A seed crystal with the desired crystal orientation is dipped into the melt until it begins to melt itself. The seed is then withdrawn from the melt so that the neck is formed by gradually reducing the diameter; this is the most delicate step. During the entire crystal growth process, inert gas (usually argon) flows downward through the pulling chamber in order to carry off reaction products such as SiO and CO.
4. By gradually increasing the crystal diameter, the conical part and shoulder are grown. The diameter is increased up to the target diameter by decreasing the pulling rate and/or the melt temperature.
5. Finally, the cylindrical part of the body with a constant diameter is grown by controlling the pulling rate and the melt temperature while compensating for the drop in the melt level as the crystal grows. The pulling rate is generally reduced toward the tail end of a growing crystal, mainly due to increasing heat radiation from the crucible wall as the melt level drops and exposes more crucible wall to the growing crystal. Near the end of the growth process, but before the crucible is completely drained of molten silicon, the crystal diameter must be gradually reduced to form an end-cone in order to minimize thermal shock, which can cause slip dislocations at the tail end. When the diameter becomes small enough, the crystal can be separated from the melt without the generation of dislocations.

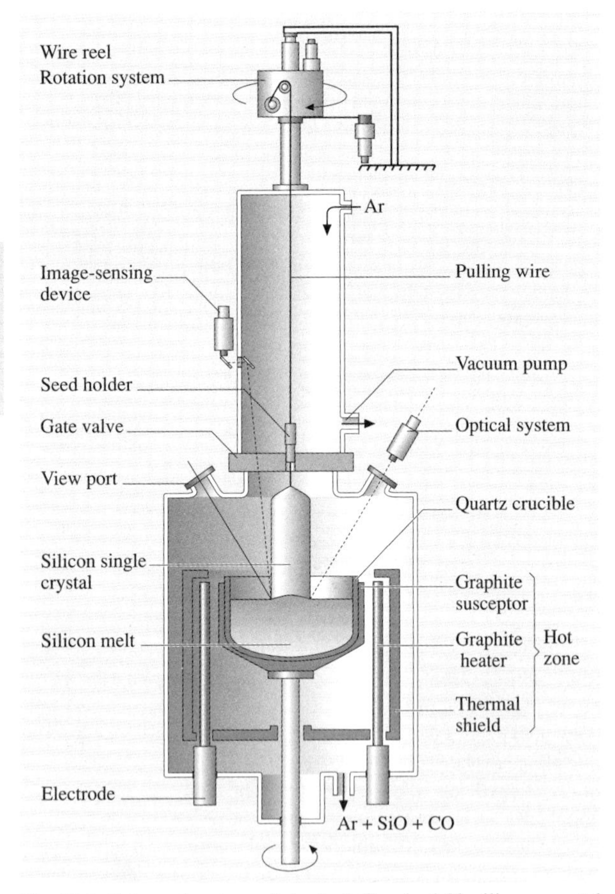

Fig. 13.6 Schematic view of typical Czochralski silicon crystal growing system. (After *Shimura* [13.1])

Figure 13.7 shows the seed-end part of an as-grown CZ silicon crystal. Although a seed-corn, which is the transition region from the seed to the cylindrical part, is usually formed to be rather flat for economic reasons, a more tapered shape might be desirable from a crystal quality point of view. The shoulder part and its vicinity should not be used for device fabrication because this part is considered a transition region

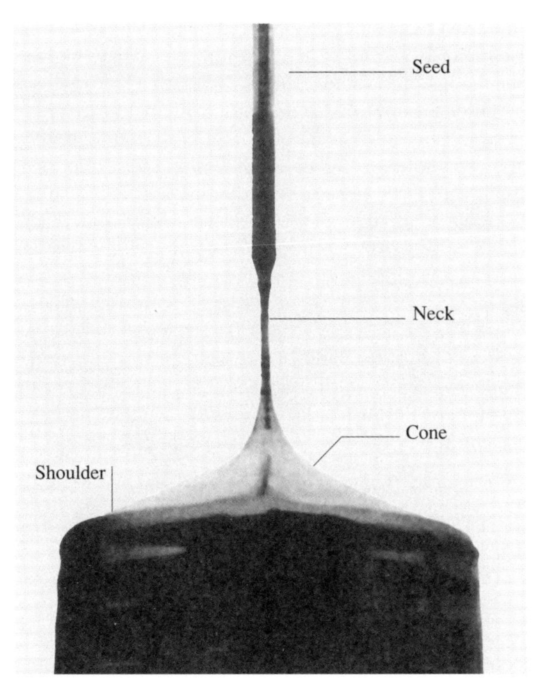

Fig. 13.7 Seed-end part of as-grown Czochralski silicon crystal

in many senses and it exhibits inhomogeneous crystal characteristics due to the abrupt change in growth conditions.

Figure 13.8 shows an extra-large as-grown CZ silicon crystal ingot 400 mm in diameter and 1800 mm in length grown by the Super Silicon Crystal Research Institute Corporation in Japan [13.2].

Influence of Spatial Location in a Grown Crystal

As Fig. 13.9 clearly depicts, each portion of a CZ crystal is grown at a different time with different growth conditions [13.19]. Thus, it is important to understand that each portion has a different set of crystal characteristics and a different thermal history due to its different position along the crystal length. For example, the seed-end portion has a longer thermal history, ranging from the melting point of 1420 °C to around 400 °C in a puller, while the tail-end portion has a shorter history and is cooled down rather rapidly from the melting point. Ultimately, each silicon wafer prepared from a different portion of a grown crystal could exhibit different physico-chemical characteristics depending on its location in the ingot. In fact, it has been reported that the oxygen precipitation behavior exhibits the greatest location dependence, which, in turn, affects the generation of bulk defects [13.20].

Also, a nonuniform distribution of both crystal defects and impurities occurs across the transverse section of a flat wafer prepared from a CZ crystal silicon melt

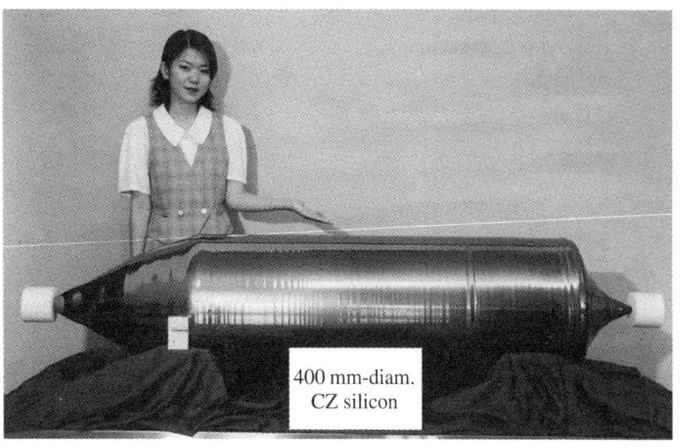

Fig. 13.8 Extra-large as-grown Czochralski silicon ingot 400 mm in diameter and 1800 mm in length. (Courtesy of Super Silicon Crystal Research Institute Corporation, Japan)

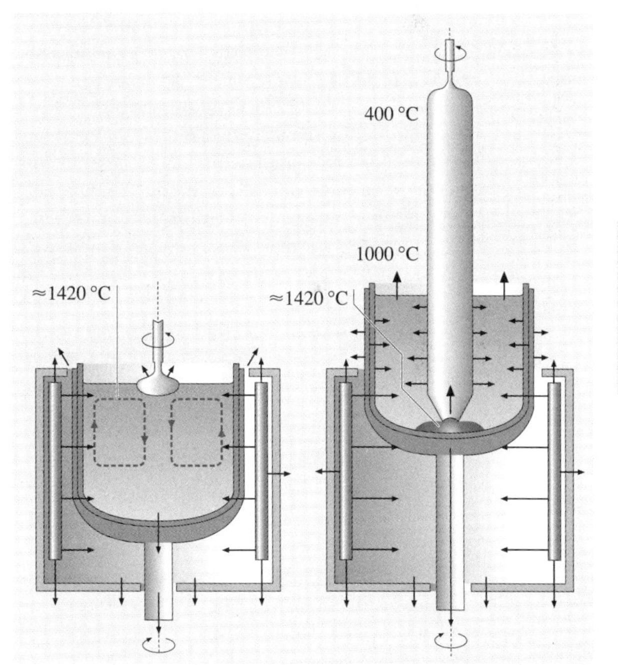

Fig. 13.9 Thermal environment during Czochralski crystal growth at initial and final stages. *Arrows* indicate approximate directions of heat flow. (After *Zulehner* and *Huber* [13.19])

crystallized or solidified successively at the crystal–melt interface, which is generally curved in the CZ crystal growth process. Such inhomogeneities can be observed as *striations*, which are discussed later.

13.3.3 Impurities in Czochralski Silicon

The properties of the silicon semiconductors used in electronic devices are very sensitive to impurities. Because of this sensitivity, the electrical/electronic properties of silicon can be precisely controlled by adding a small amount of dopant. In addition to this dopant sensitivity, contamination by impurities (particularly transition metals) negatively affects the properties of silicon and results in the serious degradation of device performance. Moreover, oxygen is incorporated at levels of tens of atoms per million into CZ silicon crystals due to the reaction between the silicon melt and the quartz crucible. Regardless of how much oxygen is in the crystal, the characteristics of silicon crystals are greatly affected by the concentration and the behavior of oxygen [13.21]. In addition, carbon is also incorporated into CZ silicon crystals either from polysilicon raw materials or during the growth process, due to the graphite parts used in the CZ pulling equipment. Although the concentration of carbon in commercial CZ silicon crystals is normally less than 0.1 ppma, carbon is an impurity that greatly affects the behavior of oxygen [13.22, 23]. Also, nitrogen-doped CZ silicon crystals [13.24, 25] have recently attracted much attention due to their high microscopic crystal quality, which may meet the requirements for state-of-the-art electronic devices [13.26, 27].

Impurity Inhomogeneity

During crystallization from a melt, various impurities (including dopants) contained in the melt are incorporated into the growing crystal. The impurity concentration of the solid phase generally differs from that of the liquid phase due to a phenomenon known as *segregation*.

Segregation. The equilibrium segregation behavior associated with the solidification of multicomponent systems can be determined from the corresponding phase diagram of a binary system with a *solute* (the impurity) and a *solvent* (the host material) as components.

The ratio of the solubility of impurity A in solid silicon $[C_A]_s$ to that in liquid silicon $[C_A]_L$

$$k_0 = [C_A]_s/[C_A]_L \tag{13.6}$$

is referred to as the *equilibrium segregation coefficient*. The impurity solubility in liquid silicon is always higher than that in solid silicon; that is, $k_0 < 1$.

The equilibrium segregation coefficient k_0 is only applicable to solidification at negligibly slow growth rates. For finite or higher solidification rates, impurity atoms with $k_0 < 1$ are rejected by the advancing solid at a greater rate than they can diffuse into the melt. In the CZ crystal growth process, segregation takes place at the start of solidification at a given seed–melt interface, and the rejected impurity atoms begin to accumulate in the melt layer near the growth interface and diffuse in the direction of the bulk of the melt. In this situation, an *effective segregation coefficient* k_{eff} can be defined at any moment during CZ crystal growth, and the impurity concentration $[C]_s$ in a CZ crystal can be derived by

$$[C]_s = k_{eff}[C_0](1-g)^{k_{eff}-1}\,, \tag{13.7}$$

where $[C_0]$ is the initial impurity concentration in the melt and g is the fraction solidified.

Consequently, it is clear that a macroscopic longitudinal variation in the impurity level, which causes a variation in resistivity due to the variation in the dopant concentration, is inherent to the CZ batch growth process; this is due to the segregation phenomenon. Moreover, the longitudinal distribution of impurities is influenced by changes in the magnitude and the nature of melt convection that occur as the melt aspect ratio is decreased during crystal growth.

Striations. In most crystal growth processes, there are transients in the parameters such as instantaneous microscopic growth rate and the diffusion boundary layer thickness which result in variations in the effective segregation coefficient k_{eff}. These variations give rise to microscopic compositional inhomogeneities in the form of *striations* parallel to the crystal–melt interface. Striations can be easily delineated with several techniques, such as preferential chemical etching and X-ray topography. Figure 13.10 shows the striations revealed by chemical etching in the shoulder part of a longitudinal cross-section of a CZ silicon crystal. The gradual change in the shape of the growth interface is also clearly observed.

Striations are physically caused by the segregation of impurities and also point defects; however, the striations are practically caused by temperature fluctuations near the crystal–melt interface, induced by unstable thermal convection in the melt and crystal rotation in an asymmetric thermal environment. In addition, mechanical vibrations due to poor pulling control mechanisms

Fig. 13.10 Growth striations, revealed by chemical etching, in a shoulder of Czochralski silicon

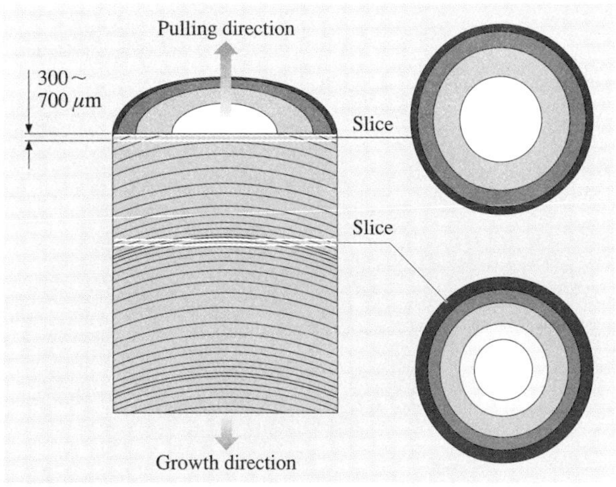

Fig. 13.11 Schematic illustration of Czochralski crystal cross-section containing a curved crystal–melt interface and planar wafers sliced into different portions. (After *Shimura* [13.1])

in the growth equipment can also cause temperature fluctuations.

Figure 13.11 schematically illustrates a CZ-grown crystal cross-section containing a curved crystal–melt interface, which results in inhomogeneities on the surface of a slice. As each planar wafer is sliced, it contains different portions of several curved striations. Different "phonograph rings", referred to as *swirl*, can then occur in each wafer, which can be observed across the wafer using the techniques mentioned above.

Doping

In order to obtain the desired resistivity, a certain amount of dopant (either donor or acceptor atoms) is added to a silicon melt according to the resistivity–concentration relation. It is common practice to add dopants in the form of highly doped silicon particles or chunks of about 0.01 Ω cm resistivity, which are called the dopant fixture, since the amount of pure dopant needed is unmanageably small, except for heavily doped silicon materials (n^+ or p^+ silicon).

The criteria for selecting a dopant for a semiconductor material are that it has the following properties; (1) suitable energy levels, (2) high solubility, (3) suitable or low diffusivity, and (4) low vapor pressure. A high diffusivity or high vapor pressure leads to undesirable diffusion or vaporization of dopants, which results in unstable device operation and difficulties in achieving precise resistivity control. A solubility that is too small limits the resistivity that can be obtained. In addition to those criteria, the chemical properties (the toxicity for example) must be considered. A further consideration from the viewpoint of crystal growth is that the dopant has a segregation coefficient that is close to unity in order to make the resistivity as uniform as possible from the seed-end to the tail-end of the CZ crystal ingot. Consequently, phosphorus (P) and boron (B) are the most commonly used donor and acceptor dopants for silicon, respectively. For n^+ silicon, in which donor atoms are heavily doped, antimony (Sb) is usually used instead of phosphorus because of its smaller diffusivity, in spite of its small segregation coefficient and high vapor pressure, which lead to large variations in concentration in both the axial and the radial directions.

Oxygen and Carbon

As shown schematically in Figs. 13.3b and 13.6, a quartz (SiO_2) crucible and graphite heating elements are used in the CZ-Si crystal growth method. The surface of the crucible that contacts the silicon melt is gradually dissolved due to the reaction

$$SiO_2 + Si \rightarrow 2SiO\,. \tag{13.8}$$

This reaction enriches the silicon melt with oxygen. Most of the oxygen atoms evaporate from the melt surface as volatile silicon mono-oxide (SiO), but some of them are incorporated into a silicon crystal through the crystal–melt interface.

However, the carbon in CZ silicon crystals originates mainly from the polycrystalline starting material. Levels

Part B | 13.3

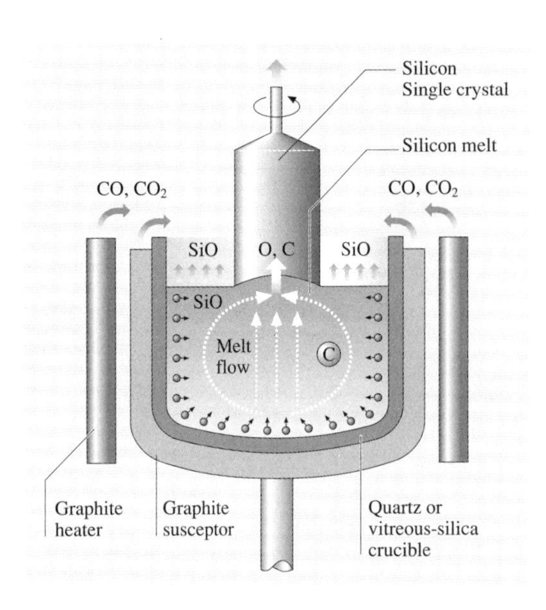

Fig. 13.12 Incorporation of oxygen and carbon into Czochralski silicon crystal. (After *Shimura* [13.1])

of carbon ranging from 0.1 to 1 ppma, depending on the manufacturer, are found in the polysilicon. Sources of carbon in polysilicon are assumed to be mainly carbon-containing impurities found in the trichlorosilane used in the production of polysilicon. Graphite parts in CZ pulling equipment can also contribute to carbon contamination by reacting with oxygen, which is always present during the ambient growth. The resulting products of CO and CO_2 are dissolved into the silicon melt and account for the carbon impurities in silicon crystals. Thus, oxygen and carbon are the two major nondoping impurities that are incorporated into CZ silicon crystals in the way shown schematically in Fig. 13.12. The behavior of these impurities in silicon, which affect a number of the properties of CZ silicon crystals, has been the subject of intensive study since the late 1950s [13.21].

13.4 New Crystal Growth Methods

Silicon crystals used for microelectronic device fabrication must meet a variety of requirements set by device manufacturers. In addition to the requirements for silicon *wafers*, the following crystallographic demands have become more common due to high-yield and high-performance microelectronic device manufacturing.

1. large diameter
2. low or controlled defect density
3. uniform and low radial resistivity gradient
4. optimum initial oxygen concentration and its precipitation.

It is clear that silicon crystal manufacturers must not only meet the above requirements but also produce those crystals economically and with high manufacturing yields. The main concerns of silicon crystal growers are the crystallographic perfection and the axial distribution of dopants in CZ silicon. In order to overcome some problems with the conventional CZ crystal growth method, several new crystal growth methods have been developed.

13.4.1 Czochralski Growth with an Applied Magnetic Field (MCZ)

The melt convection flow in the crucible strongly affects the crystal quality of CZ silicon. In particular, unfavorable growth striations are induced by unsteady melt convection resulting in temperature fluctuations at the growth interface. The ability of a magnetic field to inhibit thermal convection in electrically conducting fluid was first applied to the crystal growth of indium antimonide via the horizontal boat technique [13.28] and the horizontal zone-melting technique [13.29]. Through these investigations, it was confirmed that a magnetic field of sufficient strength can suppress the temperature fluctuations that accompany melt convection, and can dramatically reduce growth striations.

The effect of the magnetic field on growth striations is explained by its ability to decrease the turbulent thermal convection of a melt and in turn decrease the temperature fluctuations at the crystal–melt interface. The fluid flow damping caused by the magnetic field is due to the induced magnetomotive force when the flow is orthogonal to the magnetic flux lines, which results in an increase in the effective kinematic viscosity of the conducting melt.

Silicon crystal growth by the magnetic field applied CZ (MCZ) method was reported for the first time in 1980 [13.30]. Originally MCZ was intended for the growth of CZ silicon crystals that contain low oxygen concentrations and therefore have high resistivities with low radial variations. In other words, MCZ silicon was expected to replace the FZ silicon almost exclusively used for power device fabrication. Since then, various magnetic field configurations, in terms of the magnetic field direction (horizontal or vertical) and the type of magnets used (normal conductive or superconductive), have been developed [13.31]. MCZ silicon produced with a wide range of desired oxygen concentrations (from low to high) has been of great interest for different device applications. The value of MCZ silicon lies in its high quality and its ability to control the oxygen concentration over a wide range, which cannot be achieved using the conventional CZ method [13.32], as well as its enhanced growth rate [13.33].

As far as the crystal quality is concerned, there is no doubt that the MCZ method provides the silicon crystals most favorable to the semiconductor device industry. The production cost of MCZ silicon may be higher than that of conventional CZ silicon because the MCZ method consumes more electrical power and requires additional equipment and operating space for the electromagnets; however, taking into account the higher growth rate of MCZ, and when superconductive magnets that need smaller space and consume less electrical power compared with conductive magnets are used, the production cost of MCZ silicon crystals may become comparable to that of conventional CZ silicon crystals. In addition, the improved crystal quality of MCZ silicon may increase production yields and lower the production cost.

13.4.2 Continuous Czochralski Method (CCZ)

Crystal production costs depend to a large extent on the cost of materials, in particular the cost of those used for quartz crucibles. In the conventional CZ process, called a *batch process*, a crystal is pulled from a single crucible charge, and the quartz crucible is used only once and is then discarded. This is because the small amount of remaining silicon cracks the crucible as it cools from a high temperature during each growth run.

One strategy for replenishing a quartz crucible with melt economically is to continuously add feed as the crystal is grown and thereby maintain the melt at a constant volume. In addition to saving crucible costs, the continuous-charging Czochralski (CCZ) method provides an ideal environment for silicon crystal growth. As already mentioned, many of the inhomogeneities in crystals grown by the conventional CZ batch process are a direct result of the unsteady kinetics arising from the change in melt volume during crystal growth. The CCZ method aims not only to reduce production costs but also to grow crystals under steady conditions. By maintaining the melt volume at a constant level, steady thermal and melt flow conditions can be achieved (see Fig. 13.9, which shows the change in thermal environments during conventional CZ growth).

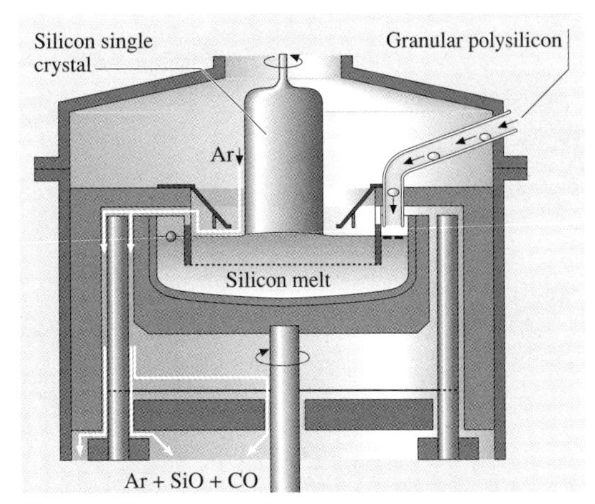

Fig. 13.13 Schematic illustration of the continuous-charging Czochralski method. (After *Zulehner* [13.34])

Continuous charging is commonly performed by polysilicon feeding, as shown in Fig. 13.13 [13.34]. This system consists of a hopper for storing the polysilicon raw material and a vibratory feeder that transfers the polysilicon to the crucible. In the crucible that contains the silicon melt, a quartz baffle is required to prevent the melt turbulance caused by feeding in the solid material around the growth interface. Free-flowing polysilicon granules such as those mentioned previously are obviously advantageous for the CCZ method.

The CCZ method certainly solves most of the problems related to inhomogeneities in crystal grown by the conventional CZ method. Moreover, the combination of MCZ and CCZ (the magnetic-field-applied continuous CZ (MCCZ) method) is expected to provide the ultimate crystal growth method, giving ideal silicon crystals for a wide variety of microelectronic applications [13.1]. Indeed, it has been used to grow high-quality silicon crystals intended for microelectronic devices [13.35].

However, it should be emphasized that the different thermal histories of different parts of the crystal (from the seed to the tail ends, as shown in Fig. 13.9) must be considered even when the crystal is grown by the ideal growth method. In order to homogenize the grown crystal or to obtain axial uniformity in the thermal history, some form of post-treatment, such as high-temperature annealing [13.36], is required for the crystal.

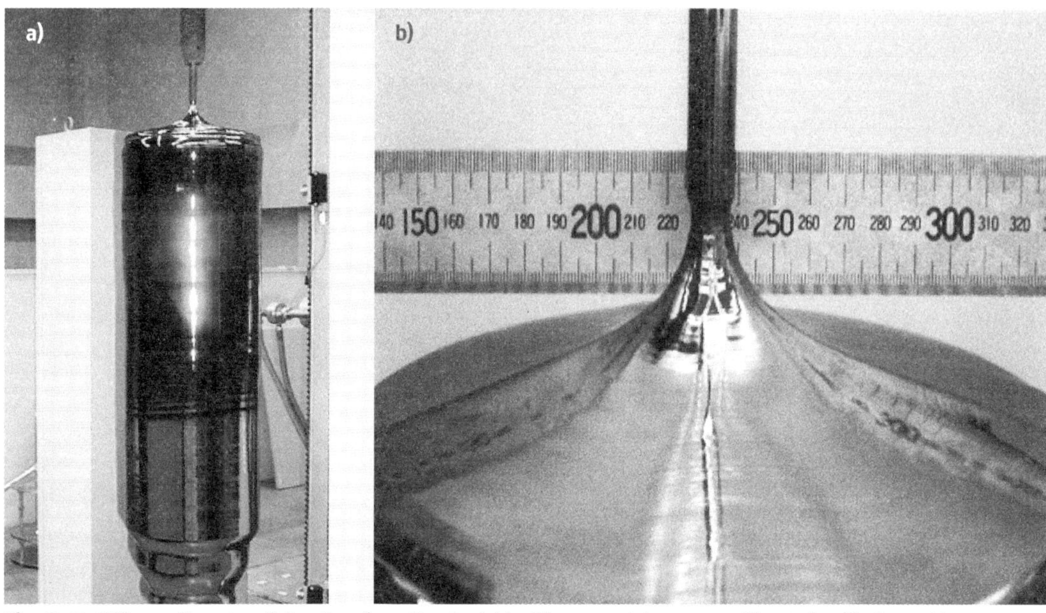

Fig. 13.14 200 mm-diameter dislocation-free Czochralski silicon crystal grown without the Dash necking process. (Courtesy of Prof. K. Hoshikawa)

13.4.3 Neckingless Growth Method

As mentioned previously, Dash's necking process (which grows a thin neck 3–5 mm in diameter, Fig. 13.7) is a critical step during CZ crystal growth because it eliminates grown-in dislocations. This technique has been the industry standard for more than 40 years. However, recent demands for large crystal diameters (> 300 mm, weighing over 300 kg) have resulted in the need for larger diameter necks that do not introduce dislocations into the growing crystal, since a thin neck 3–5 mm in diameter can not support such large crystals.

Large diameter seeds that are typically 170 mm long, with a minimum diameter of > 10 mm and an average of 12 mm grown from silicon melt heavily doped with boron (> 10^{19} atoms/cm^3) have been used to grow dislocation-free 200 mm-diameter CZ silicon crystals [13.37, 38]. It is estimated that large diameter necks 12 mm in diameter can CZ support crystals as heavy as 2000 kg [13.39]. Figure 13.14a shows a 200 mm-diameter dislocation-free CZ silicon crystal grown without the Dash necking process, and Fig. 13.14b shows its enlarged seed (compare with Fig. 13.7). The mechanism by which dislocations are not incorporated into the growing crystal has been primarily attributed to hardening effect of the heavy doping of boron in the silicon.

References

13.1 F. Shimura: *Semiconductor Silicon Crystal Technology* (Academic, New York 1988)

13.2 K. Takada, H. Yamagishi, H. Minami, M. Imai: *Semiconductor Silicon* (The Electrochemical Society, Pennington 1998) p. 376

13.3 W. C. Dash: J. Appl. Phys. **29**, 736 (1958)

13.4 J. R. McCormic: *Semiconductor Silicon* (The Electrochemical Society, Pennington 1986) p. 43

13.5 P. A. Taylor: Solid State Technol. **53** (1987)

13.6 W. G. Pfann: Trans. Am. Inst. Min. Metall. Eng. **194**, 747 (1952)

13.7 C. H. Theuere: U.S. Patent 3,060,123,(1962)

13.8 P. H. Keck, M. J. E. Golay: Phys. Rev. **89**, 1297 (1953)

13.9 W. Keller, A. Muhlbauer: *FLoating-Zone Silicon* (Marcel Dekker, New York 1981)

13.10 J. M. Meese: *Neutron Transmutation Doping in Semiconductors* (Plenum, New York 1979)
13.11 H. M. Liaw, C. J. Varker: *Semiconductor Silicon* (The Electrochemical Society, Pennington 1977) p. 116
13.12 E. L. Kern, L. S. Yaggy, J. A. Barker: *Semiconductor Silicon* (The Electrochemical Society, Pennington 1977) p. 52
13.13 S. M. Hu: Appl. Phys. Lett **31**, 53 (1977)
13.14 K. Sumino, H. Harada, I. Yonenaga: Jpn. J. Appl. Phys. **19**, L49 (1980)
13.15 K. Sumino, I. Yonenaga, A. Yusa: Jpn. J. Appl. Phys. **19**, L763 (1980)
13.16 T. Abe, K. Kikuchi, S. Shirai: *Semiconductor Silicon* (The Electrochemical Society, Pennington 1981) p. 54
13.17 J. Czochralski: Z. Phys. Chem **92**, 219 (1918)
13.18 G. K. Teal, J. B. Little: Phys. Rev. **78**, 647 (1950)
13.19 W. Zulehner, D. Hibber: *Crystals 8: Silicon, Chemical Etching* (Springer, Berlin, Heidelberg 1982) p. 1
13.20 H. Tsuya, F. Shimura, K. Ogawa, T. Kawamura: J. Electrochem. Soc. **129**, 374 (1982)
13.21 F. Shimura: *Oxygen In Silicon* (Academic, New York 1994) pp. 106, 371
13.22 S. Kishino, Y. Matsushita, M. Kanamori: Appl. Phys. Lett **35**, 213 (1979)
13.23 F. Shimura: J. Appl. Phys **59**, 3251 (1986)
13.24 H. D. Chiou, J. Moody, R. Sandfort, F. Shimura: *VLSI Science and Technology* (The Electrochemical Society, Pennington 1984) p. 208
13.25 F. Shimura, R. S. Hocket: Appl. Phys. Lett **48**, 224 (1986)
13.26 A. Huber, M. Kapser, J. Grabmeier, U. Lambert, W. v. Ammon, R. Pech: *Semiconductor Silicon* (The Electrochemical Society, Pennington 2002) p. 280
13.27 G. A. Rozgonyi: *Semiconductor Silicon* (The Electrochemical Society, Pennington 2002) p. 149
13.28 H. P. Utech, M. C. Flemings: J. Appl. Phys. **37**, 2021 (1966)
13.29 H. A. Chedzey, D. T. Hurtle: Nature **210**, 933 (1966)
13.30 K. Hoshi, T. Suzuki, Y. Okubo, N. Isawa: *Ext. Abstr. Electrochem. Soc. 157th Meeting* (The Electrochemical Society, Pennington 1980) p. 811
13.31 M. Ohwa, T. Higuchi, E. Toji, M. Watanabe, K. Homma, S. Takasu: *Semiconductor Silicon* (The Electrochemical Society, Pennington 1986) p. 117
13.32 M. Futagami, K. Hoshi, N. Isawa, T. Suzuki, Y. Okubo, Y. Kato, Y. Okamoto: *Semiconductor Silicon* (The Electrochemical Society, Pennington 1986) p. 939
13.33 T. Suzuki, N. Isawa, K. Hoshi, Y. Kato, Y. Okubo: *Semiconductor Silicon* (The Electrochemical Society, Pennington 1986) p. 142
13.34 W. Zulehner: *Semiconductor Silicon* (The Electrochemical Society, Pennington 1990) p. 30
13.35 Y. Arai, M. Kida, N. Ono, K. Abe, N. Machida, H. Futuya, K. Sahira: *Semiconductor Silicon* (The Electrochemical Society, Pennington 1994) p. 180
13.36 F. Shimura: *VLSI Science and Technology* (The Electrochemical Society, Pennington 1982) p. 17
13.37 S. Chandrasekhar, K. M. Kim: *Semiconductor Silicon* (The Electrochemical Society, Pennington 1998) p. 411
13.38 K. Hoshikawa, X. Huang, T. Taishi, T. Kajigaya, T. Iino: Jpn. J. Appl. Phys **38**, L1369 (1999)
13.39 K. M. Kim, P. Smetana: J. Cryst. Growth **100**, 527 (1989)

14. Epitaxial Crystal Growth: Methods and Materials

The epitaxial growth of thin films of material for a wide range of applications in electronics and optoelectronics is a critical activity in many industries. The original growth technique used, in most instances, was liquid-phase epitaxy (LPE), as this was the simplest and often the cheapest route to producing device-quality layers. These days, while some production processes are still based on LPE, most research into and (increasingly) much of the production of electronic and optoelectronic devices now centers on metalorganic chemical vapor deposition (MOCVD) and molecular beam epitaxy (MBE). These techniques are more versatile than LPE (although the equipment is more expensive), and they can readily produce multilayer structures with atomic-layer control, which has become more and more important in the type of nanoscale engineering used to produce device structures in as-grown multilayers. This chapter covers these three basic techniques, including some of their more common variants, and outlines the relative advantages and disadvantages of each. Some examples of growth in various important systems are also outlined for each of the three techniques.

This chapter outlines the three major epitaxial growth processes used to produce layers of material for electronic, optical and optoelectronic applications. These are liquid-phase epitaxy (LPE), metalorganic chemical vapor deposition (MOCVD) and molecular beam epitaxy (MBE). We will also consider their main variants. All three techniques have advantages and disadvantages when applied to particular systems, and these will be highlighted where appropriate in the following sections.

14.1 Liquid-Phase Epitaxy (LPE)

14.1.1 Introduction and Background

Liquid-phase epitaxy (LPE) is a mature technology and has unique features that mean that it is still applicable for use in niche applications within certain device technologies. It has given way in many areas, however, to various vapor-phase epitaxy techniques, such as metalorganic vapor phase, molecular beam and atomic layer epitaxies (MOVPE, MBE, ALE), see Sects. 14.2 and 14.3. When selecting an epitaxial growth technology for a particular material system and/or device application, the choice needs to take into account the basic principles of thermodynamics, kinetics, surface energies, and so on, as well as practical issues of reproducibility, scalability, process control, instrumentation, safety and capital equipment costs. A systematic comparison of the various epitaxy techniques suggests that no single technique can best satisfy the needs of all of the material/device combinations needed in microelectronics, optoelectronics, solar cells, thermophotovoltaics, thermoelectrics, semiconductor electrochemical devices, magnetic devices and microelectromechanical systems. LPE is still a good choice for many of these application areas (M. Mauk, private communication, 2004).

14.1.2 History and Status

LPE is basically a high-temperature solution growth technique [14.1] in which a thin layer of the required material is deposited onto a suitable substrate. Homoepitaxy is defined as growth of a layer of the same composition as the substrate, whereas heteroepitaxy is the growth of a layer of markedly different composition. A suitable substrate material would have the same crystal structure as the layer, have as close a match in terms of lattice parameters as possible and be chemically compatible with the solution and the layer. *Nelson* [14.2] is commonly thought to have developed the first LPE systems, in this case for producing multilayer compound semiconducting structures. In the following decades a large technology base was established for III–V compound semiconductor lasers, LEDs, photodiodes and solar cells. LPE has been applied to the growth of Si, Ge, SiGe alloys, SiC, GaAs, InP, GaP, GaSb, InAs, InSb (and their ternary and quaternary alloys), GaN, ZnSe, CdHgTe, HgZnTe and HgMnTe. It has also been used to produce a diverse range of oxide/fluoride compounds, such as high-temperature superconductors, garnets, para- and ferroelectrics and for various other crystals for optics and magnetics. The early promise of garnet materials for making 'bubble' memories was not fully realised as standard semiconductor memory was more commercially viable. Dipping LPE is still used to make magneto-optical isolators by epitaxially growing garnet layers on gadolinium gallium garnet substrates.

It is probably true to say that most of these systems were first studied using LPE, where it was used in the demonstration, development and commercialization of many device types, including GaAs solar cells, III–V LEDs and laser diodes, GaAs-based Gunn-effect and other microwave devices and various IR detectors based on InSb and on CdHgTe. Nevertheless, LPE does not appear in the research literature as often as, say, MOVPE, MBE and ALE in reference to work in these systems. However, it is still used extensively in industrial applications, including III–V LEDs, particularly those based on AlGaAs and GaP alloys, where it is ideally suited to the small die areas, the high luminescence efficiencies and the relatively simple device structures needed, and IR detectors based on CdHgTe.

Realistic industrial production data is difficult to obtain, but *Moon* [14.3] noted that the large majority of optoelectronic devices were still being grown by LPE at that time, amounting to $\approx 4000\,m^2$ per year. He also estimated that despite the loss of market share to more advanced techniques, the total demand for LPE material was still increasing at $\approx 10\%$ per year. LPE was discontinued for many applications because of its perceived limitations in regard to control of layer thickness, alloy compositions, doping, interface smoothness and difficulties in growing certain combinations of interest for heterostructure devices. LPE is normally dismissed for the production of superlattices, quantum wells, strained-layer structures and heterojunctions with large lattice mismatches of chemical dissimilarities. It also suffers from a reputation for poor reproducibility, problems with scaling up in size or throughput, and difficulties in achieving abrupt interfaces between successive layers within structures.

14.1.3 Characteristics

LPE is characterized as a near-equilibrium growth process, when compared to the various vapor-phase epitaxy techniques. Heat and mass transport, surface energies, interface kinetics and growth mechanisms are different in LPE compared to those in vapor-phase epitaxy or bulk growth techniques. These features result in both advantages and disadvantages for LPE. The former include:

- High growth rates. These are typically $0.1-10\,\mu m/h$, i.e. faster than in MOVPE or MBE. This feature is useful when thick layers or "virtual substrates" are required.
- Favorable segregation of impurities into the liquid phase. This can lead to lower residual or background impurities in the epitaxial layer.
- Ability to produce very flat surfaces and excellent structural perfection (Fig. 14.1).
- Wide selection of dopants. Most solid or liquid elements can be added to a melt and incorpo-

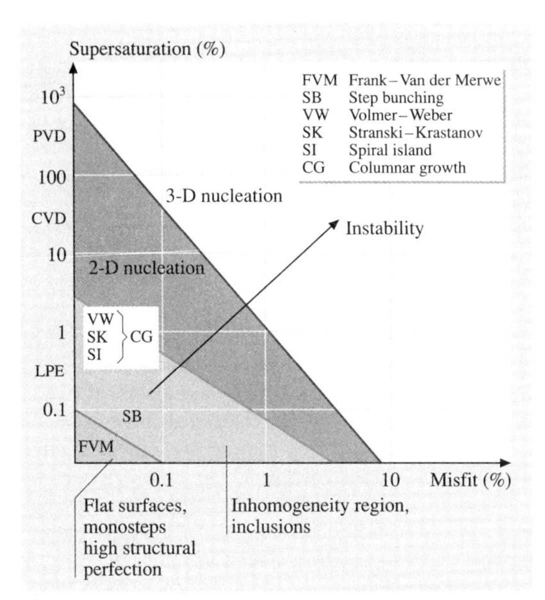

Fig. 14.1 The effects of both supersaturation (and the growth method) and misfit on the nucleation and growth regimes. Only at very low supersaturation in LPE using low-misfit substrates can really flat surfaces be expected. (After [14.4])

rated in the layer, unlike in vapor-phase growth where the development of volatile dopant precursors with suitable kinetics and sticking coefficients is a major undertaking. In this regard, there is work underway on rare-earth doping of semiconductor layers to exploit their gettering and optical properties.

- Suppression of certain types of defects. In general, LPE material has lower point defects (vacancies, interstitials, antisites) than material made by other techniques. For example, the Ga-rich conditions during GaAs LPE inhibits the formation of the As antisite defect that is responsible for the nonradiative losses in luminescent devices.
- Once the relevant phase diagram is established, growth can be made to occur over a wide range of temperatures.
- Absence of highly toxic precursors or byproducts.
- Low capital equipment and operating costs. A research LPE kit can be constructed for under $50 000.

The main consideration when designing an LPE process is to determine accurate phase equilibria (S–L and/or S–L–V) of the required system. Solution modeling, extrapolations of existing phase equilibria and semi-empirical predictions are usually sufficient to guide developments in new systems/applications. The near-equilibrium nature of LPE provides for several important growth modes, such as selective epitaxy (deposition through masks on a substrate) and epitaxial layer overgrowth (ELOG, where growth over a mask occurs), which are useful for defect reduction and new device structures. These new areas include work on the currently important growth of SiC and GaN for diode applications.

14.1.4 Apparatus and Techniques

The basic requirement is to bring the substrate and growth solution into contact while the epitaxial layer grows, and then to separate them cleanly at the end of the growth cycle. The three main embodiments of the LPE growth method are tipping, dipping and sliding boat, see Fig. 14.2.

Figure 14.3 shows the tipping furnace system used by *Nelson* [14.2]. The boat, normally graphite or silica, sits in the work tube in the center of the tilted furnace such that the substrate, held with a clamp, is held at one end of the boat with the growth solution at the other end. Once the melt has been equilibrated the temperature is

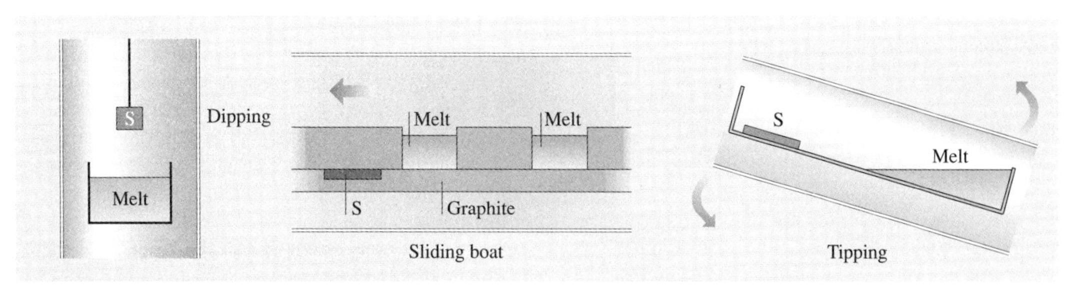

Fig. 14.2 Dipping, sliding boat and tipping LPE arrangements. (After [14.5])

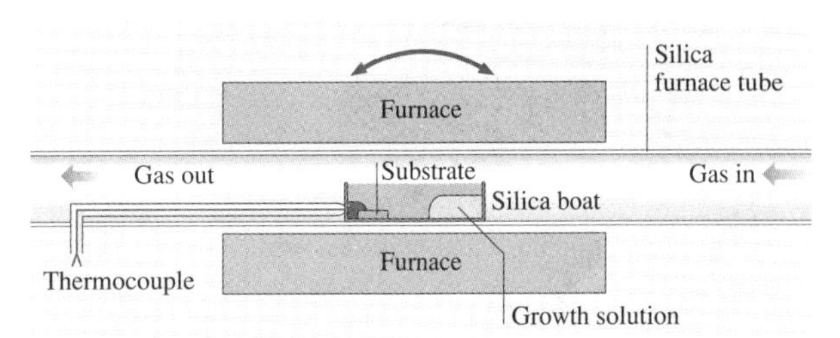

Fig. 14.3 Tipping LPE furnace. (After [14.2])

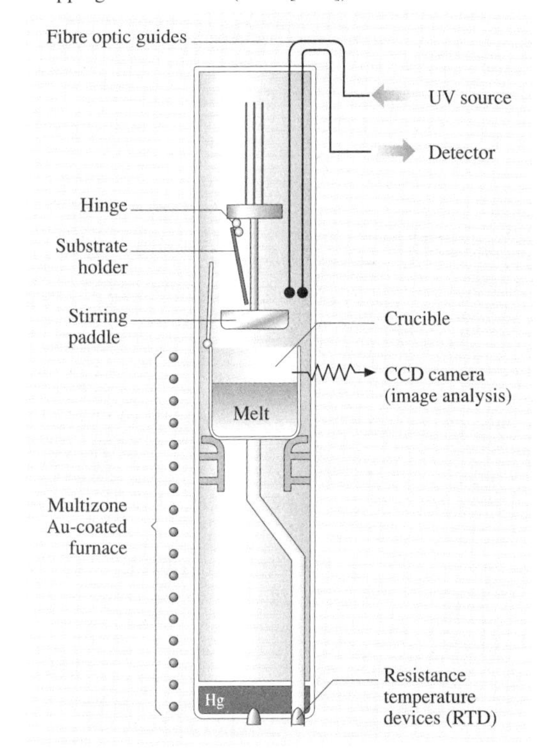

Fig. 14.4 Schematic diagram of a dipping LPE reactor showing the Te-rich melt, the mercury reservoir and positions of the sensors. (After [14.5])

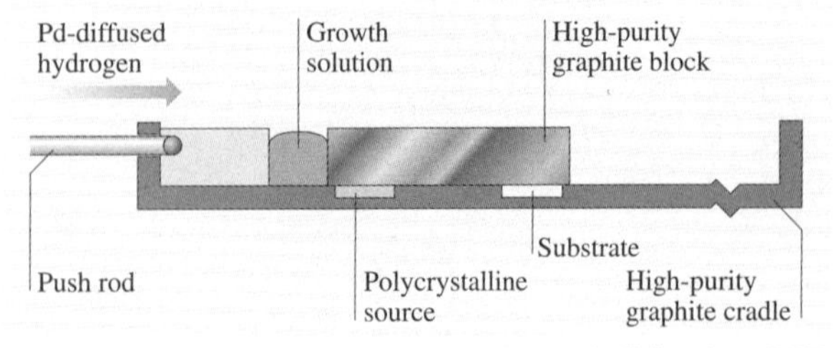

Fig. 14.5 Basic structure of graphite horizontal sliding-boat LPE. (After [14.6])

slowly reduced and the furnace is tipped to roll the solution over the substrate. After a suitable time the furnace is tipped back to the start position and the solution rolls off the grown layer. This is a relatively simple and cheap technique but has the limitations that solution removal is difficult and it is normally only suitable for single-layer growth.

Figure 14.4 shows the dipping system used for the growth of CdHgTe [14.5]. The design and operation of a system for growth from Hg-rich melts is dominated by consideration of the high vapor pressure of Hg that comprises $\geq 90\%$ of the growth solution, which led to the evolution of a vertical high-pressure furnace design with a cooled reflux region. The cylindrical melt vessel consists of a high-strength stainless steel chamber lined with quartz. Such systems are capable of containing about 10–20 kg of melt at 550 °C for several years with no degradation in melt integrity or purity. The prepared substrates are introduced into the melt through a transfer chamber or air lock. The paddle assembly can be lowered into the melt and rotated to stir the melt. In general, the high-purity melt components are introduced into the clean melt vessel at room temperature. The system is sealed, evacuated and pressurized. The temperature of the furnace is raised above the predicted melting point and held constant until all the solute dissolves. The use of large melts results in a near-constant saturation temperature from run to run and ensures excellent reproducibility of layer characteristics.

Figure 14.5 shows the basic structure of a graphite sliding-boat system, which has turned out to be the most popular and versatile of the three main methods [14.6]. The substrate sits in a recess in a slider supported by a base section. Growth solutions reside in wells in the upper section of the boat and can be repositioned over the substrate using a push-rod arrangement. One of the main drawbacks of this method is that of melt retention on the grown layer. Various means, such as empty wells, slots, lids on the solutions, and pistons to tilt the substrate have been tried with varying degrees of success. The critical design feature of the boat to aid wipe-off is to control the gap between the top of the grown layer and the underside of the top section. If this is too large, melt retention occurs, but if it is too small the layer may be scratched. Multilayer growth is easily possible using this sliding-boat method, providing melt retention is kept to a minimum. The thin melts lead to suppression of thermal or solutal convection, and hence reduce enhanced edge growth. Scale-up has also been achieved in this method with several substrates (up to 16) growing three-layer GaAlAs structures in a single run.

In all of the LPE methods, production of supersaturation in the growth solution drives the deposition of the layer on the substrate. This supersaturation can be produced by ramp cooling, step cooling, supercooled growth (a hybrid of the previous two techniques), two-phase growth, constant-temperature growth or transient growth [14.7]. The choice between these various means will depend on the details of the particular material system and the precise requirement for the material. An additional means of producing the required supersaturation is that of electroepitaxy, in which an electric current is passed through the interface to stimulate layer growth. It is now thought that this occurs via an electromigration process rather than via Peltier cooling. Benefits claimed for the technique include reduced surface ripple, a reduced number of certain microdefects and an ability to grow millimeter-thick layers of GaAlAs with uniform composition.

14.1.5 Group IV

Silicon and Silicon/Germanium

Ciszek [14.9] noted that high-quality Si layers have been grown on Si substrates at temperatures in the range 700–900 °C at a rate of 1 μm/min. The potential application was for solar cells, but because growth was on silicon, rather than a low-cost alternative, this is not considered to be a viable production process.

Alonso and *Winer* [14.10] grew SiGe alloys of various compositions from Si–Ge–In melts at temperatures between 640 and 900 °C. Layers were 1–5 μm thick and were used to study Raman spectra features seen in material grown by MBE. The advantage of LPE-grown material was thought to lie in the random distribution of Si and Ge atoms (no ordering is present) compared to the MBE-grown material. The authors were able to show that the Raman peaks seen in MBE-grown material were not due to ordering; rather they were due to optical phonons associated with Si–Si motion. This demonstrates the benefits that a near-equilibrium growth process can have when studying material grown by 'nonequilibrium' techniques.

Silicon Carbide

Dmitriev [14.8, 11] has described the production of high-quality 6H-SiC and 4H-SiC p-n junctions by LPE from Si melts. Layer thicknesses range from 0.2 to 100 μm with growth rates of 0.01–2 μm/min. Nitrogen is used as the donor impurity and aluminium, gallium and boron as acceptor impurity elements. The material showed high carrier mobility and low deep-center concentrations. Initial attempts used a technique where molten silicon ran from an upper section of the crucible to a lower section where the SiC substrates were held fixed. Dipping was also used in an attempt to grow material that was less stressed by Si melt solidification. Growth temperatures were 1500–1750 °C and layer thicknesses were 20–40 μm. Both of these techniques produced material that was successfully used to make blue LEDs. A new version of LPE, so-called container-free LPE (CFLPE) based on the electromagnetic crucible technique, was also developed, see Fig. 14.6. Liquid metal (molten Si) is suspended in a high-frequency electromagnetic field at 1000 °C and the substrates, SiC, are placed on top of the melt after heating to 1450 °C. A source of SiC is also placed at the bottom of the Si melt. Cooling of the solution was used to produce the epitaxial layer and the

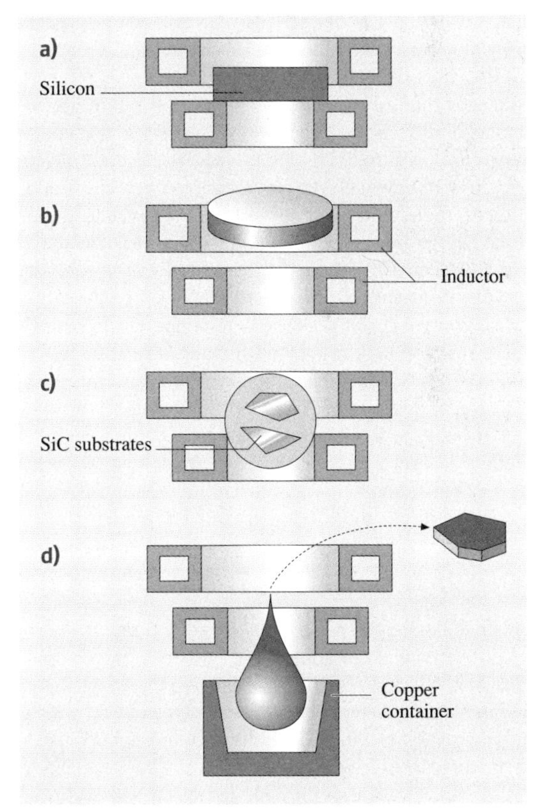

Fig. 14.6a–d Schematic of CFLPE: 1 – silicon, 2 – inductor, 3 – SiC substrate, 4 – copper container; **(a)** solid Si in copper inductor, **(b)** solid Si suspended in the inductor, **(c)** liquid Si with SiC crystals suspended in the electromagnetic field, **(d)** Si is dropped into the container. (After [14.8])

samples were then removed from the melt. The technique succeeded in producing p–n junctions by growing both layers in the same run. These formed the basis of several types of high-temperature devices (up to 500 °C), including diodes, FETs, bipolar transistors and dynistors, and optoelectronic devices such as green, blue, violet and UV LEDs. Reductions in melt temperature have been attempted by adding Sn, Ge and Ga to Si melts, with some success being reported for the latter; growth at 1100–1200 °C was obtained.

14.1.6 Group III–V

Arsenic- and Phosphorus-Based Materials

The majority of work in the area of III–V growth has been on GaAs and GaP, plus additions of As and Al. Following the earlier treatment by *Elwell* and *Scheel* [14.1], *Astles* [14.7] gave a comprehensive treatment of the LPE growth of GaAs and other III–V binaries and ternaries. He lists the advantages of LPE as: high luminescence efficiency due to the low concentration of nonradiative centers and deep levels, growth of ternary and quaternary alloys, controlled p- and n-type doping, multilayer growth with low interface recombination velocities and good reproducibility and uniformity. Disadvantages included: large areas that are required to be free of surface features (such as for photocathodes or ICs), very abrupt control of doping/composition profiles is required (as for microwave devices), accurate thickness control is required (as for microwave and quantum-well devices), and compositional grading between the substrate and the layer is inevitable. A problem associated with the use of phosphorus-containing substrates is the need to provide an overpressure source or a dummy solution to prevent phosphorus loss during the pregrowth phases.

All of the methods outlined above were attempted for the growth of GaAs and related materials. In addition, because LPE is a near-equilibrium technique that uses low supersaturation, nucleation is very sensitive to substrate lattice parameters and the growth rate is influenced by the substrate orientation. This enables localized growth in windows on the substrate surface and growth on nonplanar substrates with ribs or channels produced by preferential etching. The latter feature has been used to produce novel laser structures.

In fact, a vast array of both optoelectronic and microwave devices have been produced in LPE GaAs and related materials. The earliest were the GaAs Gunn devices and GaP/GaAsP LEDs. Later, GaAs/GaAlAs heterojunctions were produced for use in lasers, photocathodes and solar cells. Other alloy systems, such as GaInP for blue/green LEDs, GaInSb for improved Gunn devices, and GaInAs or GaAsSb for photocathodes were also studied. Later still came growth of ternaries, such as GaInAsP (lattice-matched to InP) for heterostructure optoelectronic devices. Finally, OEICs and buried heterostructure lasers were developed to exploit the potential for selective-area growth and anisotropy of growth rate.

III–V Antimonides

Commercially available substrates for epitaxy are limited in their lattice constant spread and this imposes certain constraints in terms of lattice-matched growth and miscibility gaps. Ternary and quaternary alloy substrates with adjustable lattice parameters would open up new device applications. However, bulk-grown ternary alloys suffer from segregation and stress effects. An alternative approach is to grow very thick layers ($> 50\,\mu m$) of these compounds for use as 'virtual substrates', *Mao* and *Krier* [14.12]. For III–V antimonides, where substrate and lattice-matching problems are acute, such thick layers are feasible by LPE due to the relatively fast growth rates ($1-10\,\mu m/min$). Either gradual compositional grading or growing multilayers with abrupt but incremental compositional changes between layers can by combined with either selective removal of the substrate (to produce free-standing layers) or wafer-bonding techniques, yielding an alloy layer bonded to a surrogate substrate. The challenge for these virtual substrates is to produce lattice constants that are sufficiently different from those available using binary substrates, without introducing an excessive level of defects.

Another interesting application of antimonides is that of InSb-based quantum dots, *Krier* et al. [14.13]. The potential application here is in mid-IR lasers, LEDs and detectors. In particular, there is a market for these materials as gas detectors based on IR absorption. The principle is that of rapid slider LPE, in which a thin slit of melt is wiped across the substrate producing contact times of 0.5–200 ms. This produces low-dimensional structures such as quantum wells and quantum dots. InSb quantum dots were grown on InAs substrates at 465 °C with 10 °C supercooling and a 1 ms melt–substrate contact time. Both small (4 nm high and 20 nm in diameter) and large quantum dots (12 nm high and 60 nm in diameter) are produced. Extensions to this work included growing InSb dots on GaAs and InAsSb dots. Photoluminescence and electroluminescence in the mid-IR region ($\approx 4\,\mu m$) were observed in these dots.

A Japanese group [14.14] is pioneering a technique called melt epitaxy, which can be viewed as a variant of LPE. A sliding-boat arrangement rapidly solidifies a ternary melt into a $\approx 300\,\mu m$-thick ternary slab on a binary substrate. For example, thick InGaSb and InAsSb layers were grown onto GaAs and InAs substrates, respectively. Low background doping and high electron mobilities are achieved in material that demonstrates cut-off wavelengths in the $8-12\,\mu m$ region, potentially a competitor to the more established IR detectors based on MCT (Sect. 14.1.7).

Group III Nitrides

The LPE of GaN is difficult due to the low solubility of nitrogen in molten metals at atmospheric pressure. There are reports of growth of GaN from gallium and bismuth melts, and in some instances the melt is replenished with nitrogen by introducing ammonia into the growth ambient, relying on a so-called VLS (vapor–liquid–solid) growth mechanism that essentially combines LPE with CVD (chemical vapor deposition). Another report [14.15] notes the use of Na fluxes as a solvent. *Klemenz* and *Scheel* [14.16] used a dipping mode at $900\,°C$ with sapphire, $LiGaO_2$, $LiAlO_2$ and CVD GaN on sapphire substrates.

Other Topics

Doping with rare-earth elements (Dy, Er, Hl, Nd, Pr, Yb, Y, ...) in the AlGaAs, InGaAs, InGaAsSb and InGaAsP systems can lead to impurity gettering effects that radically reduce background doping and junction saturation currents and increase carrier mobilities and minority carrier lifetimes. Such rare-earth doping in InAsSb LEDs [14.17] increases the luminescence by $10-100$ times.

There is no fundamental limit to the number of components in mixed alloy layers produced by LPE. For example, AlGaInPAs layers have been grown on GaAs by LPE [14.18]. Each additional element adds an extra degree of freedom for tailoring the properties of the layer, although more detailed phase equilibria data or models are required to determine accurate melt compositions and temperatures. However, as more constituents are added the melt becomes more dilute and more nearly approaches ideal behavior.

Traditionally, LPE melts are rich in one of the major components of the layer to be grown. However, there are certain advantages to using alternative solvents, such as bismuth, as used for GaAs. In the latter case the melt is then dilute in both arsenic and gallium and the chemical activities can be separately controlled to try to reduce point defects since the concentrations of these defects depend on the chemical potentials of the constituents. Bismuth also has lower surface tension that provides better wetting of the substrate. Solubilities can also be changed to affect growth rates or segregation of certain elements, such as Al in AlGaAs. Other solvents that might be considered include molten salts, alloys with Hg, Cd, Sb, Se, S, Au, Ag, or even perhaps some fused oxides.

Several groups have reported success with LPE growth of several less-common semiconductors, such as InTlAsSb, InBiSb and GaMnAs [14.14]. The drive for this work is for low-bandgap material for use in detectors to rival those made in MCT (Sect. 14.1.7).

The low supersaturation of LPE makes selective modes of epitaxy feasible. A substrate can be masked (using, say, SIO_2, Si_3N_4, TiN) and patterned with openings that serve as sites for preferential nucleation. In epitaxial lateral overgrowth (ELO), the selectively seeded material overgrows the mask. This technique has been used for defect filtering, stress reduction, substrate isolation and buried mirrors and electrodes [14.14]. ELO is difficult with vapor-phase methods; aspect ratios (width to thickness of selectively grown material) are small, whereas they can be 100 in LPE. This could have potential for light-emitting diodes [14.14]. Another interesting application of selective LPE is the growth of pyramidal AlGaAs microtips for scanning near-field optical microscopy.

LPE growth of heterostructures with high lattice mismatch has also been attempted, for example of InSb on GaAs [14.14] and AlGaAs on GaP [14.14]. This can be assisted by growing a buffer layer by CVD, as in the LPE of AlGaAs on GaAs-coated (by MOCVD or MBE) silicon substrates. Defect-density reductions of ≈ 2 orders of magnitude can be achieved relative to the GaAs buffer layer grown by MOCVD or MBE.

Another variant of the basic LPE process is that of liquid-phase electroepitaxy (LPEE), where application of an electric current through the growth interface can enhance growth rates for producing thick ternary layers [14.14]. Selective LPEE on patterned, tungsten-masked GaAs substrates can produce inverted pyramid-shaped crystals that can be used to make very high efficiency LEDs [14.14].

Mauk et al. [14.19] have reported on a massive scaling up of the LPE growth of thick ($> 50\,\mu m$) AlGaAs on 75 mm-diameter GaAs substrates. The method produces a two orders of magnitude improvement in areal throughput compared to conventional horizontal sliding boat systems and has applications for LEDs, thermophotovoltaic devices, solar cells and detectors. A large rectangular aluminium chamber is used instead

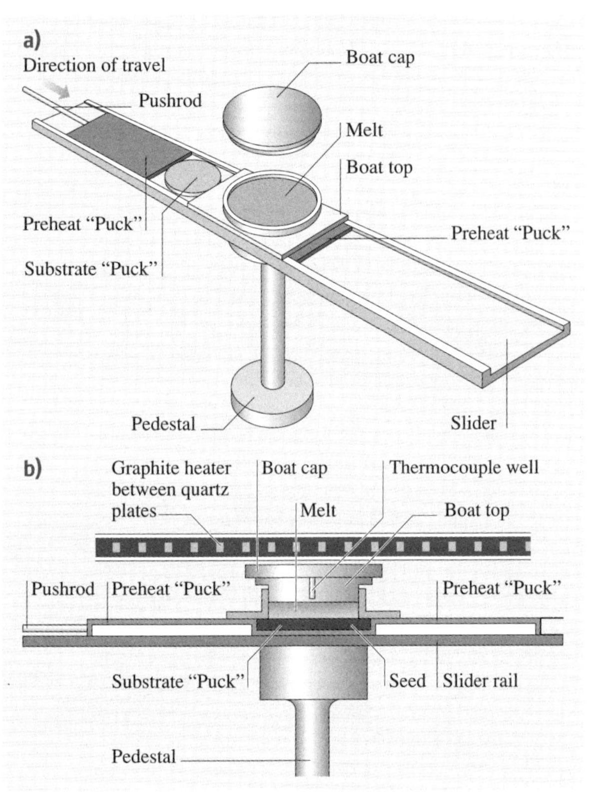

Fig. 14.7a,b Schematic of three-inch LPE apparatus, (**a**) perspective view, (**b**) side view. (After [14.19])

of a silica tube (Fig. 14.7). A modified sliding-boat arrangement is used with a top heater mounted above the boat cap, in conjunction with a heat exchanger/pedestal that acts as a cold finger to impose a vertical temperature gradient, which is the driving force for growth. These $> 50\,\mu$m-thick layers were then produced as free-standing layers bonded to glass.

14.1.7 Group II–VI

Widegap Compounds

While vapor-phase methods are normally used, LPE has been used to grow some widegap Zn-based II–VI compounds [14.20]. This work was driven by the view at that time that blue LEDs could be made economically in ZnSe. The aim was to produce p–n junctions directly by LPE via growth at 950–650 °C, much lower than the bulk crystal growth, to reduce the number of defects. The problems included the high Se vapor pressure, necessitating a closed-tube approach, and the need to maintain the ZnSe substrates in the upper portion of the vertically held melt during deposition. Nevertheless, 10–20 μm-thick layers were grown in 2 h. Growth at 950 °C produced smoother surfaces, but the layers contained more deep levels and impurities compared with those grown at the lower temperatures. Addition of a separate Zn vapor pressure source improved the properties, showing p-type conductivity, and doping with Au, Na, and Li was also attempted. All of these produced p-type material but there was no n-type material reported. More recently, the same group [14.21] reported growth of p-type ZnSe doped with Na_2Se from which p–n junctions were fabricated after Ga diffusion from a Zn solution, to produce the n-type layer. Blue light was emitted at a wavelength of 471 nm.

Astles [14.22] has reviewed the work done on LPE of CdTe-based compounds. Most studies have been carried out from Te-rich solutions in the temperature range 500–900 °C. Layers are p-type as-grown or n-type if doped with In or Al. Growth rates are typically 0.5 μm/°C at 500 °C. Growth from Bi-rich melts was also studied and this was found to improve melt wipe-off and surface morphology. Buffer layer growth of CdZnTe layers was used by *Pelliciari* et al. [14.23] as impurity barrier layers. Both CdMnTe [14.24] and HgCdMnTe [14.25] have also been grown by LPE. The latter compound was used to produce mesa diodes for room-temperature 1.3–1.8 μm applications.

Mercury Cadmium Telluride (MCT)

The situation regarding LPE of MCT was reviewed by *Capper* et al. [14.5]. LPE has emerged as the predominant materials growth technology for the fabrication of both first- and second-generation MCT IR focal plane arrays (FPAs). The technology has advanced to the point where material can now be routinely grown for high-performance photoconductive (PC), photovoltaic (PV) and laser detector devices covering the entire 2–18 μm spectral region. Two different technical approaches have been pursued with almost equal success: growth from Hg solutions and growth from Te solutions. One major advantage of the Hg-solution technology is its ability to produce layers of excellent surface morphology due to the ease of melt decanting. Two additional unique characteristics have now been widely recognized as essential for the fabrication of high-performance double-layer heterojunction (DLHJ) detectors by LPE: low liquidus temperature (< 400 °C), which makes a cap-layer growth step feasible, and ease of incorporating both p-type and n-type temperature-stable impurity dopants, such as As, Sb and In, during growth.

Figure 14.4 shows the dipping system used for the growth of CdHgTe [14.5]. A typical growth procedure begins by lowering the paddle plus substrates into the melt and allowing thermal equilibrium to be reached while stirring. After reaching equilibrium, a programmed ramp reduces the melt temperature to the required level at which point the shutters are opened and the substrates are exposed to the melt. Upon completion, the paddle is withdrawn into the transfer chamber and the isolation valve is closed. Large melts allow the production of layers of up to 30 cm^2 with excellent compositional and thickness uniformity and allow dopant impurities to be accurately weighed for incorporation into layers and to maintain stable electrical characteristics over a long period of time. Four layers (30 cm^2 each) with a total area of 120 cm^2 can be grown in a single run [14.27]. *Norton* et al. [14.28] also scaled up for the growth of cap layers from Hg-rich solutions, each reactor capable of growth on four 24 cm^2 base layers per run.

While layers grown from Hg-rich solutions are easily doped with group VB elements with high solubility, layers grown from tellurium-rich solutions are not. Group VB dopants have low solubility and are not 100% active electrically. Group IIIB elements, indium in particular, are easily incorporated from both solutions. Indium doping from tellurium-rich melts, however, has one advantage in that the segregation coefficient is near unity.

Astles [14.22] has reviewed the experimental data of Te-rich LPE growth at 460 to 550 °C. As an example, to compare growth parameters for Te solutions with those for Hg solutions, consider the growth of LWIR MCT ($x = 0.2$) at 500 °C from both Te and Hg solutions. The x_{Cd} for Te-rich solutions is 8.3×10^{-3}, while x_{Cd} for Hg-rich solutions is 2.6×10^{-4}. This is one of the difficulties encountered in LPE growth from Hg-rich solutions. Use of large melts, however, overcomes the Cd depletion problem. MCT epitaxial layers of the desired thickness (> 10 μm) and of uniform composition through the thickness can be grown.

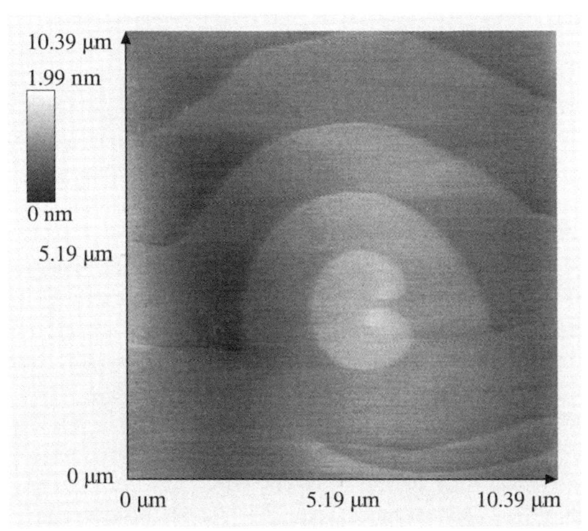

Fig. 14.8 AFM image of a Frank–Read growth spiral. (After [14.26])

A typical sliding-boat arrangement has been described by *Capper* et al. [14.29]. The LPE boat is made from purified, high-density, small-grain, electronic-grade graphite. The precompounded growth solution is placed in a growth well, and crushed HgTe is placed in a blind well. The HgTe acts as a buffering source for the volatile elements to help stabilize the growth solution composition and also to deliver an overpressure to the annealing well to control the metal vacancy level during annealing. These wells are connected to the annealing well with a gas channel plate that controls the movement of the volatile elements within the boat. The growth solution and HgTe overpressure source are made from high-purity elements. The control over the impurity levels in the major constituent elements is a crucial part of the control of the overall process.

At the start of the growth cycle the slider is positioned so that the substrate is under the annealing well. The loaded boat is placed in the reactor tube and the furnace is pre-heated to 520 °C and then moved over the boat. The boat heats rapidly, and after a solution melting and equilibration period, the furnace is cooled rapidly by ≈ 20 °C, and then a slow cooling ramp (2 or 3 °C/h) is initiated. When the boat reaches the required growth start temperature, the slider is moved so that the substrate is positioned under the molten solution. The ramp continues until the required film thickness has been grown, after which the slider is returned to the starting position and the furnace temperature reduced rapidly to an annealing temperature. Following the anneal, the furnace is moved back to its starting position and the system is allowed to cool.

Surface morphology is controlled at two levels: microtexture and long-range variation. The microtexture is a result of misalignment of the substrate crystal plane with the growing surface. Deviations > 0.1° from the <111>B plane lead to significant surface texture. Growth on accurately orientated substrates gives a specular surface on which atomic-scale growth features can be seen using atomic force microscopy (AFM). Figure 14.8 shows a classical Frank–Read site on an as-grown LPE layer surface.

The approach to forming p-on-n DLHJ structures by LPE is virtually universal. LPE from Hg-rich solution is used to grow the As- or Sb-doped p-type cap layers. The In-doped n-type base layers are grown by various Te-melt LPE techniques including tipping, sliding, and dipping. The trend appears to be in favor of the p-on-n DLHJ structures, as passivation is more controllable than that of the n-on-p structures [14.30]. A bias-selectable two-color (LWIR/MWIR) detector structure was first fabricated by growing three LPE layers from Hg-rich melts in sequence on a bulk CdZnTe substrate, *Casselman* et al. [14.31].

Other Narrowgap II–IV Compounds

HgZnTe was first proposed as an alternative detector material to MCT due to its superior hardness and its high energies for Hg vacancy formation and dislocation formation [14.32]. *Rogalski* [14.33] reviewed the LPE growth of HgZnTe and noted that Te-rich growth is favored due to the low solubility of Zn in Hg and the high Hg partial pressure. He also commented that the same factors apply to the growth of HgMnTe. *Becla* et al. [14.24] grew HgMnTe in a two-temperature, closed-tube tipping arrangement at 550–670 °C onto CdMnTe bulk substrates and CdMnTe LPE layers previously grown on CdTe substrates. Phase diagram data were also presented and the value of k_{Mn} was quoted as 2.5–3. *Rogalski* [14.33] also reviewed the status of PC and PV detectors in both HgZnTe and HgMnTe.

14.1.8 Atomically Flat Surfaces

Chernov and *Scheel* [14.34] have argued that far from the perceived drawback of LPE of producing rough surfaces, it may be uniquely suited to providing atomically flat, singular surfaces over distances of several micrometers. These surfaces would have applications in surface physics, catalysis and improved homogeneity of layers and superlattices of semiconductors and superconductors.

In support of this view, Fig. 14.8 shows an AFM image of a Frank–Read growth spiral on the surface of an MCT layer grown by LPE in this author's laboratory [14.26].

14.1.9 Conclusions

LPE was generally the first epitaxial technique applied to most systems of interest in micro- and optoelectronics. It is now generally a mature technology, with large fractions of several optoelectronic, IR detectors and other device types being made in LPE material, although some developments are still taking place. LPE has several advantages over the various vapor-phase epitaxial techniques, such as high growth rates, favorable impurity segregation, ability to produce flat faces, suppression of certain defects, absence of toxic materials, and low cost. There is much less emphasis on LPE in the current literature than on the vapor-phase methods, but LPE continues to seek out and develop in several niche markets where vapor-phase techniques are not suitable.

14.2 Metalorganic Chemical Vapor Deposition (MOCVD)

14.2.1 Introduction and Background

The technique of MOCVD was first introduced in the late 1960s for the deposition of compound semiconductors from the vapor phase. The pioneers of the technique, *Manasevit* and *Simpson* [14.35] were interested in a method for depositing optoelectronic semiconductors such as GaAs onto different substrates such as spinel and sapphire. The near-equilibrium techniques such as LPE and chloride VPE were not suitable for nucleation onto a surface chemically very different to the compound being deposited. These pioneers found that if they used combinations of an alkyl organometallic for the Group III element and a hydride for the Group V element, then films of GaAs could be deposited onto a variety of different surfaces. Thus, the technique of MOCVD was born, but it wasn't until the late 1980s that MOCVD became a production technique of any significance. This success depended on painstaking work improving the impurity of the organometallic precursors and hydrides. By this time the effort was on high-quality epitaxial layers on lattice-matched substrates, in contrast with the early work. The high-quality epitaxial nature of the films was emphasized by changing the name of the growth method to metalorganic vapor phase epitaxy (MOVPE) or organometallic VPE (OMVPE). All of these variants of the name can be found in the literature and in most cases they can be used interchangeably. However, MOCVD can also include polycrystalline growth that cannot be described as epitaxy. The early niche applications of MOVPE were with GaAs photocathodes, GaAs HBT lasers and

GaInAsP lasers and detectors for 1.3 μm optical fiber communications.

The characteristics of MOCVD that have taken it from a research curiosity to production have been in the simplicity of delivery of the reactive vapors and the versatility of compositions, dopants and layer thicknesses. These basic attributes have enabled the same basic technique to be used for narrow bandgap semiconductors such as the infrared detector materials $Cd_xHg_{1-x}Te$ and GaInSb and now for wide bandgap semiconductors such as GaN and ZnO. Indeed, the success of GaInN in the 1990s for high-brightness blue LEDs has now led to this being the most popular material produced by MOCVD. The early strength of MOCVD was its ability to grow onto different substrates but this was later abandoned in favor of the more conventional homoepitaxy; however, the nitrides rely on heteroepitaxy onto sapphire and SiC substrates, bringing MOCVD back to its roots with the early work of Manasevit. This versatility with substrate materials presents MOCVD with the ultimate challenge of mating high-performance optoelectronic materials with silicon substrates in order to combine the best of optoelectronic and electronic performance.

This section of the chapter will cover the key elements of the MOCVD process from the physical characteristics of the precursors through reactor design to getting the right materials properties for high-performance devices.

14.2.2 Basic Reaction Kinetics

The precursors for III–V MOCVD are generally a simply alkyl for the Group III source and a hydride for the Group V source. Both have the essential properties of being volatile in a suitable carrier gas stream (usually hydrogen) and being chemically stable at ambient temperature. These precursors are normally mixed outside the reaction chamber, introduced into the reaction chamber through a suitable injector arrangement and directed onto a hot substrate. This is shown schematically in Fig. 14.9. The details of reactor design will be discussed later in this chapter. The reaction of the precursors to yield the III–V compound on the substrate can occur either in the hot vapor above the surface or on the hot surface. The stoichiometric reaction for GaAs growth is given as

$$(CH_3)_3Ga + AsH_3 \rightarrow GaAs + CH_4 \,. \qquad (14.1)$$

This reaction has been the most widely studied of all the MOCVD reactions and was one of the original processes reported by *Manasevit* and *Simpson* [14.35]. One

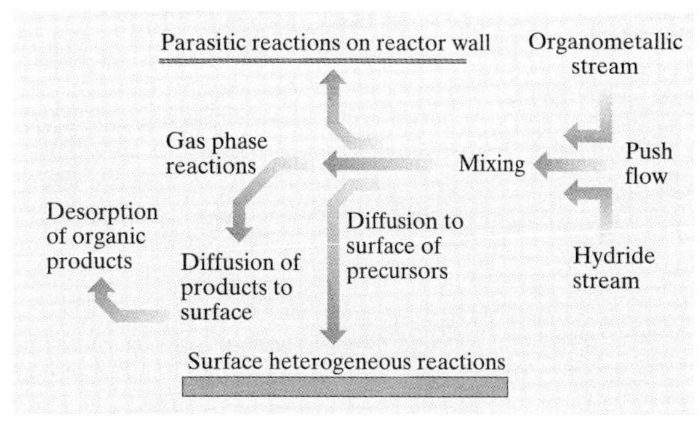

Fig. 14.9 Schematic of MOCVD process from mixing of gas streams to reaction on the substrate surface

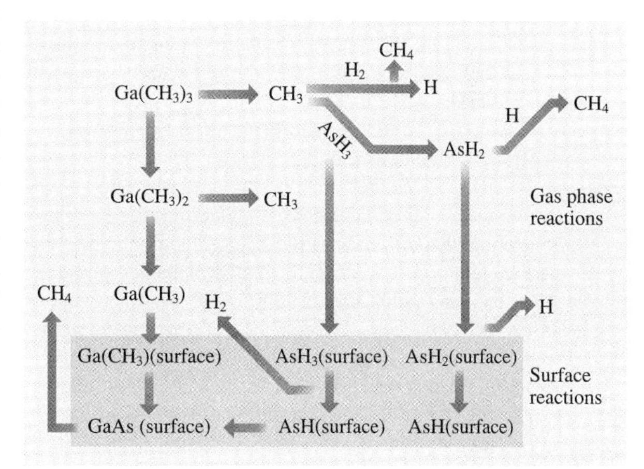

Fig. 14.10 Reaction scheme for epitaxial deposition of GaAs showing the most important vapor phase and surface reaction steps

reason that it works so well is that the hydrogen required to satisfy the $\cdot CH_3$ radical bond is supplied from the arsine hydride, and at normal growth temperatures (around 700 °C) avoids unsatisfied carbon bonds that could lead to carbon incorporation. In reality this very simple picture covers a complexity of reaction steps that have to take place, that have been discussed in great detail by *Chernov* [14.36]. However, some of the important reaction steps will be described here as an introduction to the kinetics of GaAs MOCVD.

The schematic shown in Fig. 14.10 gives some of the important reaction steps that have been identified for the reaction of GaAs. This gives some insight into the complexity of the reaction kinetics and it is worth remembering that this is a relatively straightforward reaction for MOCVD. Fortunately, one does not have to understand every step in the process before attempting to grow a layer, and this goes some way to explaining why MOCVD has developed along very empirical lines. An understanding of the reaction kinetics does, however, enable some of the problems that are associated with MOCVD to be understood, particularly when these relatively simple precursors are replaced by more complex precursors.

It can be seen from Fig. 14.10 that the reaction process is started by gas-phase homolysis of TMGa [$(CH_3)_3Ga$] to yield dimethylgallium and methyl radicals ($\cdot CH_3$). There are two important roles that the methyl radicals can take and this is generally important in all MOCVD processes for deposition of III–V semiconductors:

1. Methyl radicals can react with the ambient hydrogen carrier gas to yield stable methane and hydrogen radicals.
2. Methyl radicals can react with the arsine (AsH_3) to yield stable methane and AsH_2.

Both of these steps can initiate the decomposition of arsine either through either a methyl radical or a hydrogen radical removing a hydrogen atom from the arsine; these reaction steps can be seen in Fig. 14.10. There are some important consequences of these initial reaction steps, and one you may have already spotted is that the hydrogen carrier gas is not included in the stoichiometric reaction (14.1) but can play a part in the reaction process. Another consequence is that although it is highly unlikely that the two (or more) precursors have the same

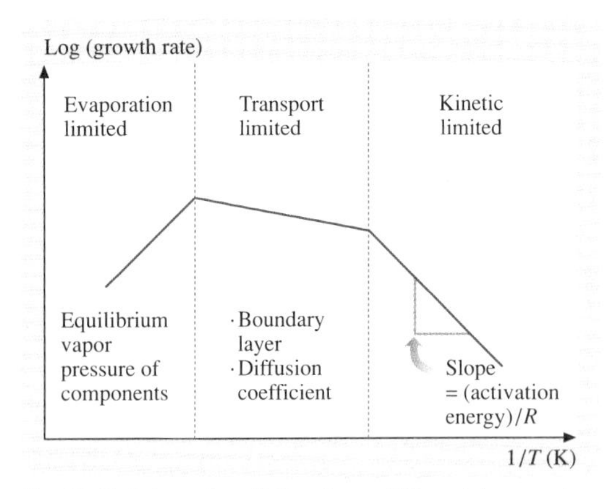

Fig. 14.11 Schematic of the three different temperature regimes for growth of a compound semiconductor by MOCVD

thermal decomposition characteristics, they can be used together to react at the same temperature through radical reaction mechanisms. The latter has been illustrated in many studies of different systems using ersatz reaction tube experiments, and further information on these can be found in *Stringfellow* [14.37]. For example, in the case of GaAs, the pyrolysis of TMGa will occur at 500 °C, but for arsine decomposition, a temperature of around 700 °C is required.

The stepwise removal of ligands from TMGa and arsine can occur in the vapor or on the surface. Figure 14.10 shows both alternatives and the dominant path will depend on both the surface temperature and vapor temperature above the substrate. A general rule is that more of the reaction process will occur on the surface at lower temperatures when the vapor reaction rate will be slower and the surface adsorption will be more efficient. It is also important to note that there are two different adsorption sites on the surface: the Ga sites that will take the anion species and the As sites that will take the cation species.

Part of the early success of GaAs and the alloy GaAlAs was due to the absence of unwanted vapor-phase reactions that could lead to the formation of polymer chains that would contaminate the layer. This was not the case with InP where the liquid alkyl source TEIn was used instead of the solid TMIn. This led to a mist in the reactor resulting from the formation of an adduct between the precursors that subsequently polymerized. This was overcome through the use of low-pressure reactors, and this has remained a feature of most production reactors today. However, the preferred precursor is TMIn and the problems of uncertain supply from a solid source has been overcome by a variety of different innovative methods. One example is to form a liquid with a stable amine, and the source then will behave in the same way as any other liquid precursor (*Frigo* et al. [14.38]).

The details of the reaction kinetics will change with substrate temperature, but as the temperature is increased a point will be reached where the rate of epitaxial growth will no longer be determined by the overall reaction rate but will be determined by the supply of precursors to the substrate. This will be reflected by a depletion of the precursor concentration immediately above the surface and a gradient in precursor concentration towards the undepleted free stream. The limitation on the rate of epitaxial growth then becomes the rate of diffusion through the depleted boundary layer to the substrate. This is called transport-limited growth and is characterized by high growth rates and only a weak dependence of growth rate on substrate temperature. This is shown schematically in Fig. 14.11. The plot is of ln(growth rate) versus $1/T$ because of the expected Arrhenius relationship in the rate constants. This really only applies to the low-temperature (kinetic) regime. Here the growth rate can be expressed as

$$\text{Rate} = A \exp -(E_a/RT) , \quad (14.2)$$

where A is a constant and E_a is the activation energy. It is unlikely that E_a can be attributed to the activation energy for a single reaction step, but it is still useful for characterizing the kinetics when different precursors are being tested.

In the transport-limited regime there will be a small dependence on temperature due to the increase in diffusion rate with temperature, and this is illustrated in Fig. 14.11. Most MOCVD growth processes will take place in the transport-limited regime where it is easier to control growth rate. However, there are a number of growth processes that will occur at lower temperatures in order to control the properties such as native defect concentrations of the epitaxial films. This is generally the case with II–VI semiconductors, but can also apply to the formation of thermodynamically unstable III–V alloys.

In the high-temperature regime, the growth rate decreases with temperature, as the equilibrium vapor pressure of the constituent elements in the film will increase and give desorption rates similar to the deposition rate, leading to significant loss of material through evaporation to the gas stream.

14.2.3 Precursors

The choice of precursors is not confined to simple alkyls and hydrides but can extend to almost any volatile organometallic as a carrier for the elemental components of a film. In the case of II–VI semiconductors it is usual to use an alkyl for both the Group II and the Group VI elements. Hydrides have been used as Se and S sources but prereaction makes it difficult to control the growth process and in particular can make it difficult to incorporate dopants. The use of combined precursor sources has been extensively researched but is not in common use for epitaxial device-quality material. One reason for this is the difficulty in controlling the precursor ratio that is needed to control the stoichiometry of the material.

The important properties of precursors, and their selection, can be generalized and provides a basis for optimizing the MOCVD process. These properties can be summarized as follows:

1. Saturated vapor pressure (SVP) should be in the range of 1–10 mbar in the temperature range 0–20 °C.
2. Stable for long periods at room temperature.
3. Will react efficiently at the desired growth temperature.
4. The reaction produces stable leaving groups.
5. Avoids unwanted side reactions such as polymerization.

According to the Clausius–Clapeyron equation, the SVP of a liquid is given by an exponential relationship:

$$\text{SVP} = \exp(-\Delta G/RT) , \quad (14.3)$$

where ΔG is the change in Gibbs free energy on evaporation, R is the gas constant and T the temperature of the liquid in the bubbler. This can be expressed as the heat of evaporation ΔH and the entropy for evaporation ΔS, where $\Delta G = \Delta H - T\Delta S$; this gives the familiar form of the SVP equation:

$$\begin{aligned}\text{SVP} &= \exp(-\Delta H/RT) + \Delta S/R \\ &= \exp(\Delta S/R)\exp(-\Delta H/RT) . \quad (14.4)\end{aligned}$$

This is of the form:

$$\log_e(\text{SVP}) = A - B/T , \quad (14.5)$$

where A and B are constants given by $A = \Delta S/R$ and $B = \Delta H/R$. Manufacturers of the precursors will generally give the SVP data in the form of the constants A

Table 14.1 List of precursors with vapor pressure constants derived according to (14.6)

Precursor	A	B	SVP at 20 °C (mm Hg)
TMGa	8.07	1703	182
TEGa	8.08	2162	5.0
TMAl	8.22	2134	8.7
TEAl	9.0	2361	0.02
TMIn	10.52	3014	1.7
TEIn	8.94	2815	1.2
Solution TMIn	10.52	2014	1.7
DMZn	7.80	1560	300
DEZn	8.28	2109	12
DMCd	7.76	1850	28.2
DES	8.184	1907	47
DMSe	9.872	2224	
DESe	8.20	2020	
DMTe	7.97	1865	40.6
DIPTe	8.29	2309	2.6

and B in (14.5). It can also be given in the form

$$\log_{10}(\text{SVP}) = A' + B'/T . \quad (14.6)$$

To convert the constants in (14.6) to (14.5), just multiply by ln10. Some examples of the SVP constants, along with the calculated SVP at 20 °C, for a number of typical precursors are shown in Table 14.1.

14.2.4 Reactor Cells

The design of reactor cells has formed a very important part of the development of MOVPE and has been crucial in scaling laboratory processes to large-scale production. The original research reactors fell into one of two groups, either the vertical reactor or horizontal reactor. These reactor designs are shown schematically in Fig. 14.12. The substrate is placed onto a graphite susceptor that is heated by either RF coupling via a coil surrounding the reactor, a resistance heater underneath the susceptor, or lamps placed underneath the susceptor. The reactor wall can be water-cooled or gas-cooled to minimize reaction and deposition onto these surfaces. Either of these reactor cells could be operated at atmospheric or reduced pressure. For reduced-pressure operation the reactor cell pressure would be typically a tenth of an atmosphere but a wide range of different pressures have been successfully used. Reduced pressure will increase the gas velocity and help to overcome the effects of free convection from a hot substrate. The forced convection parameter that is often quoted is the Reynolds number and is proportional to gas velocity. A high Reynolds number will ensure streamline flow, while at low Reynolds number the buoyancy effects of the hot substrate will take over and the gas flow will be dominated by free convection (characterized by the Grashof number) and become disorganized with recirculation cells. In transport-limited growth the exact nature of the gas flow will determine the uniformity of deposition and can also affect the defect concentration in the films due to particulates and reaction products being swept back across the growing surface. It is not normal to achieve the very high flow velocities and Reynolds numbers associated with turbulent flow in an MOVPE chamber, but the disorganized flow due to free convection is often (wrongly) referred to as 'turbulent flow'.

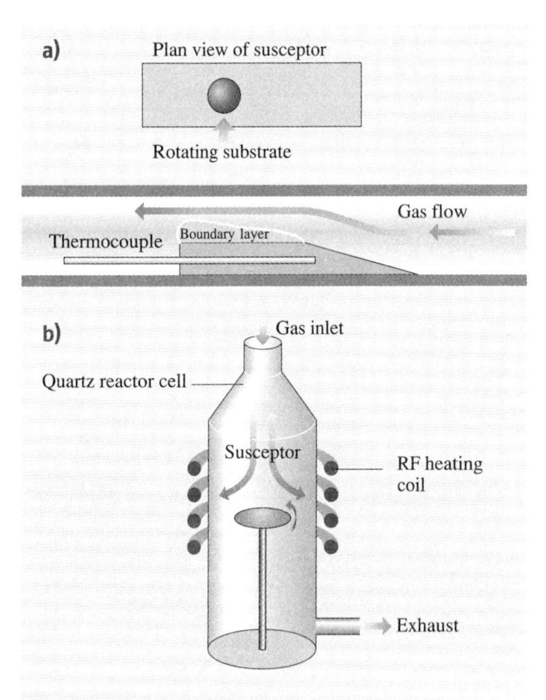

Fig. 14.12a,b Schematic of (a) horizontal and (b) vertical reactor cells

Another reason for using high flow velocities is to overcome the effects of depletion of the precursor concentration at the downstream end of the deposition region. For transport-limited growth the growth rate is limited by the rate of diffusion from the free stream to the substrate. This region is called the boundary layer and increases in thickness going downstream from the leading edge of the susceptor, as shown in Fig. 14.13. Some horizontal reactors are designed with a tilt in the susceptor so that the free cross-sectional area decreases and hence the flow velocity increases going downstream. This helps to flatten the boundary layer and ensure better uniformity.

Maintaining a boundary layer has a cost: the high utilization of expensive precursors and gases, as the

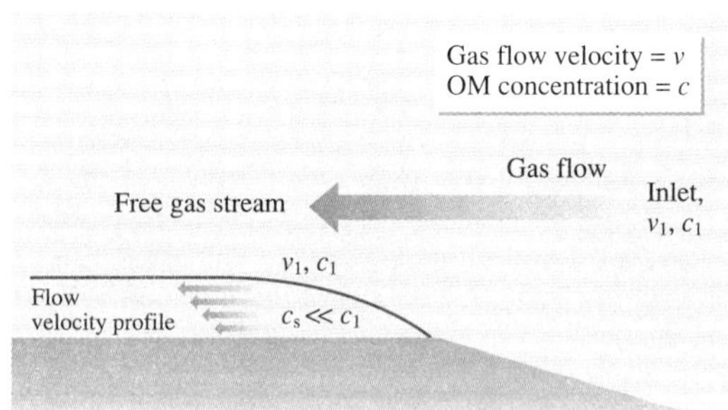

Fig. 14.13 Schematic of a boundary layer in a horizontal reactor

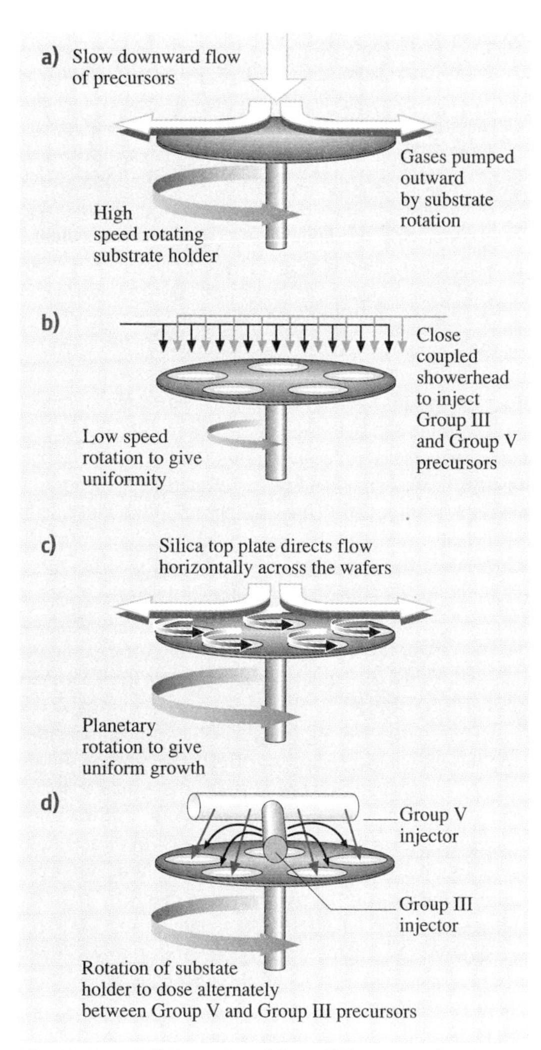

Fig. 14.14a–d Schematic of four different production reactor designs: (**a**) the Emcore (Veeco) Turbo-disc, (**b**) the Thomas Swan showerhead, (**c**) the Aixtron planetary, and (**d**) the EMF Vector flow reactor

flow throughput is typically very high and the proportion of precursors reacted in the region of the substrates is very low. This became a more serious problem when reactors were scaled to hold multiple wafers for production. These problems have been resolved with different solutions in the vertical and horizontal reactor configurations with the Emcore (now Veeco) turbo-disc reactor, the Thomas Swan showerhead reactor and the Aixtron planetary reactor. A fourth production-scale reactor design has recently been introduced by EMF, called the Titan, and works on the principle of atomic layer deposition (ALD), where the Group II and Group V gases are fed over the substrate surface separately. A common feature to all of these production reactor designs is that the substrates are rotated so that the concentrations of precursors and reactants arriving at the substrate do not have to be uniform across the surface, as a portion of the substrate will alternately experience high and low concentrations that will average out. Each of these reactor designs is shown schematically in Fig. 14.14.

The turbo-disc reactor, shown schematically in Fig. 14.14a, is a vertical reactor configuration but the boundary layer is kept to a narrow region above the susceptor by high-speed rotation that pumps the gas radially outwards due to viscous drag. The rotation speeds are up to 2000 rpm in order to create this lateral flow of the constituents above the substrate. This is continuously replenished from the slower downward gas stream, resulting in excellent uniformity of deposition across the wafers and a high utilization of the reactant gases (*Tompa* et al. [14.39]). The reactor pressure is typically around 100 mbar.

The showerhead reactor is another vertical reactor arrangement but it takes a different approach to overcoming free-convection currents and poor uniformity [14.40]. The precursors are introduced through a water-cooled showerhead placed just above the susceptor. The susceptor is rotated but typically at much lower speeds than for the turbo-disc reactor. The precursor distribution can be balanced across the width of the reactor to give a uniform supply of precursors.

The planetary reactor is a horizontal flow arrangement where the reactants enter at the center of rotation of the susceptor and flow outwards. This is an example of a fully developed flow where depletion of the reactants is occurring as the gases move away from the center and this will be accentuated by a decrease in the mean flow velocity as the gases move outwards [14.41]. This would normally give very poor uniformity but the planetary rotation mechanism will rotate each wafer on the platen so it will sample alternately high and low concentrations, giving uniform deposition. This approach has the advantage of high utilization of the precursors and the ability to extend the design to very large reaction chambers for multiple wafers, with the Aixtron 3000 reactor holding 95 2 inch-diameter wafers.

The fourth approach to multiple wafer deposition is the EMF Ltd vector flow epitaxy (VFE), which introduces the Group II and Group V precursors separately over a rotating susceptor platen, as shown in Fig. 14.14d.

The rotation of the platen will direct the gases across the wafers and out through separate exhausts, thus keeping the gases separate in the reactor chamber. This has the advantage of alternately dosing the surface with Group III and Group V precursors to grow the film from atomic layers, which in turn prevents prereaction between the precursors and maintains excellent film uniformity over the growth surface. This could be particularly important for compounds of nitrides and oxides where reduced pressure is normally required to avoid significant prereaction. The advantage of the ALD approach is that the reaction chamber can be operated at atmospheric pressure, which simplifies the operation of the system.

14.2.5 III–V MOCVD

This section will consider the range of III–V materials grown by MOCVD and the precursors used. Most of the III–V semiconductors can be grown from organometallics of the Group III element and hydrides of the Group V element. Exceptions to this will be noted where appropriate.

Arsenides and Phosphides

The most commonly studied alloy system is $Al_{1-x}Ga_xAs$, which is used for LEDs and laser diodes from the near-infrared to the red part of the visible spectrum. This is a well-behaved alloy system with only a small change in lattice parameter over the entire composition range and it covers a range of bandgaps from 1.435 eV for GaAs to 2.16 eV for AlAs. One problem with this alloy is the sensitivity of aluminium to oxygen, which makes it extremely difficult to grow high-quality AlAs. Just 1 ppm of oxygen contamination will result in $10^{20}\,cm^{-3}$ incorporation of oxygen into $Al_{0.30}Ga_{0.7}As$ [14.42]. In addition to the normal MOCVD precautions of using ultrahigh-purity hydrogen carrier gas and ensuring that the moisture in the system is removed, the hydrides and organometallics also need to have extremely low oxygen contents. Precursor manufacturers have tended to keep to the simple alkyl precursors but to find innovative ways of reducing the alkoxide concentrations.

Alternative Group V precursors have been sought due to the high toxicity of arsine and phosphine. These hydride sources also suffer from the fact that they are stored in high-pressure cylinders and any leakage could result in the escape of large quantities of toxic gas. Alternative alkyl Group V sources have been extensively researched but only two precursors have proved to be suitable for high-quality epitaxial growth, tertiarybutylarsine (TBAs) and tertiarybutylphosphine (TBP). These precursors only have one of the hydrogen ligands replaced with an alkyl substituent but they are liquid at room temperature rather than high-pressure gases. In the reactor chamber the likely reaction path is to form the hydride by a process called beta-hydrogen elimination. This entails one of the hydrogen atoms from the methyl groups satisfying the bond to As (or P) with a butene leaving group as shown below [14.43]:

$$C_4H_9AsH_2 \rightarrow C_4H_8 + AsH_3 \,. \qquad (14.7)$$

This process is more likely to dominate at the normal growth temperature for transport-limited growth and it effectively yields the arsine precursor that can then react in the normal way. In the search for alternative alkyl precursors this proved to be an important factor, as the fully substituted alkyl arsenic sources tended to incorporate large concentrations of carbon, degrading the electrical properties of the film. The importance of the Group V hydride was discussed in Sect. 14.2.2 and it can be understood why TBA and TBP (for the phosphorus alloys) have proved to be good alternatives to the hydrides. However, it is fair to say that these have never been widely utilized due to much higher cost than the hydrides and poor availability.

An alternative for improved safety has been investigated more recently and relies on the same principle of reducing the toxic gas pressure in the event of a system leak. This alternative stores the hydride in a reversible adsorption system [14.44]. The adsorption system keeps the hydride at sub-atmospheric pressure and requires pumping to draw off the hydride when needed, making it inherently safer. One major advantage to this system, in addition to the inherent improvement in safety, is that the precursors and hence the precursor chemistry are unchanged in the reactor cell.

Other alloys commonly grown using MOCVD include $In_{0.5}Ga_{0.5}P$, which has a band gap of approximately 2 eV and is lattice-matched to GaAs. The quaternary alloy GaInAsP enables lattice-matching to InP substrates while controlling the bandgap in the 1.3 μm and 1.55 μm bands used for long-range fiber-optic telecommunications.

Antimonides

The antimonides cover an important range of bandgaps from the near-infrared to the mid-infrared bands, up to 5 or 6 μm. These compounds and alloys can be used in infrared detectors, thermophotovoltaic (TPV) devices and high-speed transistors. The growth of the antimonides is more complex than for the arsenides and phosphides because the hydride, stibine, is not very stable at room temperature so the use of alkyl precursors has been a more natural choice. Another factor that has influenced the growth of the antimonides is the lower thermodynamic stability and decomposition of substrates such as InSb above about 400 °C. It is also desirable to grow the films at a much lower temperature than for the arsenides and phosphides in order to keep the native defect concentration low for controlled n-type and p-type doping. However, the antimonides have an advantage in that they do not incorporate carbon as readily as in the arsenides and phosphides and there is greater flexibility over the choice of antimony precursors [14.45]. The easiest choice is to use trimethylantimony (TMSb) or to use larger alkyl groups such as triethyl and triisopropyl to reduce the reaction temperature as required. It is also possible to reduce the reaction temperature by replacing TMGa with triethylgallium (TEGa) [14.46]. This approach has been particularly advantageous when growing aluminium-containing alloys where carbon incorporation can be a problem, but is reduced using TEGa. This serves to illustrate the flexibility of MOCVD and has given more scope for the design of precursors, not envisaged in the early days of MOCVD.

Nitrides

The nitrides, mainly GaN and the alloy GaInN have brought MOCVD into prominence as a manufacturing technology with the success of high-brightness blue LEDs making large-screen full-color LED displays a reality. The precursors used for the nitrides are standard with the methyl alkyls (TMGa and TMIn) for the Group III elements and ammonia for the nitrogen. The key technological barriers to obtaining device-quality GaN were to overcome the problems associated with heteroepitaxy onto a non-lattice-matched substrate, sapphire or SiC and to control p-type doping [14.47]. Some research has been carried out with homoepitaxy onto GaN substrates, but the very high pressures needed for bulk crystal growth of GaN will restrict the sizes of substrates available.

The heteroepitaxial problems have been overcome with a two-stage growth. High-quality GaN requires growth temperatures in excess of 1000 °C, much higher than is needed for the arsenides and phosphides. At these temperatures, nucleation onto sapphire is poor and large faceted islands grow before complete coalescence of the film occurs. This not only leads to very poor surface morphology but a high dislocation density where the islands coalesce. Two-stage growth overcomes this by growing a uniform nucleation (or buffer) layer onto the sapphire at 600 °C. This is then heated to normal growth temperature, where a thicker GaN film is then grown. The whole nucleation process can be monitored in situ using laser reflectometry and an example is shown in Fig. 14.15 [14.48]. It can be seen that approximately 50 to 100 nm of GaN is grown as a smooth layer but during heating this changes to a rough layer. A remarkable part of the process is that during the high-temperature growth, shown in Fig. 14.15 by interference oscillations, there is a recovery in the surface morphology, resulting in smooth, device-quality layers. In some cases the GaN buffer layer is substituted with an AlN layer, but in all cases a two-stage growth process is required. The growth of device layers is achieved by controlling the band gap through growth of the alloys GaInN or GaAlN. The band gap of GaN is 3.4 eV, which is in the UV, so the color of the LED is determined by the alloy composition where increasing the In content will reduce the band gap and push the emission wavelength from blue to green. However, this is not an easy alloy to form due to the different stabilities of the GaN and InN bonds [14.49]. In fact, it cannot be grown to any useful In content at temperatures above 1000 °C, so the temperature must be reduced to around 800 °C. Even with this compromise, the different lattice parameters of GaN and InN result in poor solubility and In contents of more than 40% are not practical. In principle it is possible to prepare LEDs of any color by just changing the In content in the alloy, but in practice the high-brightness diodes can only be prepared from nitrides covering the blue to green portion of the spectrum. The growth of Al-containing alloys is

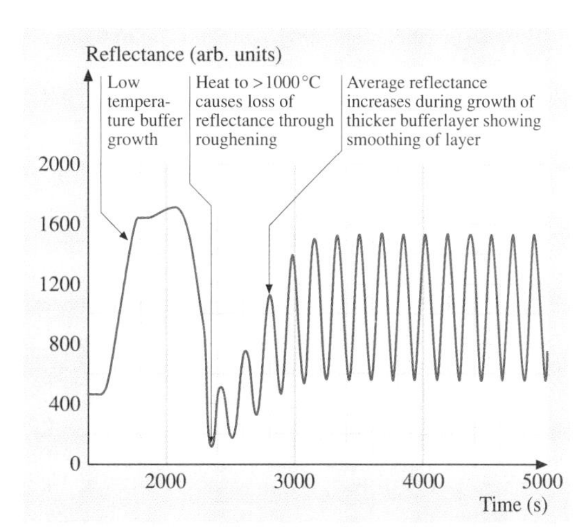

Fig. 14.15 Laser reflectometry of GaN nucleation and growth

desirable in order to achieve wider band gaps for applications such as UV LEDs and solar-blind detectors. The problems are similar to those of the In and Ga alloys, but in the case of Al and Ga the stability of AlN is much greater than that of GaN, which will tend to form AlN clusters. The growth of different alloy compositions, particularly the growth of higher In content GaInN, has stimulated some research on alternative precursors for nitrogen, as ammonia pyrolysis is not very efficient at temperatures below 800 °C. One of the favorite candidates is dimethylhydrazine, which will react readily with TMGa at temperatures down to 400 °C.

The reaction of the ammonia with the Group III alkyls to form adducts that can then polymerize is a problem associated with the high growth temperature. This requires special care over the introduction of the precursors, the control of gas flows and wall temperatures. A failure to adequately control these parasitic reactions will lead to poor growth efficiency, higher defect concentration in the GaN layer and poor dopant control. The dopants used for n-type and p-type GaN are Si from silane and Mg from dicyclopentadienylmagnesium. The n-type doping has proved to be fairly straightforward, but Mg doping results in the formation of Mg–H bonds that passivate the acceptor state. This problem was solved by annealing the epitaxial films after growth to remove the hydrogen. This is possible due to the thermal stability of GaN and the high mobility of hydrogen in the lattice. A further problem with p-type doping is that the Mg acceptor has an ionization energy of between 160 and 250 meV and only about 10% of the chemically introduced Mg is ionized at room temperature.

Despite the materials challenges of GaN and its alloys, MOCVD has enabled the production of a wide range of devices based on these alloys over the past decade, from high-power transistors to laser diodes. Both of these examples have required improvements in material quality and a reduction in the relatively high dislocation densities. In fact, the potential for nitrides is enormous as the quaternary GaInNAs can be tuned to around 1 eV with just 4% nitrogen and is a challenger to the use of InP-based materials for 1.3 μm telecommunications lasers.

14.2.6 II–VI MOCVD

The MOCVD of II–VI semiconductors is carried out at much lower temperatures than for their III–V counterparts and this has stimulated a wide range of research on alternative precursors, growth kinetics and energy-assisted growth techniques such as photoassisted growth. The basic principles are the same as for III–V MOCVD and, in general, the same reaction chambers can be used but the lower growth temperatures have led to the development of new precursors, particularly for the Group VI elements. Hydrides are, in general, not used now but early work on ZnSe and ZnS used hydrogen selenide and hydrogen sulfide [14.50]. A strong prereaction occurred between the hydrides and dimethylzinc that could result in deposition at room temperature, but as with III–V MOCVD, prereactions can make it difficult to control the defect chemistry and the doping. These II–VI compounds and their alloys have been investigated as blue emitter materials with similar bandgaps to GaInN. Alternatively, ZnTe is a potential green emitter and the narrower bandgap tellurides are used for infrared detectors. In fact, the only commercial application of II–VI MOCVD has been for the fabrication of HgCdTe alloys for infrared detectors. However, the processes used are quite different to standard MOCVD and require different designs of reactor cells, as will be shown in the next section.

MOCVD of HgCdTe

HgCdTe is one of the few direct bandgap semiconductors suitable for infrared detection in the important 10 μm band. The alloy has only a 0.3% mismatch over the entire composition range and will cover the entire infrared spectrum from the near-infrared with CdTe to the far-infrared (HgTe is a semimetal so there is no lower limit to the band gap). The main difficulty with growing HgCdTe by MOCVD has been the very high equilibrium vapor pressure of Hg over the alloy even at relatively low temperatures. For example, MBE has to be carried out at temperatures below 200 °C. A further difficulty created by the instability of HgTe is that the tellurium-rich phase boundary, which represents the minimum Hg pressure required to achieve growth, has a high concentration of doubly ionized metal vacancies that make the material p-type. At typical MOCVD growth temperatures for HgCdTe, 350 to 400 °C, the equilibrium vapor pressure for Hg would have to be close to the saturated vapor pressure for liquid Hg in order to keep the metal vacancy concentration below the impurity background. This is clearly not realistic in MOCVD as the walls of the reaction chamber would have to be heated to the same temperature as the substrate to avoid mercury condensation, and this, in turn, would cause pyrolysis of the precursors before they arrived at the substrate. Fortunately, it is possible to grow HgCdTe film on the tellurium-rich phase boundary where the Hg

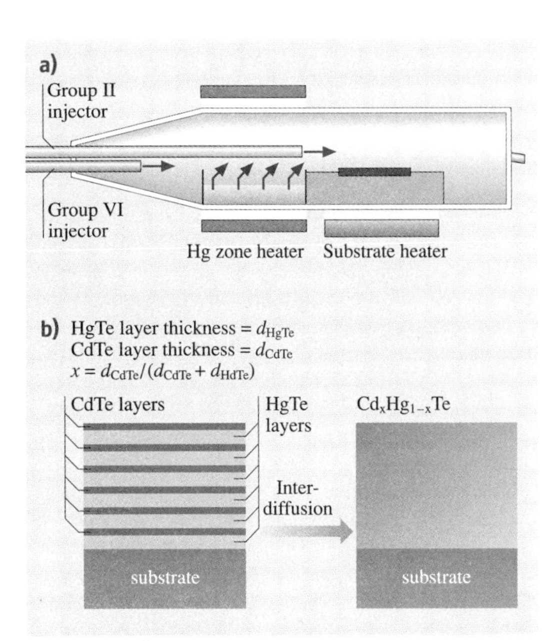

Fig. 14.16 (**a**) Schematic of MOCVD cell for HgCdTe growth, (**b**) schematic of IMP

source and reactor walls can be held at around 250 °C. At this temperature the pyrolysis of the Cd source, dimethylcadmium (DMCd), and the tellurium source, diisopropyltellurium (DIPTe), are sufficiently slow to minimize the premature reaction.

A schematic of an MOCVD reactor suitable for the growth of HgCdTe is shown in Fig. 14.16. The main features are the heating of the reactor wall and the internal source of liquid Hg. This is the only MOCVD process that uses an elemental source and is only possible because of the high vapor pressure of liquid Hg. In fact, a higher partial pressure can be achieved using the elemental source than an organometallic source, and this is the only metal where this is true. Another feature of this reactor cell is the injection of DMCd through the Hg source to avoid a radical exchange reaction between DMCd and Hg [14.51].

The alloy composition needs to be precisely controlled in order to control the detection wavelength of the infrared detector. For a 10 μm detector the proportion of Cd on the metal sub-lattice is 21% and it needs to be controlled to better than 0.5%. The reaction rates for CdTe and HgTe require different optimum flow rates, which makes simultaneous alloy control very difficult. This problem was overcome using the interdiffused multilayer process (IMP), where alternate layers of HgTe and CdTe were grown and the flow rate optimized for each [14.52]. This process relied on very rapid Cd/Hg interdiffusion in the HgCdTe alloy, which enables complete homogenization of the alloy at the growth temperature. The composition is now simply controlled by the relative thicknesses of HgTe and CdTe, as shown schematically in Fig. 14.16b.

Sulfides and Selenides

Low-temperature growth of ZnSe (below 400 °C) has been achieved using ditertiarybutylselenide (DTBSe) with dimethylzinc (DMZn) or the adduct DMZn.TEN. The amine, triethylamine (TEN), was first introduced to suppress gas-phase reactions with the hydrides H_2Se and H_2S. However, this adduct has the additional advantage of reducing the saturated vapor pressure of DMZn and making it easier to manage. The significance of keeping the growth temperature low is to avoid deep-level native defects that act as trapping sites for donors and acceptors. The p-type doping of ZnSe with nitrogen was problematic due to hydrogen passivating the dopant. Incorporation of nitrogen from a variety of precursor sources such as amines and azides up to concentrations of $10^{18}\,cm^{-3}$ could be readily achieved but the active dopant concentration, in general, remained below $10^{15}\,cm^{-3}$. This was in contrast with the success of MBE growth of ZnSe that was doped from a nitrogen-plasma source. *Fujita* and *Fujita* [14.53] overcame this problem by ex situ annealing in a similar manner to GaN but the weaker lattice gives less scope for this treatment compared with GaN.

An alternative approach for low-temperature growth of ZnSe was to use short-wavelength light and UV wavelengths to illuminate the growing surface and promote surface photocatalytic reactions [14.54]. It was shown by *Irvine* et al. [14.55] that the reaction kinetics did not depend on the Se precursor, giving similar growth rates for DMSe, DESe and DIPSe, but depended on a hydrogen radical reaction that was initiated by a surface decomposition of the Group II precursor. Although this was effective for growing epitaxial films of ZnSe at temperatures well below 400 °C, it was clear that hydrogen incorporation was a natural consequence of the reaction mechanism.

MOCVD of Group II Oxides

A recent resurgence of interest in ZnO and related materials such as ZnMgO and ZnCdO has arisen because of the success of GaInN as a blue emitter and the potential for further developments with UV laser diodes and a solid state replacement for domestic lighting. GaInN white light LEDs already exist but the efficiency of the phosphors would improve if they were excited with UV rather than blue photons. A further potential advantage of ZnO is that large ZnO single-crystal substrates can be grown by the hydrothermal method and would eventually avoid the defect problems associated with heteroepitaxy that have slowed progress with GaN.

All the early work on ZnO MOCVD used oxygen or water vapor as the oxygen source. These react strongly at room temperature with DMZn and DEZn. Although reasonably good quality ZnO films have been deposited with this approach, it is unlikely that it will lead to high-quality epitaxial growth or good doping control. Essentially, prereaction in all of the III–V and II–VI semiconductors has been a barrier to obtaining device-quality material. The favored alternative oxygen precursors are the alcohols: isopropanol and tertiarybutanol. For higher temperatures, N_2O is a suitable precursor. In general, for epitaxial growth on sapphire or ZnO substrates it is necessary to grow at temperatures above 600 °C, but for polycrystalline transparent conducting oxides (TCOs) these precursors can react at temperatures as low as 300 °C. It is possible to readily dope ZnO n-type using TMAl, but as with ZnSe it has been difficult to achieve p-type doping. Some encouraging results have been obtained using ammonia [14.56], but this work is still at an early stage of development and must be solved before electroluminescent devices can be made. This is proving to be another class of materials where the versatility of MOCVD has a lot of potential for innovative solutions.

14.2.7 Conclusions

This section of the chapter has covered the basic principles of MOCVD and reviewed the range of III–V and II–VI semiconductors that can be grown in this way. This can be contrasted with LPE and MBE, where each method will have its own strengths and weaknesses for a particular material or application. The strength and the weakness of MOCVD is in its complexity. With the right precursors it is possible to deposit almost any inorganic material, but in many cases the reaction mechanisms are not well understood and the development is empirical, with the researcher spoilt by a very wide choice. This is not to deny the very considerable successes that have led to major industries in compound semiconductors that has been epitomized in the past 10 years by the productionization of GaN and the plethora of large LED displays that would not have been possible without MOCVD. Without the pioneering work of *Manasevit* and *Simpson*, who demonstrated the potential to grow so many of these materials in the early years, and the fortuitous ease with which GaAs/AlGaAs could be grown, we might not have tried so hard with the more difficult materials and hopefully we will see many more innovations in the future with MOCVD.

14.3 Molecular Beam Epitaxy (MBE)

14.3.1 Introduction and Background

MBE is conceptually a very simple route to epitaxial growth, in spite of the technology required, and it is this simplicity that makes MBE such a powerful technique. It can be thought of as a refined form of vacuum evaporation, in which neutral atomic and molecular beams from elemental effusion sources impinge with thermal velocities on a heated substrate under ultrahigh vacuum (UHV). Because there are no interactions within or between the beams, only the beam fluxes and the surface reactions influence growth, giving unparalleled control and reproducibility. Using MBE, complex structures can be grown atomic layer by atomic layer, with precise control over thickness, alloy composition and intentional impurity (doping) level. UHV confers two further advantages: cleanliness, because the partial pressures of impurities are so low, and compatibility with in situ analytical techniques – essential to understanding the surface reaction kinetics. The basic elements of an MBE system are shown schematically in Fig. 14.17. A number of reviews [14.57–59] and books [14.60, 61] have discussed the physics, chemistry, technology and applications of MBE.

The technique that became known as MBE evolved from surface kinetic studies of the interaction of silane (SiH_4) beams with Si [14.62] and of Ga and As_2 beams with GaAs [14.63]. *Cho* and coworkers, who first used the term molecular beam epitaxy, demonstrated that MBE was a viable technique for the growth of III–V material for devices, leading the way for a worldwide expansion of effort.

Much early MBE equipment had a single vacuum chamber for loading, deposition and analysis, which led to prolonged system pumpdown between growths. The technique required significant improvements in vacuum conditions before very high quality thin films were grown. *Cho* surrounded the effusion cells with a liquid nitrogen cryopanel to give thermal isolation and reported the MBE growth of thin films of n- and p-type GaAs for device purposes [14.64] and of GaAs/AlGaAs heterostructures [14.65]. The introduction of a substrate-exchange load lock [14.66] drastically reduced pumpdown times and reduced contamination of the deposition chamber. The installation of extensive internal liquid nitrogen-cooled cryopanels [14.67] substantially increased the pumping of oxygen containing species and permitted the growth of AlGaAs with superior quality. *Tsang* [14.68] demonstrated lasers with threshold current densities superior to those grown by LPE. Uniform growth over a 2″-diameter wafer was achieved by the introduction of a rotating substrate holder capable of 5 rpm [14.69]. Advanced forms of these features are now standard on commercial MBE systems, many of which feature a modular design. The technology of MBE is now mature, with increasing numbers of ever larger high-throughput, multiwafer production MBE machines in widespread use since the early 1990s.

MBE has been used to grow a wide range of materials, including semiconductors, superconductors, metals, oxides, nitrides and organic films. In almost all cases there is a drive to produce structures with ever smaller dimensions, whether for higher-performance devices, quantum confinement or, more recently, nanotechnology. This is longest established in III–V semiconductors, from the GaAs/AlGaAs superlattice [14.70] through quantum wells (QWs) and modulation doping to quantum wires and quantum dots [14.71]. Such low-dimensional structures form the basis of the QW lasers and p-HEMTs produced in huge volumes by MBE for optoelectronic and microwave applications. The combination of precise growth control and in situ analysis makes MBE the preeminent technology used to meet such demands.

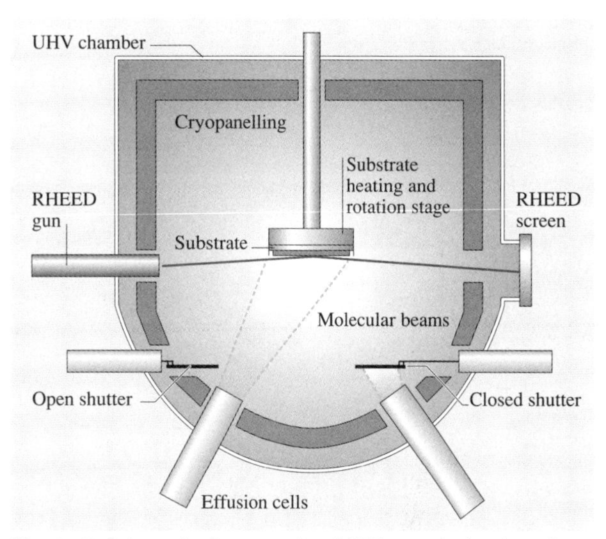

Fig. 14.17 Schematic diagram of an MBE growth chamber, showing the effusion cells and shutters, the substrate stage and the arrangement of the RHEED system

14.3.2 Reaction Mechanisms

One of the first and most important of the in situ analytical techniques to be used in MBE is reflection high-energy electron diffraction (RHEED). From an early stage [14.72] it was used to determine the surface structure of the clean substrate and growing layer. This revealed that, in general, all surfaces are reconstructed (they have a lower symmetry than the bulk), and Cho was the first to propose that the two-fold periodicity observed in the $[\bar{1}10]$ direction on the (001) surface was the result of dimerization of As atoms on the arsenic-terminated surface, which was confirmed many years later by scanning tunneling microscopy [14.73]. RHEED is a forward-scattering technique and therefore more compatible with the MBE arrangement of normally incident fluxes than the back-scattering geometry of LEED.

A further application of RHEED is the in situ measurement of growth rate. It was found [14.74, 75] that the intensity of any diffraction feature oscillated with a period corresponding to the growth of a single monolayer (ML) – a layer of Ga + As – in the [001] direction on a (001) substrate. These oscillations arise from surface morphological changes during two-dimensional (2-D) layer-by-layer growth, the Frank–van der Merwe mode, and a typical result for GaAs is shown in Fig. 14.18. The exact origin of the oscillations is still the subject of debate [14.59], but the technique was found to be applicable to many other material systems, including elemental semiconductors, metals, insulators, superconductors and even organic compounds.

Thus RHEED provided information on surface reconstruction and quantitative measurements of growth dynamics, which could be combined with those from other in situ techniques and related to theoretical treatments. In the early 1970s, the surface chemistry involved

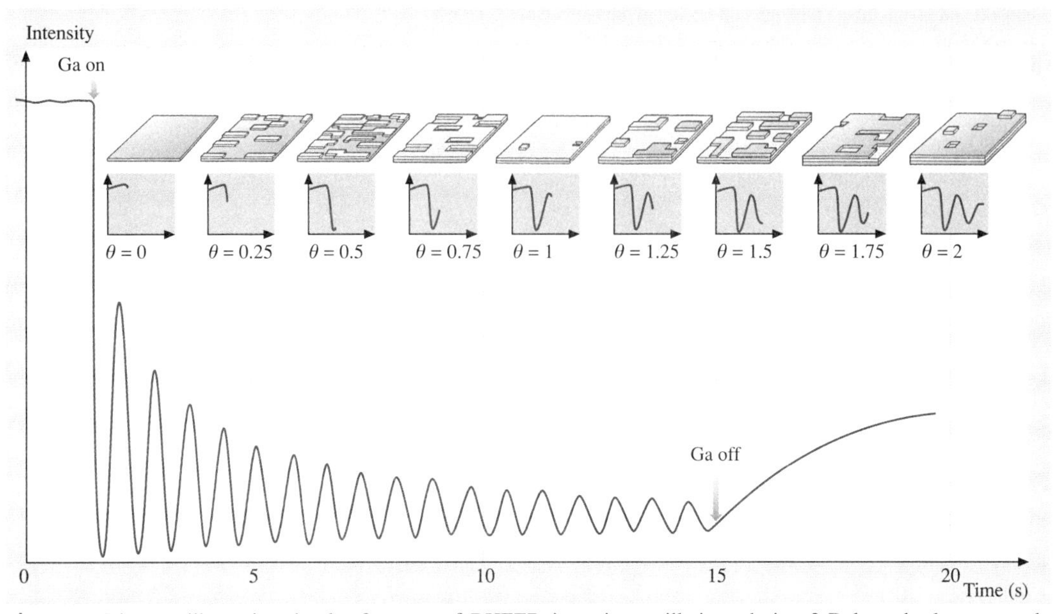

Fig. 14.18 Diagram illustrating the development of RHEED intensity oscillations during 2-D layer-by-layer growth, together with an actual experimental plot (courtesy of B. A. Joyce)

in the MBE growth of III–V compounds was studied extensively using a combination of RHEED and modulated molecular beam mass spectrometry [14.76] or temperature-programmed desorption [14.77].

Detailed information was obtained on surface reactions involving gallium and arsenic. It was found that the dissociation of GaAs results in the desorption of As_2, and not of As or As_4 [14.78], and that a significant amount of desorption takes place at MBE growth temperatures. The sublimation of elemental arsenic, as from an effusion cell, results in the formation of As_4 alone. It was also shown that growth from Ga + As_2 is a first-order reaction (Fig. 14.19), whereas growth from Ga + As_4 is second-order [14.79,80]. These results suggest that in MBE growth Ga sticks to available As sites and chemisorption of As_2 occurs on available Ga atoms. Chemisorption of As_4 occurs with two As_4 molecules interacting on adjacent Ga atoms. The sticking coefficient of As_4 is observed to be less than or equal to 0.5, whereas the sticking coefficient of As_2 can be equal to one, in agreement with this model. This implies that maximum coverage will be less than 100%, since single Ga sites cannot be occupied, and that As vacancies will be introduced into material grown using As_4. This was thought to be responsible for the higher deep-level concentrations observed, which were found to be reduced when As_2 was used or at higher growth temperatures. It is now known that above 580 °C As_4 dissociates to As_2 on the surface [14.81].

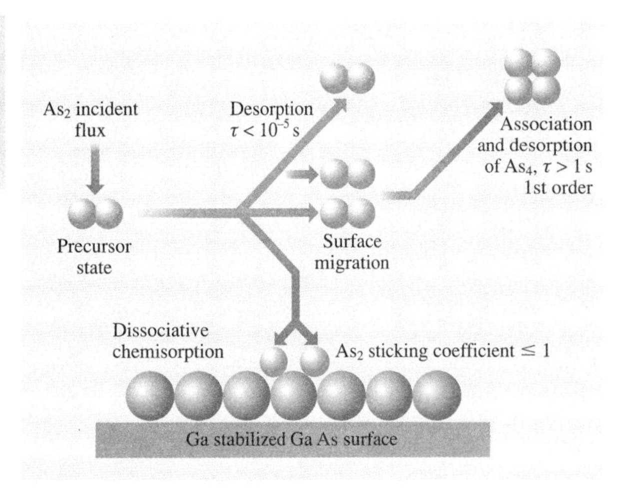

Fig. 14.19 Surface chemistry of the growth of GaAs from As_2 and Ga (courtesy of B. A. Joyce)

Detailed models were established for the growth of GaAs [14.79, 80] and AlAs [14.82]. With minor modifications these models are also valid for ternary alloys such as AlGaAs [14.83]. It was established that growth rate depends solely on the net Group III flux (incident flux minus desorbing flux), and that ternary alloy composition can be controlled by adjusting the ratio of the Group III fluxes provided the thermal stabilities of both of the binary compounds that make up the alloy are considered. The Group V element need only be supplied in excess. The situation is more complicated for alloys containing both arsenic and phosphorus, since the presence of one Group V element influences the sticking coefficient of the other. *Foxon* et al. [14.84] found that phosphorus has a much lower incorporation probability than arsenic.

More recent studies have made use of in situ STM and more powerful theoretical treatments to consider nucleation and growth at the atomistic level, but the basic models are still sound.

14.3.3 MBE Growth Systems

The UHV system required for MBE is of conventional stainless steel construction, with an ultimate or background vacuum of less than 5×10^{-11} torr achievable with a clean system after baking, and with the liquid nitrogen cryopanels filled. At such a pressure the molecules have a typical mean free path of 10^6 m and so only suffer collisions with the internal surfaces of the system. It would take several hours to build up a monolayer of impurity on the wafer surface.

Oil-free pumping is used to eliminate the possibility of contamination by hydrocarbon backstreaming; typically rough pumping is with sorption pumps and UHV pumping is with ion pumps and titanium sublimation pumps. Diffusion pumps or turbomolecular pumps can be used, however, provided suitable cold traps are fitted, and such pumps are required for the higher gas loads involved in GSMBE and CBE. A two- or three-stage substrate entry load-lock and preparation chamber isolated by gate valves is used to minimize the exposure of the system to air. In modular systems, further deposition and analysis chambers may be added to the system and samples are transferred via the preparation chamber under UHV conditions.

The growth chamber substrate stage is surrounded by a large liquid nitrogen-cooled cryopanel, which has a high pumping speed for H_2O, CO, O_2 and other condensable species. This is arranged so that the heated (500–700 °C) substrate is not directly exposed to thermal sources other than the molecular beams themselves, and impurities emanating from any other source can only reach the substrate after suffering at least one collision with, and probable adsorption by, a surface at liquid nitrogen temperatures. The cryopanel also reduces contamination arising from outgassing from the walls of the chamber that are exposed to radiation from the effusion cells.

Control over the composition and doping levels of the epitaxial layers is achieved by precise temperature control of the effusion cells and the use of fast-acting shutters in front of these cells. In most systems, the growth rate is about 1 monolayer per second, and the shutter operation time of 0.1 s thus corresponds to less than a monolayer of growth. A growth rate of one monolayer per second closely approximates one micrometer per hour for GaAs/AlGaAs.

Although often referred to as K-cells, the solid-source effusion cells used in MBE growth have a large orifice so as to obtain a high flux at reasonable temperatures. A true Knudsen cell has a very small outlet orifice compared to the evaporating surface, so that an equilibrium vapor pressure, typically 10^{-3} torr, is maintained within the cell. The diameter of this orifice is less than one tenth of the molecular mean free path, which is typically several centimeters. Under these conditions, a near approximation to ideal Knudsen effusion is obtained from the cell, giving molecular flow with an approximately cosine distribution. The flux from such a cell can be calculated quite accurately, but a high temperature is required to produce a reasonable growth rate; for example a gallium Knudsen cell would need to be at 1500 °C to produce the same flux as an open-ended effusion cell at 1000 °C. The lower operating temperature helps to reduce impurities in the flux and puts a lower thermal load on the system. The beam from an open-ended cell may not be calculable with any degree of accuracy but it is highly reproducible. Once calibrated via growth rate, normally by in situ measurement, the flux can be monitored using an ion gauge located on the substrate stage.

A number of effusion cells can be fitted to the growth chamber, generally in a ring facing towards the substrate with the axis of each cell at an angle of 20–25° to the substrate normal. Simple geometrical considerations therefore dictate the best possible uniformity that can be achieved with a stationary substrate [14.85]. Associated with each cell is a fast-action refractory metal shutter with either pneumatic or solenoid operation.

Uniformity of growth rate for a binary compound can be achieved by rotating the substrate at speeds as low as a few rpm, but compositional uniformity of ternary or quaternary alloys requires rotation of the substrate at speeds of up to 120 rpm (normally rotation is timed so that one rotation corresponds to the growth of one monolayer). The mechanical requirements for a rotating substrate stage in a UHV system are quite demanding, as no conventional lubricants can be used on the bearings or feedthroughs and yet lifetime must exceed several million rotations. Magnetic rotary feedthroughs have largely replaced the earlier bellows type.

The need to rotate the substrate to give uniformity also leads to complications in substrate temperature measurement. The substrate is heated by radiation from a set of resistively heated tantalum foils behind the substrate holder, and both the heater and the thermocouple are stationary. Without direct contact between the thermocouple and the wafer the "indicated" thermocouple temperature will be very different from the "actual" substrate temperature. Some form of calibration can be obtained by using a pyrometer, although problems with window coating, emissivity changes and substrate transparency below the bandgap impose limits on the accuracy of such measurements. Alternatively, a number of "absolute" temperature measurements can be obtained by observing transitions in the RHEED pattern, which occur at reasonably fixed temperatures. However, such transitions occur in the lower temperature range and extrapolation to higher growth temperatures is not completely reliable. If the substrate is indium bonded to a molybdenum block (using the surface tension of the indium to hold the substrate), then inconsistencies in wetting can lead to variations in temperature across the substrate. Most modern systems and all production machines use "indium-free" mounting, which avoids these problems. However, the substrate is transparent to much of the IR radiation from the heater, putting a higher thermal load on the system.

The substrate preparation techniques used prior to MBE growth are very important, as impurities on the surface provide nucleation sites for defects. Historically, various chemical clean and etch processes were used, but wafers are now usually supplied "epi-ready", with a volatile oxide film on the surface that protects the surface from contamination and can be thermally removed within the UHV chamber. RHEED is used to confirm the cleanliness of the surface prior to growth.

Historically, one of the major problems in MBE was the presence of macroscopic defects, with a typical density of 10^3–10^5 cm^{-2}, although densities below 300 cm^{-2} were reported for ultraclean systems [14.86]. Defects are generally divided into two types; small hillocks or pits and oval defects. Such defects are a serious obstacle to the growth of material for integrated circuits, and considerable effort was devoted to the problem. Oval defects are microtwin defects originating at a local imperfection, oriented in the (110) direction and typically 1 to 10 μm in length [14.87]. There are several possible sources of these defects, including foreign impurities on the substrate surface due to inadequate substrate preparation or to oxides from within the system, and possibly from the arsenic charge or the condensate on the cryopanels. The fact that oval defects were not seen when graphite crucibles were used but were common with PBN crucibles suggests that gallium oxide from the gallium melt is a major source of such defects since oxides would be reduced by the graphite. *Chai* and *Chow* [14.88] demonstrated a significant reduction in defects by careful charging of the gallium source and prolonged baking of the system. The irregular hillocks and pits seen in MBE-grown material were probably produced by microdroplets of gallium spitting from the effusion cells on to the substrate surface [14.89]. Gallium spitting can be caused by droplets of Ga that condense at the mouth of the effusion cell, fall back into the melt and explode, ejecting droplets of liquid Ga, or by turbulence in the Ga melt due to uneven heating that causes a sudden release of vapor and droplets.

Continuous developments in the design of Group III effusion cells for solid-source MBE have largely eliminated the problem of macroscopic defects. Large-area Ta foil K-cell heaters have reduced the uneven heating of the PBN crucible; the use of a "hot-lipped" or two-temperature Group III cell, designed with a high-temperature front end to eliminate the condensation of gallium metal at the mouth of the cell, significantly reduced the spitting of microdroplets. Combined with careful procedures and the use of an arsenic cracker cell, defect densities as low as 10 cm^{-2} have been reported [14.90].

There have also been developments in the design of Group V cells. A conventional arsenic effusion cell produces a flux of As_4, but the use of a thermal cracker to produce an As_2 flux resulted in the growth of GaAs with better optical properties and lower deep-level concentrations [14.91]. In the case of phosphorus, growth from P_2 was strongly preferred to that from P_4 for several reasons [14.92], and this was one reason behind the development of GSMBE described below. However, the use of phosphine requires suitable pumping and safety systems. The Group V cracker cell has two distinct zones. The first comprises the As or P reservoir and produces a controlled flux of the tetramer; this passes through the second – high-temperature – zone, where dissociation to the dimer occurs. Commercial high-capacity cracker cells, some including a valve between the two zones to allow fast switching of Group V flux, have been developed for arsenic and phosphorus.

Commercial MBE systems have increased throughput with multiwafer substrate holders, cassette loading and UHV storage and preparation chambers linked to the growth chamber with automated transfer, while increased capacity effusion cells have reduced the downtime required for charging. In some cases additional analytical and processing chambers have been added to permit all-UHV processing of the device structure.

14.3.4 Gas Sources in MBE

A number of MBE hybrids were developed that combined the advantages of UHV deposition and external gas or metalorganic sources to produce a versatile technique that has some advantages over MBE and MOVPE. These techniques utilize the growth chambers developed for MBE and pumping systems with a high continuous throughput, typically liquid nitrogen-trapped diffusion pumps or turbomolecular pumps. Layers are deposited from molecular beams of the precursor materials introduced via gas source cells that are essentially very fine leak controllers. As in MBE, there are no interactions within or between beams and the precise control of beams using fast-acting gas-line valves is therefore translated into precise control of the species arriving at the substrate. Shutters are not generally required; atomically sharp interfaces and monolayer structures can be defined as a consequence of submonolayer valve switching times.

Panish [14.93] investigated the use of cracked arsine and phosphine in the epitaxial growth of GaAs and InP, later extending this work to the growth of GaInAsP, and suggested the name gas source MBE (GSMBE). The major advantage of gaseous Group V sources was that the cracker cells produced controllable fluxes of the dimers As_2 and P_2, giving improved control of the As:P ratio. The MBE growth of high-quality GaAs from cracked arsine and elemental gallium was demonstrated at the same time by *Calawa* [14.94].

The extension to gaseous Group III sources was made by *Veuhoff* et al. [14.95], who investigated the MOCVD of GaAs in a simple MBE system using trimethylgallium (TMGa) and uncracked arsine. Further study showed that cracking of arsine at the substrate surface was negligible, leading to the conclusion that unintentional cracking of the arsine had taken place in the inlet system. The acronym metalorganic MBE (MOMBE) was used to describe GaAs growth from TMGa and cracked arsine in a modified commercial MBE system [14.96].

The growth of device-quality GaAs, InGaAs and InP from alkyl sources of both Group III and Group V elements was demonstrated by *Tsang* [14.97], who used the alternative acronym chemical beam epitaxy (CBE). The use of Group V alkyls, which had much poorer purity than the hydrides, was undertaken for safety reasons [14.98]. Material quality was improved when cracked arsine and phosphine were used [14.99]. RHEED observations indicated that reconstructed semiconductor surfaces could be produced prior to growth, as for MBE [14.100], and GaAs/AlGaAs quantum well structures were demonstrated that were comparable with those grown by MBE or MOVPE [14.101].

Almost all of this work was with III–Vs, where MOMBE/CBE was seen to have several significant advantages over MBE while retaining many of its strengths, including in situ diagnostics. The use of vapor Group III sources would avoid the morphological defects associated with effusion cells, and higher growth rate and greater throughput could be achieved. Both Group III and Group V sources were external, allowing for easy replacement without the need to break vacuum. Flux control with mass flow controllers (MFCs) and valves would improve control over changes in composition or doping level, since flow could be changed faster than effusion cell temperature. Abrupt changes could thus be achieved that would require switching between two preset effusion cells in MBE (a problem when the number of cells was limited by geometry). It also offered improved long-term flux stability and greater precursor flexibility. As this was still a molecular beam technique, precise control over layer growth and abrupt interfaces would be retained, without any of the gas phase reactions, boundary layer problems or depletion of reagents associated with MOVPE. Other advantages included improved InP quality using a P_2 flux, lower growth temperatures and selective-area epitaxy.

There was, however, a price to pay in system complexity, with the need for gas handling and high-volume pumping arrangements added to the expensive UHV growth chamber. These would have been acceptable if CBE had demonstrated clear advantages, but there were a number of other issues. The standard Al and Ga precursors used in MOVPE (trimethylaluminium and trimethylgallium) produced strongly p-type material when used in CBE, due to the incorporation of C as an acceptor. Triethylgallium proved to be a viable Ga source, but alloy growth was more complicated; no universally acceptable Al source was found, while InGaAs growth was found to be strongly temperature-dependent. The surface chemistry associated with metalorganic sources proved complex and the temperature dependence of surface reactions not only restricted growth conditions, but also had a serious impact on uniformity and reproducibility, particularly for quaternary alloys such as GaInAsP [14.102]. The lack of suitable gaseous dopant sources, particularly for Si, was a further handicap [14.103], but the deliberate use of C for p-type doping proved a success and this was transferred to MBE and MOVPE. Carbon diffuses significantly less than the 'standard' MBE and MOVPE dopants, Be and Zn, respectively [14.104], and proved an ideal dopant for thin highly doped layers such as the base region in heterojunction bipolar transistors (HBTs) and for p-type Bragg reflector stacks in vertical cavity surface-emitting laser structures (VCSELs).

GSMBE remains important, not in the III–V field where, with some exceptions [14.105], the development of high-capacity Group V cracker cells provided an easier route to an As_2 and P_2 flux, but in the III–nitride field. There are two major routes to nitride MBE: active nitrogen can be supplied by cracking N_2 in an RF or ECR plasma cell, or ammonia can be injected and allowed to dissociate on the substrate surface. In contrast, CBE has not demonstrated a sufficient advantage over its parent technologies to be commercially successful, particularly as both MOVPE and MBE have continued to develop as production techniques.

14.3.5 Growth of III–V Materials by MBE

Although they were amongst the earliest materials to be grown by MBE, GaAs-based alloys retain great importance, with MBE supplying materials for the mass production of optoelectronic and microwave devices and leading research into new structures and devices.

GaAs/AlGaAs

AlGaAs is an ideal material for heterostructures, since AlAs has a greater bandgap than GaAs and the two have negligible mismatch ($\approx 0.001\%$). The growth rate in MBE depends on the net Group III flux, with one micrometer per hour corresponding to a flux of 6.25×10^{14} Ga (or Al) atoms $cm^{-2}s^{-1}$. At low growth temperatures, all incident Group III atoms are incorporated into the growing film, together with sufficient arsenic atoms to maintain stoichiometry, and excess arsenic atoms are desorbed. However, III–V compounds are thermally unstable at high temperatures. Above ≈ 600 K [14.79] arsenic is preferentially desorbed, so an excess arsenic flux is required to maintain stoichiometry. At higher temperatures, loss of the Group III element becomes significant, so that the growth rate is less than would be expected for the incident flux. This is particularly important for the growth of AlGaAs, where growth temperatures above 650 °C are generally used to give the best optical properties. The Ga flux must be significantly increased above that used at lower temperatures in order to maintain the required composition of the alloy.

Typical growth temperatures for MBE of GaAs are in the range 580–650 °C and material with high purity and low deep-level concentrations has been obtained in this temperature range [14.106, 107]. The commonly used dopants, Be (p-type) and Si (n-type), show excellent incorporation behavior and electrical activity at these temperatures and at moderate doping levels. As was noted above, for highly doped layers Be has largely been replaced by C, which diffuses somewhat more slowly. At doping levels above $\approx 5 \times 10^{17}$, Si occupies both Ga (donor) and As (acceptor) sites, producing electrically compensated material with a consequent reduction in mobility. It is still predominantly a donor, however, and is the best available n-type dopant. The electrical properties of GaAs also depend on the As/Ga flux ratio, since this influences the site occupancy of dopants. The optimum As/Ga ratio is that which just maintains As-stabilized growth conditions, which can be determined using RHEED observations of surface reconstruction.

MBE-grown GaAs is normally p-type, the dominant impurity being carbon [14.108]. The carbon concentration was found to correlate with CO partial pressure during growth [14.109] – CO is a common background species in UHV, being synthesized at hot filaments. The lowest acceptor levels commonly achieved are of the order of $5 \times 10^{13}\,cm^{-3}$, and such layers can be lightly doped to give n-type material with high mobilities. However, very high purity GaAs has been produced by adjusting the operating conditions for an arsenic cracker cell [14.110], which suggests that carbon contamination originates from hydrocarbons in the As charge. Unintentionally doped GaAs was n-type with a total impurity concentration of $< 5 \times 10^{13}\,cm^{-3}$ and a peak mobility of $4 \times 10^5\,cm^2\,V^{-1}\,s^{-1}$ at 40 K, the highest reported for n-type GaAs. The 77 K mobility of $> 200\,000\,cm^2\,V^{-1}\,s^{-1}$ is comparable with that for the highest purity GaAs grown by LPE [14.111].

MBE is capable of the growth of very high-quality material for structures whose physical dimensions are comparable to the wavelength of an electron (or hole) so that quantum size effects are important. Such structures have typical layer thicknesses from 100 Å down to

2.8 Å, the thickness of a monolayer of GaAs, and have doping and composition profiles defined on an atomic scale. These low-dimensional structures have become very important in III–V device technology, one example being the modulation-doped heterojunction.

The precise control of growth that is possible is demonstrated by the very high mobilities obtained for modulation-doped structures grown by MBE. The modulation-doped GaAs/AlGaAs heterojunction has a band structure that causes carriers from the highly doped AlGaAs to be injected into a thin undoped region in the GaAs (Fig. 14.20a). Since the carriers are then separated from the donor atoms that normally scatter them, and confined in a quantum well as a two-dimensional electron gas (2DEG), very high mobilities can be achieved. Through suitable design of the structure and the use of lightly doped AlGaAs, a GaAs/AlGaAs 2DEG structure was produced with a peak mobility of $1.1\times10^{7}\,cm^{2}\,V^{-1}\,s^{-1}$ at 1.3 K and a 4 K mobility of $4.5\times10^{6}\,cm^{2}\,V^{-1}\,s^{-1}$ and with a sheet carrier concentration of $10^{12}\,cm^{-2}$ [14.112]. GaAs layers grown in the same machine typically exhibited 77 K mobilities of the order of $100\,000\,cm^{2}\,V^{-1}\,s^{-1}$. Similar results can be obtained by delta doping the AlGaAs, that is by confining the Si donor atoms to a single monolayer in the AlGaAs separated by a few nm from the well, a technique with a wide range of applications [14.113].

The major application of this structure for device purposes is the high electron mobility transistor (HEMT), also known as the modulation-doped FET (MODFET), an FET in which the carriers are confined to the two-dimensional layer (Fig. 14.20b). The main advantage of the HEMT is not the increase in mobility, which is modest at room temperature for a practical device, but the very low noise when it is operated as a microwave amplifier, due to the reduction in impurity scattering. The HEMT is an essential component of many microwave systems, including mobile phones and satellite TV receivers.

MBE became used as a production technology for GaAs-based devices because of the excellent uniformity and reproducibility possible, and because the extensive UHV load-lock system of a production machine permits a large number of runs to be undertaken on a continuous basis. This was demonstrated as early as 1991 by a number of manufacturers producing both HEMTs and GaAs/AlGaAs lasers [14.114]. By 1994 some 5 million MQW lasers for compact disc applications were produced per month by MBE. Other devices included HEMTs and InAs Hall sensors. A comparison of production costs made at this time [14.115] showed that capital costs for MBE were very similar to those for MOVPE, which required costly safety systems for the hydride gases, while MBE costs per wafer were somewhat less than for MOVPE. The growth in demand for HEMTs for mobile phones drove a further expansion in production MBE, although MOVPE remained dominant in InP-based optoelectronic devices.

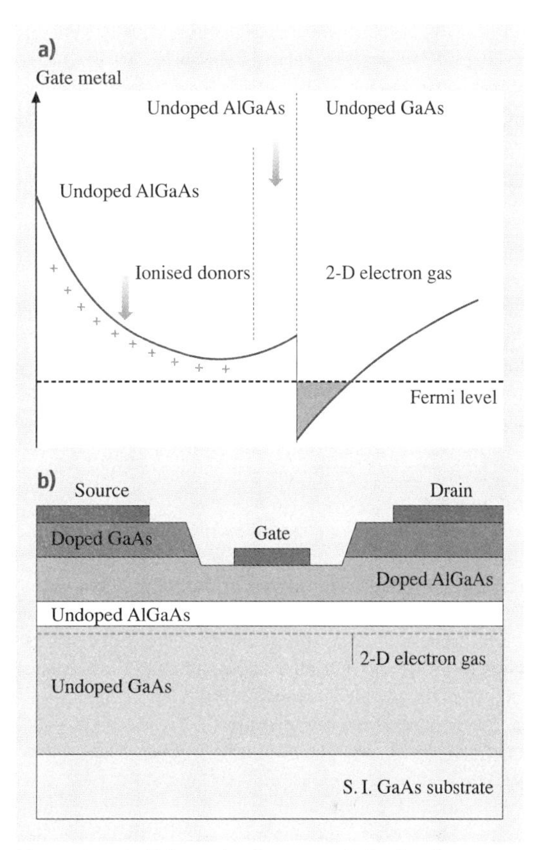

Fig. 14.20a,b Schematic diagrams of the band structure (**a**) and physical arrangement (**b**) of the high electron mobility transistor (HEMT)/modulation-doped FET

InGaAs

InGaAs is generally grown at lower temperatures than GaAs/AlGaAs because of the higher rate of In desorption. Both InGaAs and the quaternary InGaAsP can be lattice-matched to InP substrates; heterostructures in these materials form the basis of the 1.3 μm and 1.55 μm optoelectronic devices used in telecommunications. InAs has a lattice mismatch of $\approx 7\%$ with GaAs; thin films of InGaAs can be deposited pseudomorphically but as the In content is increased the critical thickness falls, so that little more than a ML of InAs can be grown in this way. Thicker films will be relaxed by the formation of dislocations, limiting their usefulness for devices. The pseudomorphic HEMT (or pHEMT) replaces the GaAs channel of the conventional HEMT with a strained InGaAs channel in order to take advantage of the greater carrier confinement and superior electron transport properties. The maximum In content of the channel is limited by the need to prevent relaxation.

Under certain growth conditions InAs or InGaAs islands are formed spontaneously. It is generally accepted that the growth of InAs on GaAs (001) follows a version of the Stranski–Krastanov mode, which implies that following the deposition of ≈ 1.7 ML of InAs in a 2-D pseudomorphic form (sometimes referred to as the wetting layer), coherent 3-D growth is initiated by a very small increment (≤ 0.1 ML) of deposited material to relax the elastic strain introduced by the lattice mismatch. The QDs rapidly reach a saturation number density, which is both temperature- and In flux-dependent, with a comparatively narrow size (volume) distribution. The actual process is rather more complicated, not least as a result of alloying with the GaAs substrate, and is the subject of much current research [14.59].

These islands can be embedded in a layer of GaAs to form self-assembled quantum dots (QDs), which have become a topic of immense interest due to the potential application of QDs in a wide range of devices, especially lasers. The volume fraction of QDs in an active layer can be increased by building up a 3-D array; the strain field induced around each dot influences not only the inter-dot spacing but also the capping layer growth, so that subsequent layers of dots are aligned ([14.71] and references therein). QD lasers offer a route to long-wavelength emission from GaAs-based devices [14.116].

Group III Nitrides

The growth of Group III nitrides has been dominated by MOVPE since the demonstration of a high-brightness blue-emitting InGaN-AlGaN double heterostructure LED by *Nichia* [14.117] and the subsequent development of other optoelectronic devices, including laser diodes also emitting in the blue [14.118]. MBE has made a significant contribution to more fundamental studies and to the growth of nitrides for high-power and microwave devices. Once again the wide range of in situ diagnostic techniques available has been important.

For heteroepitaxial growth on the most commonly used substrates, sapphire and silicon carbide, several parameters strongly influence the quality of material produced. These include substrate cleaning, initial nitridation, the nucleation and coalescence of islands involved in the low-temperature growth of a buffer layer, and subsequent annealing at a higher temperature. The polarity (nature of the outermost layer of atoms) of {0001}-oriented hexagonal structure films also has a crucial influence on material quality, but both N- or Ga-polarity can occur with MBE growth on sapphire substrates [14.119]. Under typical growth conditions with MOVPE, however, Ga-polarity material is exclusively produced.

Several of these problems can be resolved using GaN templates obtained by growing thick layers onto suitable substrates using MOVPE and then exploiting the advantages of MBE to produce the functional layer on the GaN template. These advantages include well-controlled layer-by-layer growth and lower growth temperatures than those used for MOVPE, so that InGaN phase separation and In desorption are less problematic and precise quantum wells can be grown. No post-growth thermal annealing is required to activate the p-type dopant. In this way films have been produced with smooth surface morphology and high performance, although MBE was still limited to low-power LEDs until the recent demonstration of laser diodes [14.120].

In the use of active nitrogen from plasma sources, the III/V flux ratio at the substrate during growth is also a critical parameter. GaN layers grown with a low III/V flux ratio (N-stable growth) display a faceted surface morphology and a tilted columnar structure with a high density of stacking faults. Smooth surfaces are only obtained under Ga-rich conditions, where not only is there a dramatic reduction in surface roughness, but significant improvements in structural and electrical properties are also observed. This is, of course, the exact opposite of the growth of most III–V compounds, such as GaAs. In the case of nitrides, it is thought that Ga-rich conditions (close to the point where Ga droplets are formed) promote step flow growth, whereas N-stable growth promotes the nucleation of new islands. In contrast, growth from NH_3 is smoother under N-rich conditions [14.121].

Group III–V Nitrides

The "dilute nitrides" are III–V–N materials such as GaAsN and GaInNAs, where the N concentration is $\leq 2\%$. Replacing a small fraction of As atoms with smaller N atoms reduces both the lattice constant and the bandgap. Adjusting the composition of GaInNAs allows the bandgap, band alignment, lattice constant and strain to be tailored in a material that can be lattice-matched with GaAs [14.122]. This offers strong carrier confinement and thermal stability compared to InP-based devices operating at 1.3 μm and 1.55 μm, and allows GaAs VCSEL technology to be exploited at these wavelengths [14.123].

These materials are grown in a metastable regime at a low growth temperature because of the miscibility gap in the alloys, so a less stable precursor than NH_3 is needed. A nitrogen plasma source provides active N without the incorporation of hydrogen during growth associated with hydride sources, thus avoiding the deleterious formation of N–H bonds. Material with excellent crystallinity and strong PL at 1.3 μm can be obtained by optimizing growth conditions and using post-growth rapid thermal annealing [14.124]. A number of challenges remain, particularly in the higher N material required for longer wavelengths, including the limited solubility of N in GaAs and nonradiative defects caused by ion damage from the N plasma source. These challenges may be met by using GaInNAsSb; the addition of Sb significantly improves the epitaxial growth and the material properties, and enhanced luminescence is obtained at wavelengths longer than 1.3 μm [14.125].

14.3.6 Conclusions

MBE, historically seen as centered on GaAs-based electronic devices, has broadened its scope dramatically in both materials and devices. In addition to the materials described above, MBE has been used to grow epitaxial films of a wide range of semiconductors, including other III–V materials such as InGaAsP/InP and GaAsSb/InAsSb; silicon and silicon/germanium; II–VI materials such as ZnSe; dilute magnetic semiconductors such as GaAs:Mn [14.126] and other magnetic materials. It has also been used for the growth of metals, including epitaxial contacts for devices, oxides [14.127] and organic films [14.128]. Two clear advantages possessed by MBE are the wide range of analytical techniques compatible with a UHV system and the precise control of growth to less than a monolayer, which give it unrivaled ability to grow quantum dots and other nanostructures.

References

14.1 D. Elwell, H. J. Scheel: *Crystal Growth from High-Temperature Solutions* (Academic, New York 1975)
14.2 H. Nelson: RCA Rev. **24**, 603 (1963)
14.3 R. L. Moon: J. Cryst. Growth **170**, 1 (1997)
14.4 H. J. Scheel: *The Technology of Crystal Growth and Epitaxy*, ed. by H. J. Scheel, T. Fukuda (Wiley, Chichester 2003)
14.5 P. Capper, T. Tung, L. Colombo: *Narrow-Gap II–VI Compounds for Optoelectronic and Electromagnetic Applications*, ed. by P. Capper (Chapman & Hall, London 1997)
14.6 M. B. Panish, I. Hayashi, S. Sumski: Appl. Phys. Lett. **16**, 326 (1970)
14.7 M. G. Astles: *Liquid Phase Epitaxial Growth of III–V Compound Semiconductor Materials and their Device Applications* (IOP, Bristol 1990)
14.8 V. A. Dmitriev: Physica B **185**, 440 (1993)
14.9 T. Ciszek: *The Technology of Crystal Growth and Epitaxy*, ed. by H. J. Scheel, T. Fukuda (Wiley, Chichester 2003)
14.10 M. I. Alonso, K. Winer: Phys. Rev. B **39**, 10056 (1989)
14.11 V. A. Dmitriev: *Properties of Silicon Carbide*, EMIS Datareview Series, ed. by G. L. Harris (IEE, London 1995) p. 214
14.12 Y. Mao, A. Krier: Mater. Res. Soc. Symp. Proc **450**, 49 (1997)
14.13 A. Krier, Z. Labadi, A. Manniche: J. Phys. D: Appl. Phys. **32**, 2587 (1999)
14.14 M. Mauk: private communication (2004)
14.15 H. Yamane, M. Shimada, T. Sekiguchi, F. J. DiSalvo: J. Cryst. Growth **186**, 8 (1998)
14.16 C. Klemenz, H. J. Scheel: J. Cryst Growth **211**, 62 (2000)
14.17 A. Krier, H. H. Gao, V. V. Sherstinov: IEE Proc. Optoelectron **147**, 217 (2000)
14.18 E. R. Rubstov, V. V. Kuznetsov, O. A. Lebedev: Inorg. Mater. **34**, 422 (1998)
14.19 M. G. Mauk, Z. A. Shellenbarger, P. E. Sims, W. Bloothoofd, J. B. McNeely, S. R. Collins, P. I. Rabinowitz, R. B. Hall, L. C. DiNetta, A. M. Barnett: J. Cryst Growth **211**, 411 (2000)
14.20 J.-i. Nishizawa, K. Suto: *Widegap II–VI Compounds for Optoelectronic Applications*, ed. by H. E. Ruda (Chapman & Hall, London 1992)
14.21 F. Sakurai, M. Motozawa, K. Suto, J.-i. Nishizawa: J. Cryst Growth **172**, 75 (1997)
14.22 M. G. Astles: *Properties of Narrow Gap Cadmium-Based Compounds*, EMIS Datareview series, ed. by P. Capper (IEE, London 1994) pp. 13, 380
14.23 B. Pelliciari, J. P. Chamonal, G. L. Destefanis, L. D. Cioccio: Proc. SPIE **865**, 22 (1987)
14.24 P. Belca, P. A. Wolff, R. L. Aggarwal, S. Y. Yuen: J. Vac. Sci. Technol. A **3**, 116 (1985)
14.25 S. H. Shin, J. Pasko, D. Lo: Mater. Res. Soc. Symp. Proc. **89**, 267 (1987)
14.26 A. Wasenczuk, A. F. M. Willoughby, P. Mackett, E. S. O'Keefe, P. Capper, C. D. Maxey: J. Cryst. Growth **159**, 1090 (1996)

14.27 T. Tung, L.V. DeArmond, R.F. Herald: Proc. SPIE **1735**, 109–134 (1992)
14.28 P.W. Norton, P. LoVecchio, G.N. Pultz: Proc. SPIE **2228**, 73 (1994)
14.29 P. Capper, J. Gower, C. Maxey, E. O'Keefe, J. Harris, L. Bartlett, S. Dean: *Growth and Processing of Electronic Materials*, ed. by N. McN. Alford (IOM Communications, London 1998)
14.30 C. C. Wang: J. Vac. Sci. Technol. B **9**, 1740 (1991)
14.31 T.N. Casselman, G.R. Chapman, K. Kosai, et al.: *U.S. Workshop on Physics and Chemistry of MCT and other II–VI compounds*, Dallas, TX (Oct. 1991)
14.32 R. S. Patrick, A.-B. Chen, A. Sher, M. A. Berding: J. Vac. Sci. Technol. A **6**, 2643 (1988)
14.33 A. Rogalski: *New Ternary Alloy Systems for Infrared Detectors* (SPIE, Bellingham 1994)
14.34 A. A. Chernov, H. J. Scheel: J. Cryst. Growth **149**, 187 (1996)
14.35 H. M. Manasevit, W. I. Simpson: J. Electrochem. Soc. **116**, 1725 (1969)
14.36 A. A. Chernov: Kinetic processes in vapor phase growth. In: *Handbook of Crystal Growth*, ed. by D. T. J. Hurle (Elsevier, Amsterdam 1994)
14.37 G. B. Stringfellow: J. Cryst. Growth **115**, 1 (1991)
14.38 D. M. Frigo, W. W. van Berkel, W. A. H. Maassen, G. P. M. van Mier, J.H. Wilkie, A. W. Gal: J. Cryst. Growth **124**, 99 (1992)
14.39 S. Tompa, M. A. McKee, C. Beckham, P. A. Zwadzki, J. M. Colabella, P. D. Reinert, K. Capuder, R. A. Stall, P. E. Norris: J. Cryst. Growth **93**, 220 (1988)
14.40 X. Zhang, I. Moerman, C. Sys, P. Demeester, J. A. Crawley, E. J. Thrush: J. Cryst. Growth **170**, 83 (1997)
14.41 P. M. Frijlink, J. L. Nicolas, P. Suchet: J. Cryst. Growth **107**, 166 (1991)
14.42 D. W. Kisker, J. N. Miller, G. B. Stringfellow: Appl. Phys. Lett. **40**, 614 (1982)
14.43 C. A. Larson, N. I. Buchan, S. H. Li, G. B. Stringfellow: J. Cryst. Growth **93**, 15 (1988)
14.44 M. W. Raynor, V. H. Houlding, H. H. Funke, R. Frye, J. A. Dietz: J. Cryst. Growth **248**, 77–81 (2003)
14.45 R. M. Biefeld, R. W. Gedgridge Jr.: J. Cryst. Growth **124**, 150 (1992)
14.46 C. A. Wang, S. Salim, K. F. Jensen, A. C. Jones: J. Cryst. Growth **170**, 55 (1997)
14.47 S. Nakamura: Jpn. J. Appl. Phys. **30**, 1620 (1991)
14.48 A. Stafford, S. J. C. Irvine, K. Jacobs. Bougrioua, I. Moerman, E. J. Thrush, L. Considine: J. Cryst. Growth **221**, 142 (2000)
14.49 S. Keller, S. P. DenBaars: J. Cryst. Growth **248**, 479 (2003)
14.50 B. Cockayne, P. J. Wright: J. Cryst. Growth **68**, 223 (1984)
14.51 W. Bell, J. Stevenson, D. J. Cole-Hamilton, J. E. Hails: Polyhedron **13**, 1253 (1994)
14.52 J. Tunnicliffe, S. J. C. Irvine, O. D. Dosser, J. B. Mullin: J. Cryst. Growth **68**, 245 (1984)
14.53 S. Fujita, S. Fujita: J. Cryst. Growth **145**, 552 (1994)
14.54 S. Fujita, A. Tababe, T. Sakamoto, M. Isemura, S. Fujita: J. Cryst. Growth **93**, 259 (1988)
14.55 S. J. C. Irvine, M. U. Ahmed, P. Prete: J. Electron. Mater. **27**, 763 (1988)
14.56 J. Wang, G. Du, B. Zhao, X. Yang, Y. Zhang, Y. Ma, D. Liu, Y. Chang, H. Wang, H. Yang, S. Yang: J. Cryst. Growth **255**, 293 (2003)
14.57 A. Y. Cho: J. Cryst. Growth **150**, 1 (1995)
14.58 C. T. Foxon: J. Cryst. Growth **251**, 1–8 (2003)
14.59 B. A. Joyce, T. B. Joyce: J. Cryst. Growth **264**, 605 (2004)
14.60 A. Y. Cho: *Molecular Beam Epitaxy* (AIP, New York 1994)
14.61 E. H. C. Parker: *The Technology and Physics of Molecular Beam Epitaxy* (Plenum, New York 1985)
14.62 B. A. Joyce, R. R. Bradley: Philos. Mag. **14**, 289–299 (1966)
14.63 J. R. Arthur: J. Appl. Phys. **39**, 4032 (1968)
14.64 A. Y. Cho: J. Vac. Sci. Technol. **8**, 31 (1971)
14.65 A. Y. Cho: Appl. Phys. Lett. **19**, 467 (1971)
14.66 J. W. Robinson, M. Ilegems: Rev. Sci. Instrum. **49**, 205 (1978)
14.67 P. A. Barnes, A. Y. Cho: Appl. Phys. Lett. **33**, 651 (1978)
14.68 W. T. Tsang: Appl. Phys. Lett. **34**, 473 (1979)
14.69 A. Y. Cho, K. Y. Cheng: Appl. Phys. Lett. **38**, 360 (1981)
14.70 L. L. Chang, L. Esaki, W. E. Howard, R. Ludeke: J. Vac. Sci. Technol. **10**, 11 (1973)
14.71 H. Sakaki: J. Cryst. Growth **251**, 9 (2003)
14.72 A. Y. Cho: J. Appl. Phys. **41**, 2780 (1970)
14.73 M. D. Pashley, K. W. Haberern, J. M. Woodall: J. Vac. Sci. Technol. **6**, 1468 (1988)
14.74 J. J. Harris, B. A. Joyce, P. J. Dobson: Surf. Sci. **103**, L90 (1981)
14.75 J. H. Neave, B. A. Joyce, P. J. Dobson, N. Norton: Appl. Phys. **31**, 1 (1983)
14.76 C. T. Foxon, M. R. Boudry, B. A. Joyce: Surf. Sci. **44**, 69 (1974)
14.77 J. R. Arthur: Surf. Sci. **43**, 449 (1974)
14.78 C. T. Foxon, J. A. Harvey, B. A. Joyce: J. Phys. Chem. Solids **34**, 1693 (1973)
14.79 C. T. Foxon, B. A. Joyce: Surf. Sci. **50**, 434 (1975)
14.80 C. T. Foxon, B. A. Joyce: Surf. Sci. **64**, 293 (1977)
14.81 E. S. Tok, J. H. Neave, J. Zhang, B. A. Joyce, T. S. Jones: Surf. Sci. **374**, 397 (1997)
14.82 A. Y. Cho, J. R. Arthur: Prog. Solid State Chem. **10**(3), 157–191 (1975)
14.83 C. T. Foxon, B. A. Joyce: J. Cryst. Growth **44**, 75 (1978)
14.84 C. T. Foxon, B. A. Joyce, M. T. Norris: J. Cryst. Growth **49**, 132 (1980)
14.85 M. A. Herman, H. Sitter: *Molecular Beam Epitaxy*, Springer Ser. Mater. Sci., Vol. 7 (Springer, Berlin, Heidelberg 1988) p. 7
14.86 J. Saito, K. Nambu, T. Ishikawa, K. Kondo: J. Cryst. Growth **95**, 322 (1989)
14.87 M. Bafleur, A. Munoz-Yague, A. Rocher: J. Cryst. Growth **59**, 531 (1982)

14.88 Y. G. Chai, R. Chow: Appl. Phys. Lett. **38**, 796 (1981)
14.89 C. E. C. Wood, L. Rathburn, H. Ohmo, D. DeSimone: J. Cryst. Growth **51**, 299 (1981)
14.90 S. Izumi, N. Hayafuji, T. Sonoda, S. Takamiya, S. Mitsui: J. Cryst. Growth **150**, 7 (1995)
14.91 J. H. Neave, P. Blood, B. A. Joyce: Appl. Phys. Lett. **36**(4), 311 (1980)
14.92 C. R. Stanley, R. F. C. Farrow, P. W. Sullivan: *The Technology and Physics of Molecular Beam Epitaxy*, ed. by E. H. C. Parker (Plenum, New York 1985)
14.93 M. B. Panish: J. Electrochem. Soc. **127**, 2729 (1980)
14.94 A. R. Calawa: Appl. Phys. Lett. **38**(9), 701 (1981)
14.95 E. Veuhoff, W. Pletschen, P. Balk, H. Luth: J. Cryst. Growth **55**, 30 (1981)
14.96 N. Putz, E. Veuhoff, H. Heinicke, H. Luth, P. J. Balk: J. Vac. Sci. Technol. **3**(2), 671 (1985)
14.97 W. T. Tsang: Appl. Phys. Lett. **45**(11), 1234 (1984)
14.98 W. T. Tsang: J. Vac. Sci. Technol. B **3**(2), 666 (1985)
14.99 W. T. Tsang: Appl. Phys. Lett. **49**(3), 170 (1986)
14.100 T. H. Chiu, W. T. Tsang, J. E. Cunningham, A. Robertson: J. Appl. Phys. **62**(6), 2302 (1987)
14.101 W. T. Tsang, R. C. Miller: Appl. Phys. Lett. **48**(19), 1288 (1986)
14.102 J. S. Foord, C. L. Levoguer, G. J. Davies, P. J. Skevington: J. Cryst. Growth **136**, 109 (1994)
14.103 M. Weyers, J. Musolf, D. Marx, A. Kohl, P. Balk: J. Cryst. Growth **105**, 383–392 (1990)
14.104 R. J. Malik, R. N. Nottenberg, E. F. Schubert, J. F. Walker, R. W. Ryan: Appl. Phys. Lett. **53**, 2661 (1988)
14.105 F. Lelarge, J. J. Sanchez, F. Gaborit, J. L. Gentner: J. Cryst. Growth **251**, 130 (2003)
14.106 A. Y. Cho: J. Appl. Phys. **50**, 6143 (1979)
14.107 R. A. Stall, C. E. C. Wood, P. D. Kirchner, L. F. Eastman: Electron. Lett. **16**, 171 (1980)
14.108 R. Dingle, C. Weisbuch, H. L. Stormer, H. Morkoc, A. Y. Cho: Appl. Phys. Lett. **40**, 507 (1982)
14.109 G. B. Stringfellow, R. Stall, W. Koschel: Appl. Phys. Lett. **38**, 156 (1981)
14.110 C. R. Stanley, M. C. Holland, A. H. Kean, J. M. Chamberlain, R. T. Grimes, M. B. Stanaway: J. Cryst. Growth **111**, 14 (1991)
14.111 H. G. B. Hicks, D. F. Manley: Solid State Commun. **7**, 1463 (1969)
14.112 C. T. Foxon, J. J. Harris, D. Hilton, J. Hewett, C. Roberts: Semicond. Sci. Technol. **4**, 582 (1989)
14.113 K. Ploog: J. Cryst. Growth **81**, 304 (1987)
14.114 H. Tanaka, M. Mushiage: J. Cryst. Growth **111**, 1043 (1991)
14.115 J. Miller: III–Vs Rev. **4**(3), 44 (1991)
14.116 D. Bimberg, M. Grundmann, F. Heinrichsdorff, N. N. Ledentsov, V. M. Ustinov, A. R. Korsh, M. V. Maximov, Y. M. Shenyakov, B. V. Volovik, A. F. Tsatsalnokov, P. S. Kopiev, Zh. I. Alferov: Thin Solid Films **367**, 235 (2000)
14.117 S. Nakamura, T. Mukai, M. Senoh: Appl. Phys. Lett. **64**(13), 1689 (1994)
14.118 S. Nakamura, M. Senoh, S. Nagahama, N. Iwasa, T. Yamada, T. Matsushita, H. Kiyoku, Y. Sugimoto: Jpn. J. Appl. Phys. **35**, 74 (1996)
14.119 H. Morkoç: J. Mater. Sci. Mater. El. **12**, 677 (2001)
14.120 S. E. Hooper, M. Kauer, V. Bousquet, K. Johnson, J. M. Barnes, J. Heffernan: Electron. Lett. **40**(1), 33 (2004)
14.121 N. Grandjean, M. Leroux, J. Massies, M. Laügt: Jpn. J. Appl. Phys. **38**, 618 (1999)
14.122 M. Kondow, K. Uomi, A. Niwa, T. Kitatani, S. Watahiki, Y. Yazawa: Jpn. J. Appl. Phys. **35**, 1273 (1996)
14.123 H. Riechert, A. Ramakrishnan, G. Steinle: Semicond. Sci. Technol. **17**, 892 (2002)
14.124 M. Kondow, T. Kitatani: Semicond. Sci. Technol. **17**, 746 (2002)
14.125 J. S. Harris, S. R. Bank, M. A. Wistey, H. B. Yuen: IEE Proc. Optoelectron. **151**(5), 407 (2004)
14.126 H. Ohno: J. Cryst. Growth **251**, 285 (2003)
14.127 H. J. Osten, E. Bugiel, O. Kirfel, M. Czernohorsky, A. Fissel: J. Cryst. Growth **278**, 18 (2005)
14.128 F.-J. Meyer zu Heringdolf, M. C. Reuter, R. M. Tromp: Nature **412**, 517 (2001)

15. Narrow-Bandgap II–VI Semiconductors: Growth

The field of narrow-bandgap II–VI semiconductors is dominated by the compound $Hg_{1-x}Cd_xTe$ (CMT), although some Hg-based alternatives to this ternary have been suggested. The fact that CMT is still the preeminent infrared (IR) material stems, in part, from the fact that the material can be made to cover all IR regions of interest by varying the x value. In addition, the direct band transitions in this material result in large absorption coefficients, allowing quantum efficiencies to approach 100%. Long minority carrier lifetimes result in low thermal noise, allowing high-performance detectors to be made at the highest operating temperatures reported for infrared detectors of comparable wavelengths. This chapter covers the growth of CMT by various bulk growth techniques (used mainly for first-generation infrared detectors), by liquid phase epitaxy (used mainly for second-generation infrared detectors), and by metalorganic vapor phase and molecular beam epitaxies (used mainly for third-generation infrared detectors, including two-color and hyperspectral detectors). Growth on silicon substrates is also discussed.

The field of narrow-bandgap II–VI semiconductors is dominated by $Hg_{1-x}Cd_xTe$ (CMT) (although some Hg-based alternatives to this ternary have been suggested and are discussed by *Rogalski* [15.1]). The reason that CMT is still the main infrared (IR) material is at least partially because this material can be made to cover all IR regions of interest by varying the value of x appropriately. This material also has direct band transitions that yield large absorption coefficients, allowing the quantum efficiency to approach 100% [15.2,3]. Furthermore, long minority carrier lifetimes result in low thermal noise allowing high-performance detectors to be made at the highest operating temperatures reported for infrared detectors of comparable wavelengths. These three major advantages all stem from the energy band structure of the material and they apply whatever device architecture is used. It has been shown that CMT is the third most studied semiconductor after silicon and gallium arsenide. This chapter covers the growth and characterization of CMT, mainly concentrating on the x region between 0.2 and 0.4 where the majority of applications are satisfied, and some of the Hg-based alternative ternary systems first described by Rogalski. The detectors made from these materials will be described in the chapter by Baker in this Handbook Chapt. 36.

In the first section the growth of CMT by various bulk techniques is reviewed. These include solid state recrystallization (SSR), Bridgman (plus ACRT, the accelerated crucible rotation technique), and the travelling heater method (THM). Despite the major advances made over the last three decades in the various epitaxial processes (liquid phase epitaxy, LPE, metalorganic vapor phase epitaxy, MOVPE, and molecular beam epitaxy, MBE), which are discussed in subsequent sections, bulk techniques for CMT are still used for photoconductive devices and for large-size CdTe-based materials for CMT substrates (see Chapt. 16 on wide-bandgap II–VI compounds by Isshiki). CMT material grown by LPE is currently in production for large-area focal plane arrays (FPAs) for both terrestrial and space thermal imaging applications. More complex, fully doped heterostructures are being grown by MOVPE and MBE for both FPAs and novel nonequilibrium device structures. Both MOVPE and MBE growth of device structures onto buffered silicon substrates is a topic of great interest currently, and this is briefly described.

For a fuller treatment of the topic matter in this chapter, plus further details of the characterization of CMT by various optical, electrical and structural means, the reader is referred to Chapts. 1 to 11 of [15.4], Sect. A of [15.5], Chapts. 1, 2, 10, 11, and 12 of [15.6], *Reine* [15.2], *Baker* [15.3], and Chapt. 7 of [15.7].

15.1 Bulk Growth Techniques

Despite the substantial progress that has taken place over the last three decades in various epitaxial growth techniques for CMT, bulk growth methods still play a key role when preparing device-quality material and producing substrates for these epitaxial techniques. A critical aspect of these techniques is the establishment of a thorough knowledge of the relevant phase equilibria, and a brief description of these phase diagrams for CMT is now given.

15.1.1 Phase Equilibria

The understanding of two types of equilibria is of critical importance. One is the solid compound in equilibrium with the gaseous phase (vapor growth) and the other is solid–liquid–gaseous equilibria (growth from liquid/melt). These phase equilibria also aid the understanding of post-growth heat treatments, either during cool-down to room temperature or during annealing processes used to adjust the stoichiometry, and hence the electrical properties of compounds. Solid–liquid–gaseous equilibria are described by three variables: temperature (T), pressure (p) and composition (x). It is, however, far easier to understand the interrelations between these parameters using two-dimensional (2-D) projections such as $T-x$, $p-T$ and $p-x$ plots.

A great deal of work has been carried out on the various phase equilibria ([15.5] contains reviews in this area). The liquidus and solidus lines in the pseudobinary HgTe-CdTe system are shown in Fig. 15.1. The wide separation between the liquidus and solidus, leading to marked segregation between the constituent binaries, CdTe and HgTe, was instrumental in the development of all the bulk growth techniques applied to this system. In addition to the solidus–liquidus separation, high Hg partial pressures are also influential during growth and post-growth heat treatments. A full appreciation of the

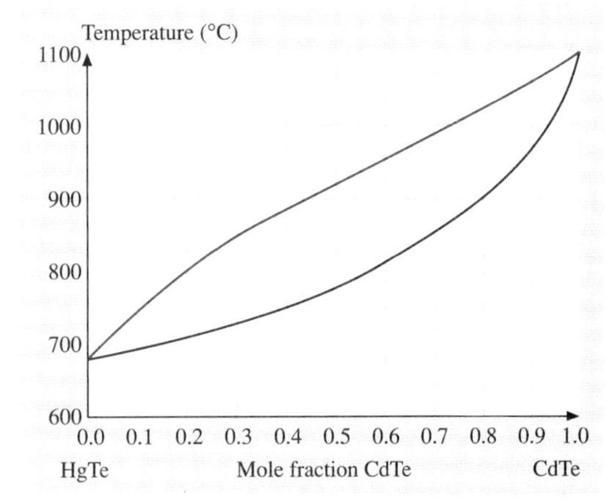

Fig. 15.1 $T-x$ phase diagram for the pseudo-binary CdTe-HgTe. (After [15.5])

$p_{Hg}-T$ diagram, shown in Fig. 15.2, is therefore essential. The curves are the partial pressures of Hg along boundaries for solid solutions of composition x where the solid solution is in equilibrium with another condensed phase as well as the vapor phase. For $x=0.1$ and $10^3/T=1.3\,K^{-1}$, CMT exists for Hg pressures between ≈ 0.1 (Te-saturated) and ≈ 7 (Hg-saturated) atm. Even at $x=0.95$ and Te-saturated conditions, Hg is the predominant vapor species. It should also be noted that no solid solution contains exactly 0.5 atomic fraction Te. These features are highly significant for controlling the native defect concentrations and hence electrical properties in CMT.

15.1.2 Crystal Growth

Growth from the melt has constituted the vast majority of work on bulk growth of CMT. CMT grown by several bulk methods is still in use, particularly for photoconductive detectors. Several historical reviews have been published [15.5, 8–10]. *Micklethwaite* [15.8] and *Kruse* [15.9] gave detailed discussions on the growth techniques and material characterization prior to 1980. Many techniques were tried, but three prime techniques survived these early attempts – SSR, Bridgman and THM – some of which are still in production, such as SSR in Germany and Bridgman in the UK. *Tennant* et al. [15.11] provided an authoritative view of the then current major issues in growth techniques. They concluded that the electrical performance had only recently been matched by LPE but pointed to the major drawbacks in bulk material of structural defects and size limitations for use in second-generation infrared detectors, such as focal plane arrays (FPAs).

Solid State Recrystallization (SSR)

This technique is used for ternary systems where there is a wide separation between solidus and liquidus, such as CMT. Other names that have been used for this process are quench anneal (QA) and cast recrystallize anneal (CRA). The term anneal is used in the first case to define a high-temperature grain-growth process, while in CRA it is a low-temperature process to adjust stoichiometry. In the basic technique, pure elements are loaded into a cleaned silica ampoule and the charge is melted and rocked to ensure complete mixing. Charges are then normally quenched rapidly, into air or oil, to room temperature in the vertical orientation. This produces a dendritic structure that is reduced/removed by the recrystallization step that occurs at temperatures just below the melting point for many days. Grain growth occurs and microinhomogeneities in composition are removed. Care must be taken in the quenching stage to avoid pipes/voids that cannot be removed by the recrystallization step.

Alternatives to the basic SSR process have included 'slush' growth [15.12], high-pressure growth [15.13], incremental quenching [15.14] and horizontal casting [15.15]. In the 'slush' process, an initial homogenous charge is prepared and then held across the liquidus–solidus gap with the lower end solid and the upper end liquid. High-pressure growth (30 atm He gas) was used in an attempt to reduce structural defects. The 'slush' technique was used in production in the USA, while the incremental quenching technique has been used to provide large-diameter feed material for THM growth by *Colombo* et al. [15.16]. Other developments were made in the basic process (e.g. [15.17, 18]) but details available in the open literature are sketchy, due to proprietary constraints. *Tregilgas* [15.19] has given a detailed review of the SSR process.

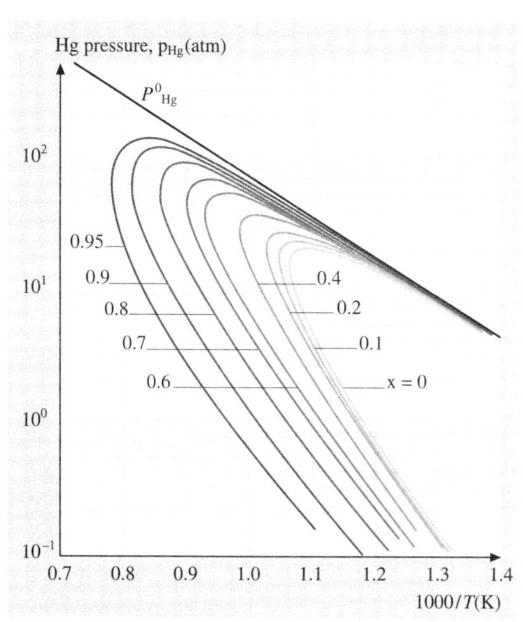

Fig. 15.2 $p-T$ diagram for $Hg_{1-x}Cd_xTe$ (CMT). (After [15.5])

Bridgman (plus ACRT)

In the standard Bridgman process, elemental Cd, Hg and Te are loaded into a clean silica ampoule, homogenized by melting/rocking, and then frozen slowly from one end in a vertical system (Fig. 15.3, [15.20]) to produce a large-grained ingot [15.21]. There is marked segregation of CdTe with respect to HgTe (due to the wide gap between the solidus and liquidus) in the axial direction, but this leads to an advantage of the Bridgman process over other techniques: that material in both the main ranges of interest (3–5 and 8–12 μm: $x = 0.3$ and 0.2, respectively) is produced in a single run. The accelerated crucible rotation technique (ACRT) in which the melt is subjected to periodic acceleration/deceleration provided a means of stirring melts contained in sealed, pressurized ampoules. However, it has been recently suggested that ACRT does not automatically promote melt mixing [15.22]. They suggest that the balance of two time-dependent fluid dynamics phenomena decides whether mixing is promoted or not.

The first report of ACRT in CMT Bridgman growth was given in [15.23]. These effects were developed and reviewed in [15.5, 21]. Crystals were produced up to

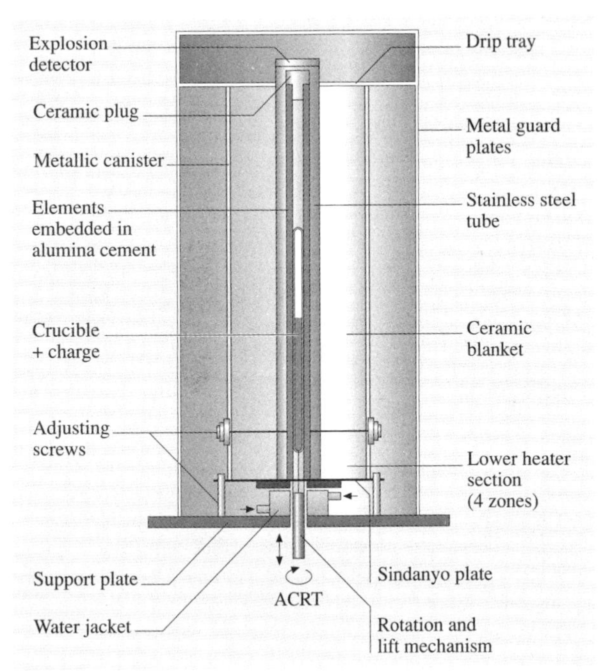

Fig. 15.3 Schematic of Bridgman growth kit for use with Hg-based compounds. (After [15.20])

20 mm in diameter and with x values of up to 0.6 in the tip regions of some crystals. Currently in this author's laboratory, crystals up to 21 mm in diameter and 200 mm in length weighing 0.5 kg are in routine production. Extensions to the basic ACRT process are enabling material of up to $x \approx 0.7$ to be produced in the first-to-freeze regions of crystals. These near-IR slices are highly transmissive and show a high degree of compositional uniformity ($\Delta x \approx 0.005$), see [15.24]. The use of ACRT produced material that enabled the fabrication of small two-dimensional arrays of photodiodes; this material was also used to demonstrate nonequilibrium detector operation and in the production of SPRITE detectors (Chapt. 11 of [15.6]).

Traveling Heater Method (THM)

Triboulet [15.25] (and his coworkers) developed this technique for CMT where diameters of up to 40 mm were accomplished and x values of up to 0.7 [15.26] for optical communication devices. *Durand* et al. [15.27] employed seeds to produce large oriented crystals, an advantage of this technique over other bulk methods. *Gille* et al. [15.28] adopted a slightly different approach. A pre-THM step was used to produce a quenched Te-rich (53–60%) CMT melt. A first THM run then provided the source ingot, after removal of the tip and tail sections, for the final THM run. The entire growth procedure took several months but gave uniform material. This group also used rotation in the horizontal growth by THM with some success [15.29]. Two groups also applied ACRT to THM and obtained enhanced material properties [15.30, 31].

15.1.3 Material Characterization

Compositional uniformity can be expressed in both the axial and the radial directions as $\Delta A(x)$ and $\Delta R(x)$, although the combination of the two clearly determines the yield of useful material. Assessment is normally by Fourier transform infrared (FTIR) spectrometry. The recrystallization process (in SSR) improves $\Delta A(x)$ from ± 0.05 to ± 0.02 [15.9] and removes the microinhomogeneity in x in both directions [15.32]. Independent control of axial and radial variations was claimed by *Nelson* et al. [15.33] to achieve ± 0.002 over 95% of slices up to 25 mm in diameter. *Colombo* et al. [15.14] quote $\Delta R(x)$ values of 0.005–0.01 in incremental-quenched material. *Su* et al. [15.34] report ± 0.008 for $\Delta A(x)$ and $\Delta R(x)$ in vertically cast material but better in horizontally cast material. *Galazka* [15.35] showed that reduced gravity gives smaller $\Delta R(x)$, and they also noted that constitutional supercooling at the first-to-freeze end is also reduced and the growth is diffusion-limited. The 'slush' process [15.8] produces $\Delta R(x)$ of ± 0.005 but only over the region of axially uniform material ≈ 3 mm in length.

Bartlett et al. [15.36] grew SSR and Bridgman crystals over a wide range of growth rates and concluded that $\Delta R(x)$ was due to the combination of a concave (to the melt) growth interface and density-driven convective flow within a boundary layer close to the interface. This, added to the normal segregation of the low melting point HgTe to the center of the concave interface, led to the observed variation. It was also concluded that crystals with simultaneous small $\Delta A(x)$ and $\Delta R(x)$ were not possible; slow growth gave low $\Delta R(x)$ while fast growth minimized $\Delta A(x)$. In Bridgman material grown with ACRT, an 80 mm length at $x = 0.21$ with both $\Delta A(x)$ and $\Delta R(x)$ within ± 0.002 was reported by *Capper* [15.21]. This was linked to a flat interface produced by strong Ekman flow in the interface region, although all Bridgman crystals have concave interfaces next to the ampoule walls [15.21]. A 20 mm-diameter ACRT crystal was grown that exhibited $\Delta R(x) = \pm 0.003$ at $x = 0.22$ [15.21].

Triboulet et al. [15.10] quoted values of ± 0.02 (over 3 cm at $x = 0.2$) and ± 0.002 for $\Delta A(x)$ and $\Delta R(x)$, respectively, for THM growth at 0.1 mm/h. Corresponding figures from *Colombo* et al. [15.16] for incremental-quenched starting material and a similar growth rate of 2 mm/d were ± 0.01 and ± 0.005. For the very slow grown material produced by *Gille* et al. [15.28], variations in composition were within experimental error for $\Delta R(x)$ while $\Delta A(x)$ was ± 0.005. *Royer* et al. [15.30] obtained improved radial uniformity with the addition of ACRT to THM, but in the work of *Bloedner* and *Gille* [15.31] both $\Delta A(x)$ and $\Delta R(x)$ were only as good as non-ACRT crystals.

With regard to electrical properties, as-grown materials from SSR, ACRT Bridgman (with $x < 0.3$) and THM are highly p-type in nature, believed to be due to metal vacancies. These types of materials can all be annealed at low temperatures, in the presence of Hg, to low n-type levels, indicating that the p-type character is due to metal vacancies. By contrast, Bridgman material is n-type as-grown [15.21] and the residual impurity donor level is found to be $< 5\times10^{14}\,\mathrm{cm^{-3}}$. *Higgins* et al. [15.18] have shown that n-type carrier concentrations can be $< 10^{14}\,\mathrm{cm^{-3}}$ at $x \approx 0.2$ in melt-grown material. A great deal of work has been done on the annealing behavior of SSR material, see *Tregilgas* [15.19]. n-type levels of $2\times10^{14}\,\mathrm{cm^{-3}}$ after Hg annealing $x = 0.2$ and higher-x material were reported by *Nguyen Duy* et al. [15.26], while *Colombo* et al. [15.16] quote $4\times10^{14}\,\mathrm{cm^{-3}}$. *Durand* et al. [15.37] note that THM growth at 600 °C results in p-type behavior, but growth at 700 °C gives n-type material. In this author's laboratory, current 20 mm-diameter ACRT material has reached mid-$10^{13}\,\mathrm{cm^{-3}}$ levels with high mobility after a normal low-temperature Hg anneal step. This is thought to reflect the improvements made in the purity of starting elements over the recent years. The basic p to n conversion process [15.19] has recently been extended by *Capper* et al. [15.38] to produce analytical expressions that can account for the temperature, donor level and composition dependencies of the junction depth.

High minority n-type carrier lifetimes have been reported in all three types of material. *Kinch* [15.39] noted values $> 1\,\mu$s (at 77 K) in $x = 0.2$ SSR crystals, while *Triboulet* et al. [15.10] quote 3 μs in THM material. *Pratt* and coworkers [15.40, 41] found high lifetimes, up to 8 μs in $x = 0.23$ Bridgman and ACRT material with n-type levels of $1{-}6\times10^{14}\,\mathrm{cm^{-3}}$, and up to 30 μs for equivalent $x = 0.3$ material (192 K).

In terms of extrinsic doping, most elements are electrically active in accordance with their position in the Periodic Table. This is true in SSR material for Group V and VII elements only after a high-temperature treatment, and is linked to the stoichiometry level at the growth temperature (in other words, those elements that substitute on Te lattice sites have to be forced onto the correct sites in Te-rich material). Group I and III elements are acceptors and donors, respectively, on the metal sites. There is evidence that some Group I elements can migrate at low temperatures to grain boundaries or to the surface of samples [15.21]. In ACRT crystals, Groups I and III are acceptors and donors, respectively, on the metal sites, as they are in Bridgman, with the exception of Au [15.21]. Groups V and VII are inactive dopants in those portions of ACRT crystals that are Te-rich as-grown ($x < 0.3$), but are active dopants for $x > 0.3$ where metal-rich conditions prevail, as found in Bridgman material. In doped material, grown by either standard Bridgman or ACRT, acceptor ionization energies were found to be lower than undoped counterparts [15.21]. Segregation of impurities in SSR is very limited due to the initial fast quench step. By contrast, Bridgman and THM benefit from marked segregation of impurities due to their relatively slow growth rate. Impurity segregation behavior was affected by ACRT [15.21], in general, segregation coefficients decrease in ACRT crystals, when compared to standard Bridgman. This segregation leads to very low levels of impurities in both THM [15.25] and Bridgman/ACRT material [15.42].

Vere [15.43] has reviewed structural properties and noted grains, subgrains, dislocations, Te precipitates and impurities on dislocations as major problems. Grain boundaries act as recombination centers and generate noise and dark current in devices. Subgrain sizes cover 50–500 μm in both SSR and Bridgman material. *Williams* and *Vere* [15.44] showed how a recrystallization step at high pressure and temperatures of > 600 °C coalesces subgrains and eventually eliminates them. Tellurium precipitation was extensively studied [15.44, 45] and was summarized in [15.44]. Precipitates were found to nucleate on dislocations during the quench from the recrystallization step but a 300 °C anneal dissolves the precipitates, generating dislocation loops that climb and lead to dislocation multiplication.

Quenching studies in Bridgman/ACRT crystals grown in flat-based ampoules [15.21] revealed not only a flat growth interface but also that the slow-grown material produced prior to quenching was single crystal. This demonstrated the power of Ekman stirring and the importance of initiating the growth of a single crystal grain. In Bridgman material, which is close to stoichiometric (Te excess < 1%), little Te precipitation occurs. Above this limit, precipitates/inclusions of 0.1–0.5 mm in size are seen. Recent work in this author's laboratory [15.24] has demonstrated flat solid–liquid interfaces in the $x \approx 0.21$ region of 20 mm-diameter ACRT crystals, mirroring the high level of compositional uniformity in these crystals.

Triboulet et al. [15.10] found no Te precipitates in their THM material, despite using a Te zone. Several groups [15.14, 46] noted an increase in dislocation density and in rocking curve widths on moving from the center to the edge of THM crystals. The addition of ACRT to the THM process led to increased growth rates with no detriment to the Te precipitation or grain structure.

We have recently [15.24, 47] extended our ACRT growth process to produce high-x material for various near-IR applications. This necessitated improving the ampoule sealing procedure to manage the higher pressures caused by the higher growth temperatures. Assessment techniques were extended beyond the normal FTIR wavelength characterization to include imaging of defects using a mid-IR camera system.

To summarize, the two main problems with all bulk material are the limitation on size – only THM has reproducibly produced material above 20 mm in diameter – and the fact that it is rarely completely single-crystal in nature. These two limitations led to the drive in the late 1970s and early 1980s to develop various epitaxial growth techniques.

15.2 Liquid-Phase Epitaxy (LPE)

The process of LPE has emerged as the predominant materials growth technology for the fabrication of second-generation CMT focal plane arrays (FPAs). Material can now be routinely grown for high-performance photoconductive, photovoltaic, and laser-detector devices over the entire 2–18 μm spectral region. This is manifested by increased size (over bulk material), enabling large-array formats to be made; reduced cycle time; better material characteristics in terms of composition uniformity, crystal quality, and electrical properties, making the material no longer the limiting factor in FPAs; and the realization of some advanced device structures, either to simplify the FPA design (back-side-illuminated detectors) or to enhance detector performance (double-layer heterostructures).

Two completely different technical approaches have been pursued: growth from Te solutions and growth from Hg solutions. One major advantage of the Hg-solution technology is its ability to produce layers of excellent surface morphology due to the ease of melt decanting. Two additional unique characteristics, essential for the fabrication of high-performance double-layer heterojunction (DLHJ) detectors by LPE, are low liquidus temperature (< 400 °C), which makes growth of the cap layer feasible, and ease of incorporating both p-type and n-type temperature-stable impurity dopants, such as As, Sb and In, during growth. While layers grown from Hg-rich solutions are easily doped with Group VB elements with high solubility [15.48], layers grown from tellurium-rich solutions are not. Group IIIB elements, in particular indium, are easily incorporated from both solutions.

A number of LPE approaches have been used to grow both thin and thick films. The principal LPE techniques used are tipping, dipping and sliding boat techniques. A schematic diagram of these three processes is shown in Fig. 15.4. The tipping and dipping techniques have been

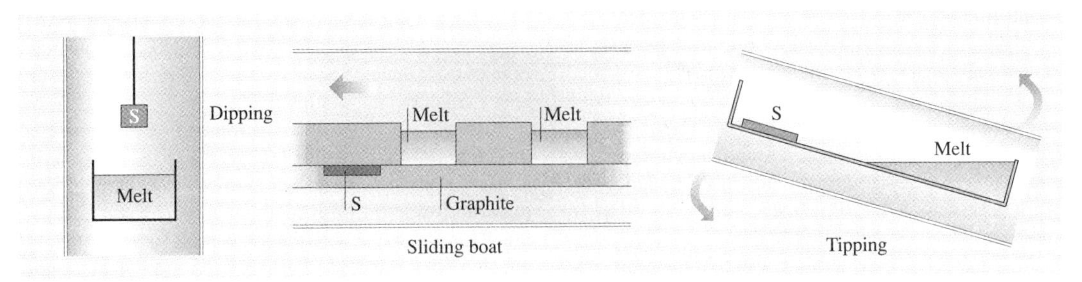

Fig. 15.4 Dipping, sliding boat and tipping LPE arrangements

implemented by using both tellurium- and mercury-rich solutions, whereas only tellurium-rich solutions have been used with the sliding boat. Both dipping and sliding boat Te-rich techniques are still in widespread use.

Extensive experimental phase diagram and thermodynamic data have been critically reviewed, along with the results calculated by the associated solution model [15.49]. As in bulk growth, full knowledge of the solid–liquid phase relation is essential for proper use of solution-growth processes. In addition, the solid–vapor and liquid–vapor phase relations are of practical importance, especially in view of the high Hg pressure in the growth process and the effect of the vapor of constituent components upon post-growth annealing and the consequent electrical properties. *Astles* [15.50] reviewed the experimental data on Te-rich LPE growth at 460 to 550 °C.

15.2.1 Hg-Rich Growth

For CMT growth from Hg-rich melts, the design and operation of a system is dominated by the consideration of the high vapor pressure of Hg, which comprises $\geq 90\%$ of the growth solution. A secondary but related factor is the requirement to minimize melt composition variation during and between growths due to solvent or solute loss. These factors led to the evolution of a vertical high-pressure furnace design with a cooled reflux region. The furnace has to provide a controllable, uniform and stable thermal source for the melt vessel, which has to be capable of maintaining at least 550 °C continuously. The cylindrical melt vessel consists of a high-strength stainless steel chamber lined with quartz. Such systems are capable of containing about 10–20 kg of melt at 550 °C for several years with no degradation in melt integrity or purity.

The system must be pressurized and leak-free to keep the Hg-rich melt from boiling or oxidizing at temperatures above 360 °C. Typical pressures range up to 200 psi and the pressurization gas may be high-purity H_2 or a less explosive reducing gas mixture containing H_2. The melts are always kept saturated and are maintained near to the growth temperature and pressure between successive runs. The prepared substrates are introduced into the melt through a transfer chamber or air lock. A high-purity graphite paddle with externally actuated shutters holds the substrates. The paddle with the shutters closed is not gas-tight but protects the substrates from undue exposure to Hg vapor/droplets, or other condensing melt components. The paddle assembly can be lowered into the melt and rotated to stir it.

In normal operation, the high-purity melt components are introduced into the clean melt vessel at room temperature and the system is sealed, evacuated and pressurized. The temperature of the furnace is raised above the predicted melting point and held constant until all of the solute dissolves. The amount of material removed from the melt during each growth run is relatively insignificant. Optimum layer smoothness occurs on polished lattice-matched CdZnTe substrates oriented close to the <111> plane. Growth begins by lowering the paddle plus substrates into the melt and allowing thermal equilibrium to be reached while stirring. A programmed ramp then reduces the melt temperature to the required level, at which point the shutters are opened and the substrates are exposed to the melt. The growth rate and layer thickness are determined mainly by the exposure temperature relative to the saturation point and the total growth range. The composition of the layer and its variation are determined mainly by the melt composition and its thermal uniformity. Large melts allow the production of layer areas of up to 30 cm^2 with excellent compositional and thickness uniformity, and allow dopant impurities to be accurately weighed for incorporation into layers and to maintain stable electrical characteristics over a long period of time. Four layers with a total area of 120 cm^2 can be grown in a single run [15.51]. *Norton* et al. [15.52] have also scaled-up for the growth of cap layers from Hg-rich solutions, with each reactor capable of growth on four 24 cm^2 base layers per run.

15.2.2 Te-Rich Growth

A number of problems encountered with bulk crystal growth techniques are solved using CMT growth from tellurium-rich solution. The most important of these is the reduction of the Hg vapor pressure over the liquid by almost three orders of magnitude at the growth temperature. Growth from Te-rich solutions is used in three embodiments: dipping, tipping and sliding boat technologies (Fig. 15.4). While the tipping process may be used for low-cost approaches, it is not as widely used as the sliding boat and the dipping techniques. A comparison of the three techniques is shown in Table 15.1.

Current dipping reactors are capable of growing in excess of 60 cm^2 per growth run and are kept at temperature for long periods, > 6 months. Melts, on the other hand, last a very long time, > 5 years. A sensor-based reactor capable of growing CMT thick layers at relatively high production volumes and with excellent

Table 15.1 Comparison of three LPE growth techniques

LPE growth technique	Advantages	Disadvantages
Sliding boat	Thickness control	Thick layers difficult to grow
	Large area	Substrate thickness/planarity control
	Double layers	High cost
Tipping	Thickness control	High cost
	Lowest investment for experimental investigation	Scale up
	Closed system/good p_{Hg} control	
Dipping	Thick layer growth	Thin-layer thickness control
	High throughput	Double layers
	Low cost	
	Large/flexible area	

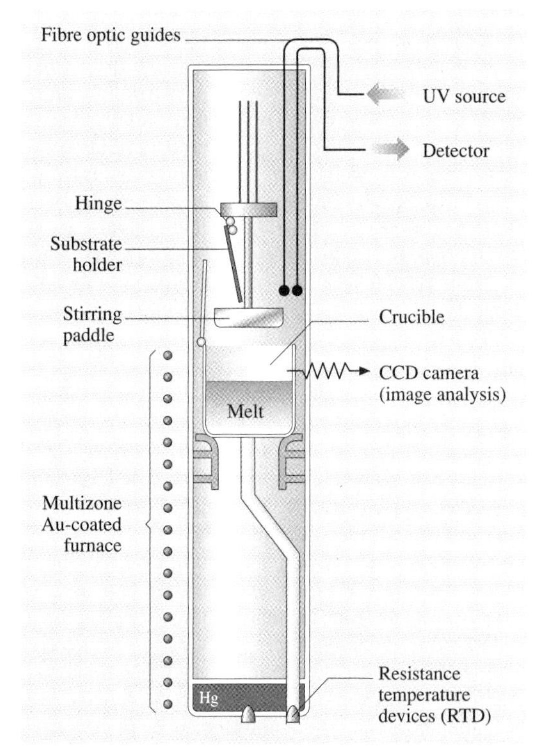

Fig. 15.5 Schematic diagram of a dipping LPE reactor showing the Te-rich melt, the mercury reservoir and positions of the sensors

reproducibility was described by *Westphal* et al. [15.53], Fig. 15.5. The reactors utilize resistance temperature devices (RTDs) for temperature control, a UV-vis spectrometer to measure the mercury partial pressure, and a video camera to aid in the observation of the liquidus temperature. The RTDs were used to control the temperature within a maximum range of $\pm 0.005\,°C$ at a temperature of $\approx 480\,°C$. The melts are about 4 kg with a liquidus temperature of $\approx 480\,°C$ for a solid CMT composition of $x = 0.225$. Accurate mercury vapor pressure control ensures constant liquidus and thus constant liquid composition.

Layers produced from Te-rich solutions are usually grown by isothermal supersaturation, programmed cooling techniques, or a combination of both. The details of the mass transport of Hg-Cd-Te solutions during LPE growth were presented by *Shaw* [15.54]. He predicted that isothermal growth of thin layers from supersaturated solutions would yield layers with uniform composition but that the growth rate would change during growth. While the growth rate is more uniform for layers grown by the programmed cooling technique, the composition gradient is higher.

The dipping growth process comprises: 1) mixing, 2) liquidus measurement, 3) meltback, and 4) growth. Because of the relatively large melt, stirring is required to mix the solution. The choice of growth method depends on the desired properties of the films. Typically, isothermal supersaturation is preferred when growing thin layers, whereas programmed cooling or a combination of both is used for thicker films. *Colombo* et al. [15.55] used a melt-tracking technique to control the composition of the solid.

Sliding-boat LPE normally consists of a graphite boat with a recess in the base to hold the substrate wafer and a movable block with wells that contain the

LPE solution and that allows the solution to be brought into contact with the substrate and then wiped off after growth. The main advantages of the technique are the efficient use of solution and the possibility of growing multilayers. The main disadvantages are the need for careful machining of the boat components in order to obtain efficient removal of the solution after growth and the need for precisely sized substrate wafers to fit into the recess in the boat. The sliding boat growth process has several variants, but essentially a polished substrate is placed into the well of a graphite slider and the Te-rich solution is placed into a well in the body of the graphite boat above the substrate and displaced horizontally from it. Normally, a separate well contains the HgTe charge to provide the Hg vapor pressure needed during growth and during cool-down to control the stoichiometry. The boat is then loaded into a silica tube that can be flushed with nitrogen/argon prior to the introduction of H_2 for the growth phase. The furnace surrounding the work tube is slid over the boat, and the temperature is increased to 10–20 °C above the relevant liquidus. At that point, a slow temperature ramp (2–3 °C/h) is initiated, and when the temperature is close to the liquidus of the melt the substrate is slid under the melt and growth commences. After the required thickness of CMT has been deposited (typical growth rates are 9–10 μm/h), the substrate is withdrawn and the temperature is decreased to an annealing temperature (to fix the p-type level in the as-grown material) before being reduced rapidly to room temperature. Layer thicknesses of 25–30 μm are normally produced for loophole diode applications [15.56].

15.2.3 Material Characteristics

Good composition uniformity, both laterally and in depth, is essential in order to obtain the required uniform device performance. Growth parameters that need to be optimized in Hg-rich LPE include the degree of supercooling and mixing of the melt, the geometrical configuration of the growth system, the melt size and the phase diagram. The standard deviation of the cut-off wavelength, for 12-spot measurements by Fourier transform infrared (FTIR) transmission at 80 K across a 30 cm^2 LWIR layer, is reported as 0.047.

Composition control and the uniformity of layers grown by dipping Te-rich LPE is one of the strengths of this process. The cut-off wavelength reproducibility is typically 10.05 ± 0.18 μm. Dipping Te-rich LPE is mainly used to grow thick films, about 100 μm, hence thickness control is not one of its advantages. Thickness control is about ±15% for layers of < 20 μm due to the relatively large amount of solidified material.

For Te-rich sliding-boat LPE, layers of ≈ 30 μm thickness can show wavelength uniformity at room temperature of 6.5 ± 0.05 μm over 90% of the area of 20×30 mm layers [15.57].

The ease of decanting the Hg-rich melt after layer growth results in smooth and specular surface morphology if a precisely oriented, lattice-matched CdZnTe substrate is used. Epitaxial growth reproduces the substructure of the substrate, especially in the case of homoepitaxy.

The dislocation density of LPE CMT and its effects on device characteristics have received much attention [15.58, 59]. The dislocation density is dominated mainly by the dislocations of the underlying substrate [15.60]. For layers grown on substrates with ZnTe ≈ 3 − 4%, the dislocations are present only in the interface region and the dislocation density is close to that of the substrate. For layers grown on substrates with ZnTe > 4.25%, dislocation generation is observed within a region of high lattice parameter gradient. The same variation of dislocation density with depth is seen for sliding-boat LPE material [15.56] (typical values are $3–7 \times 10^4\,cm^{-2}$).

For the production of heterostructure detectors with CMT epitaxial layers, it is essential that proper impurity dopants be incorporated during growth to form well-behaved and stable p–n junctions. An ideal impurity dopant should have low vapor pressure, low diffusivity, and a small impurity ionization energy. Group V and Group III dopants – As and Sb for p-type and In for n-type – are the dopants of choice. Hg-rich melts can be readily doped to produce n- and p-type layers; the solubilities of most of the useful dopants are significantly higher than in Te-rich solutions, most notably for Group V dopants, which are among the most difficult to incorporate into CMT.

Accurate determinations of dopant concentration in the solid involve the use of Hall effect measurements and secondary ion mass spectrometry (SIMS) concentration profiles. Measurements on the same sample by the two techniques are required to unequivocally substantiate the electrical activity of impurity dopants. The ease of incorporating Group I and Group III dopants into CMT, irrespective of non-stoichiometry, has been confirmed experimentally [15.5, 61].

The excess carrier lifetime is one of the most important material characteristics of CMT since it governs the device performance and frequency response. The objective is to routinely produce material with a lifetime that is limited by Auger processes, or by the radiative process in the case of the medium-wavelength infrared (MWIR) and short-wavelength infrared (SWIR) material [15.62]. It has been reported that intentionally impurity-doped LPE CMT material grown from As-doped Hg-rich melts can be obtained with relatively high minority carrier lifetimes [15.63]. The 77 K lifetimes of As-doped MWIR ($x = 0.3$) CMT layers are significantly higher than those of undoped bulk CMT and are within a factor of two of theoretical radiative lifetimes. Various annealing schedules have been proposed recently [15.38] that may lead to a reduction in Shockley–Read traps with a consequent increase in lifetime, even in undoped material. Lifetimes in In-doped MWIR CMT were also found to exhibit an inverse linear dependence on the doping concentration [15.63], with $N_d\tau$ products similar to $N_a\tau$ products observed for the As-doped material. The lifetimes of LWIR In-doped LPE material are typically limited by the Auger process at doping levels above $10^{15}\,cm^{-3}$.

The first heterojunction detectors were formed in material grown by Hg-rich LPE [15.64]. For double-layer heterojunctions (DLHJ), a second LPE (cap) layer is grown over the first (base) layer. With dopant types and layer composition controlled by the LPE growth process, this approach offers great flexibility (p-on-n or n-on-p) in junction type and in utilizing heterojunction formation between the cap and absorbing base layers to optimize detector performance. The key step in the process is to grow the cap layer doped with slow-diffusing impurities, In for an n-type cap layer, and As or Sb for a p-type cap layer.

For future large-area FPAs, Si-based substrates are being developed as a replacement for bulk CdZnTe substrates. This effort is directed at improvements in substrate size, strength, cost and reliability of hybrid FPAs, particularly during temperature cycling. These alternative substrates, which consist of epitaxial layers of CdZnTe or CdTe on GaAs/Si wafers [15.65] or directly onto Si wafers [15.66], are particularly advantageous for the production of large arrays. High-quality epitaxial CMT has been successfully grown on the Si-based substrates by the Hg-melt LPE technology for the fabrication of p-on-n DLHJ detectors. The first high-performance 128×128 MWIR and LWIR arrays were demonstrated by *Johnson* et al. [15.67]. MWIR arrays as large as 512×512 and 1024×1024 have also been produced [15.68]. A bias-selectable two-color (LWIR/MWIR) detector structure was first fabricated by growing three LPE layers from Hg-rich melts in sequence on a bulk CdZnTe substrate [15.68]. The structure forms an n-p-n triple-layer graded heterojunction (TLHJ) with two p-n junctions, one for each spectral band (color).

Destefanis et al. [15.69] have recently described their work on large-area and long linear FPAs based on Te-rich sliding-boat LPE material. Using a 15 μm pitch they were able to produce 1000×1000 MW arrays of photodiodes and 1500×2 MW and LW long linear arrays.

15.3 Metalorganic Vapor Phase Epitaxy (MOVPE)

Metalorganic vapor phase epitaxy (MOVPE) of CMT is dominated by the relatively high vapor pressures of mercury that are needed to maintain equilibrium over the growing film. This arises from the instability of HgTe compared with CdTe, and requires much lower growth temperatures than are usual for more stable compounds. The MOVPE process was developed as a vapor phase method that would provide sufficient control over growth parameters at temperatures below 400 °C, the main advantage being that the elements (although not mercury) can be transported at room temperature as volatile organometallics and react in the hot gas stream above the substrate or catalytically on the substrate surface. The first mercury chalcogenide growth by MOVPE [15.70,71] was followed by intense research activity that has brought the technology to its current state of maturity.

Although the MOVPE reaction cell conditions are far from equilibrium, an appreciation of the vapor–solid equilibrium can determine the minimum conditions needed for growth. This is particularly important with the mercury chalcogenides, where the relatively weak bonding of mercury causes a higher equilibrium vapor pressure. The equilibrium pressures of the component elements are linked and there is a range of pressures over which the solid remains in equilibrium as a single phase. At MOVPE growth temperatures, the pressure can vary by three orders of magnitude and remain in equilibrium with a single phase of HgTe. However, the Te_2 partial pressure varies across the phase field in the opposite sense to the Hg partial pressure.

The maintenance of vapor pressure equilibrium within the reactor cell does not automatically lead to the growth of an epilayer, but it does enable us to eliminate conditions that would not be suitable. For example, the supply of Hg to the substrate for growth at 400 °C must be greater than the minimum equilibrium pressure of 10^{-2} bar. This pressure is higher than would be conveniently delivered from a mercury organometallic source that would typically be less than 10^{-3} bar. All narrow-gap II–VI epilayers, except for a few examples of very low temperature growth, have used a heated mercury source that is capable of delivering the required partial pressure.

The criteria for precursors that are generally applicable to MOVPE are volatility, pyrolysis temperature, stability and volatility of organic products and purity. The latter depends on synthesis and purification routes for the precursors, but techniques such as adduct purification of the Group II organometallics have made a major contribution to epilayer purity. A good review of this topic is given by *Jones* [15.72].

The precise control over layer properties that is required for uniform detector arrays and tuning of the response wavelength band can only be achieved if the cadmium concentration is controlled to better than 1%. Improvements in reactor design, growth techniques (such as IMP – the interdiffused multi-layer process), choice of precursors and operating conditions have all contributed to successful targeting of the composition and thickness. However, these conditions are difficult to achieve on a run-to-run basis due to drifts in calibration, and they will require additional levels of control. Recent developments with in situ monitoring have been instrumental to gaining a better understanding of this technology and to reducing variance in epilayer properties. This has been achieved with better system monitoring to measure organometallic concentration and epilayer monitoring with laser reflectometry or ellipsometry [15.73].

A typical example of a horizontal reactor cell with an internal mercury source is shown in Fig. 15.6. The reactor cell wall temperature is also determined by the thermodynamics, which requires a minimum temperature in order to avoid condensation of mercury, and these requirements can be simply determined from the phase diagram. The organometallic supply shows the DMCd injected beyond the Hg source, which is the most common practice used to avoid reaction in the Hg zone.

Much of the work on precursors has been concerned with the thermal stability of the tellurium source, as this limits the growth temperature. From thermodynamic considerations, it would be desirable to grow at low temperature (below 400 °C) where the requirement on Hg equilibrium pressure will be lower. The initial demonstration of growth using MOVPE was with diethyl telluride (DETe), which requires a temperature above 400 °C for efficient pyrolysis [15.71].

Despite some high hopes of reducing the growth temperature to 200 °C, where the required equilibrium mercury vapor pressure would be as low as 10^{-7} bar, the most successful tellurium source has been diisopropyltelluride (DIPTe), which is used for growth between 350 and 400 °C. In practice, the DiPTe bubbler is held

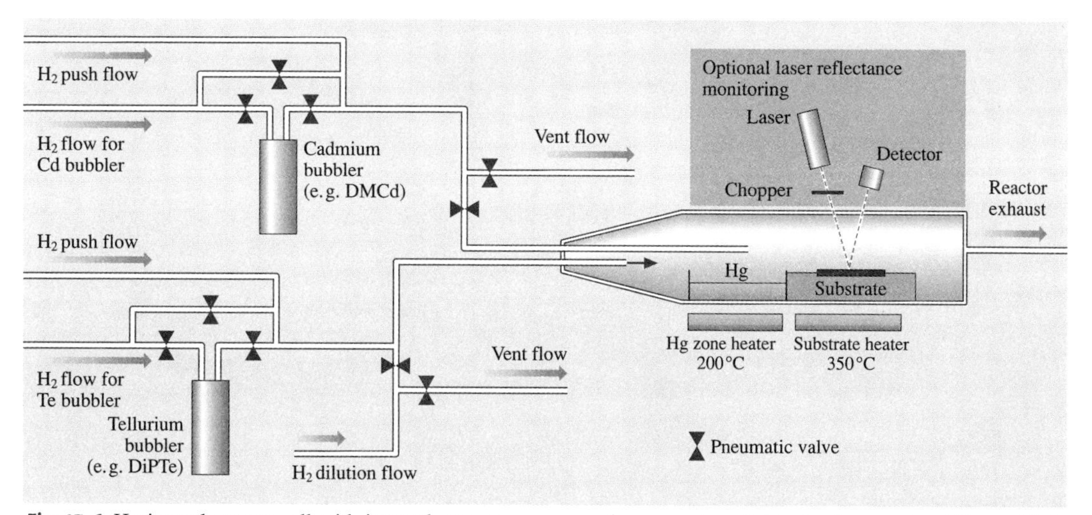

Fig. 15.6 Horizontal reactor cell with internal mercury source and gas supply suitable for the growth of $Hg_{1-x}Cd_xTe$ alloy. (After *S.J.C. Irvine*, in [15.4])

at a temperature above ambient (up to 30 °C) and the lines to the reactor are heated to avoid condensation of the precursor in the feed lines. Care must be taken, however, not to overheat the lines and cause premature pyrolysis.

Although most of the development of precursors in narrow gap II–VI semiconductors has been concerned with the tellurium source, there have been some important developments with the Group II sources. The cadmium source dimethyl cadmium (DMCd) appears to work well down to 250 °C, although this probably relies on reaction chemistry with the tellurium source.

Unlike MOVPE growth of other II–VI and III–V semiconductors, the growth of an alloy in narrow-gap II–VIs can be achieved by one of two alternative processes, direct alloy growth (DAG) and the interdiffused multilayer process (IMP). The requirement on alloy uniformity is critical, because this will determine the cut-on wavelength of a detector, and for focal plane arrays there needs to be uniformity of cut-on wavelength across the array. The DAG approach suffers from disadvantages in the thermodynamics and reaction chemistry in that the reaction rate of the tellurium precursor is very different when growing the mercury-containing binary compared with Zn, Cd, or Mn. This leads to a preferential depletion of the non-Hg Group II species as the gas flow passes over the substrate, and consequent compositional changes in the flow direction.

The interdiffused multilayer process [15.75] separates the growth of HgTe and CdTe so that the growth of the binaries can be independently optimized. The procedure entails the growth of alternate layers of HgTe and CdTe that interdiffuse during the time of growth to give a homogeneous epitaxial layer. The IMP approach relies on the very high interdiffusion coefficients in the CMT pseudobinary [15.38], and the basic principles are as follows:

1. Interdiffusion must occur at the growth temperature, and a nominal time at the end of growth for the last periods to interdiffuse must be allowed.
2. There should be no residual compositional or structural modulation attributable to the IMP oscillations.
3. Interdiffusion at heterointerfaces should be no greater than for direct alloy growth (DAG) at that temperature.
4. Flow velocity is adjusted between the two binary growths in order to optimize growth uniformity.

The alloy composition is fixed by adjusting the times for growth of the HgTe and CdTe IMP layers. Figure 15.7 shows the uniformity of wavelength over a 3 inch-diameter MOVPE layer grown on a GaAs:Si substrate [15.74]. 2-D contour maps of the cut-off wavelength of 3–5 μm wavelength slices (contour steps = 0.1 μm) showed that without rotation

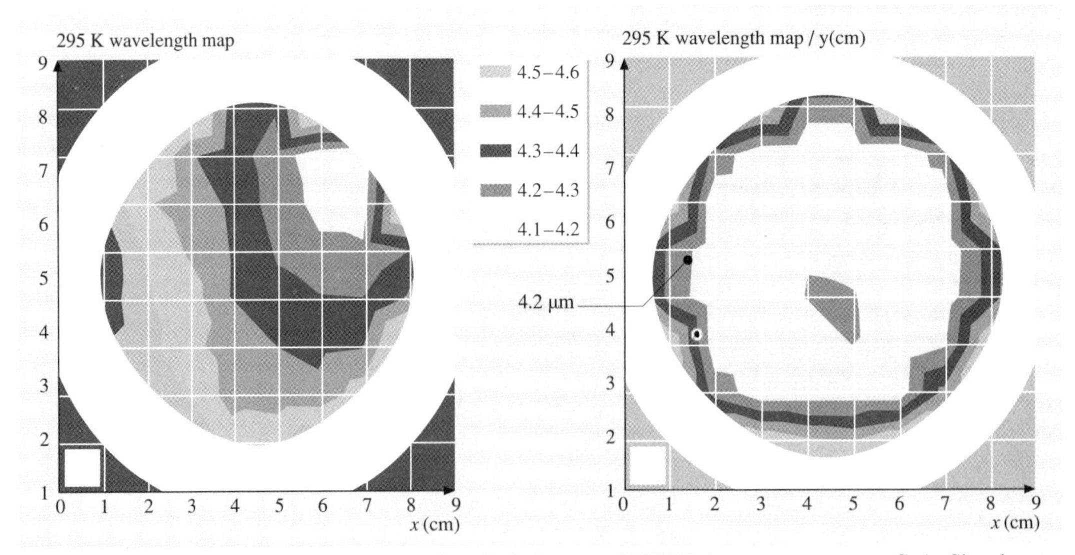

Fig. 15.7 Uniformity of wavelength over a three-inch-diameter MOVPE layer grown on a GaAs:Si substrate. (After [15.74])

(Fig. 15.7a) there is a gradual decrease in composition from upstream to downstream such that the compositional uniformity was within ± 0.022 in x. The thickness uniformity was typically $\pm 5-6\%$. However, by introducing substrate rotation, the uniformity improved dramatically such that, to within a few millimeters of the wafer edge, the composition was uniform to within ± 0.004 (Fig. 15.7b). The thickness uniformity also improved to within $\pm 2-3\%$. This was compatible with the production of twelve sites of large-format 2-D arrays (640×512 diodes on 24 μm pitch) per layer or larger numbers of smaller arrays, and was comparable with the uniformity achieved by *Edwall* [15.76]. In the past, other workers have established MOVPE reactor designs capable of large-area uniform growth of CMT on 3 inch wafers [15.76, 77] but these activities have now ceased as attention has focused on MBE growth techniques.

The usual methods for determining depth uniformity of a CMT layer are sharpness of the infrared absorption edge and SIMS depth profiles (in particular looking at the Te^{125} secondary ion). The results from these techniques indicate that the IMP structure is fully diffused for IMP periods of the order of 1000 Å and growth temperatures in the range of 350 to 400 °C unless the surface becomes faceted during growth, when microinhomogeneities may occur [15.78].

15.3.1 Substrate Type and Orientation

The search for the correct substrate material and orientation has been a major area of research in CMT because it is a limiting factor in the quality of the epilayers. Essentially, there are two categories of substrates: (i) lattice-matched II–VI substrates and (ii) non-lattice-matched 'foreign substrates'. Examples of the former are CdZnTe and CdSeTe, where the alloy compositions are tuned to the lattice parameter of the epilayer. Non-lattice-matched substrates include GaAs, Si and sapphire. The lattice mismatches can be up to 20% but, remarkably, heteroepitaxy is still obtained. The need for a ternary substrate to avoid substantial numbers of misfit dislocations has made the development of the CdTe-based substrate more complex. The small mismatch with CdTe substrates (0.2%) is sufficient to increase the dislocation density to greater than $10^6\,cm^{-2}$, comparable with some layers on CdTe-buffered GaAs, where the mismatch is 14% [15.76]. An additional problem encountered with the lattice-matched substrates is the lamella twins that form on (111) planes in Bridgman-grown crystals. It is possible to cut large (4×6 cm) (111)-oriented substrates parallel to the twins, but (100) substrates have a much lower yield. MOVPE-grown CdTe also twins on the (111), and these twins will propagate through an entire structure. However, CMT growth on the (111)B face is very smooth for layers up to 20 μm thick, which is adequate for infrared detector structures.

The majority of MOVPE growth has concentrated on orientations close to (100), normally with a misorientation to reduce the size of macrodefects, known as hillocks or pyramids. Large Te precipitates can intersect the substrate surface and nucleate macrodefects. In a detailed analysis of the frequency and shapes of defects on different misorientations, it was concluded [15.79] that the optimum orientation was (100) 3–4° towards the (111)B face. The presence of macrodefects is particularly critical for focal plane arrays, where they cause one or more defective pixels per defect. An alternative approach has been to use the (211)B orientation. In this case the surface appears to be free of macrodefects and is sufficiently misoriented from the (111) to avoid twinning. Dislocation densities of $10^5\,cm^{-2}$ have been measured in CMT grown by IMP onto CdZnTe (211)B substrates [15.76], and diffusion-limited detectors have been fabricated using this orientation [15.80].

The alternative lattice-mismatched substrates were investigated as a more producible alternative to the variable quality of the CdTe family of substrates. As the CMT arrays must be cooled during operation, there is the risk that the differential thermal contraction between the substrate and multiplexer will break some of the indium contacting columns. The ideal substrate from this point of view is of course silicon, but the initial quality of heteroepitaxy with 20% lattice mismatch was poor. One of the most successful alternative substrate technologies has been the Rockwell PACE-I (producible alternative to CdTe epitaxy) which uses c-plane sapphire with a CdTe buffer layer grown by MOVPE and a CMT detector layer grown by LPE. The sapphire substrates absorb above 6 μm and can only be used for the 3–5 μm waveband. However, even with careful substrate preparation, a buffer layer thicker than 5 μm is needed to avoid contamination of the active layer.

GaAs has been the most extensively used alternative substrate, which has been successfully used to reduce the macrodefect density to below $10\,cm^{-2}$, and X-ray rocking curve widths below 100 arcs have been obtained [15.81]. Due to the large lattice mismatch, the layer nucleates with rafts of misfit dislocations that relieve any strain. The main cause of X-ray rocking curve broadening is the tilt associated with a mosaic structure that arises from the initial island growth.

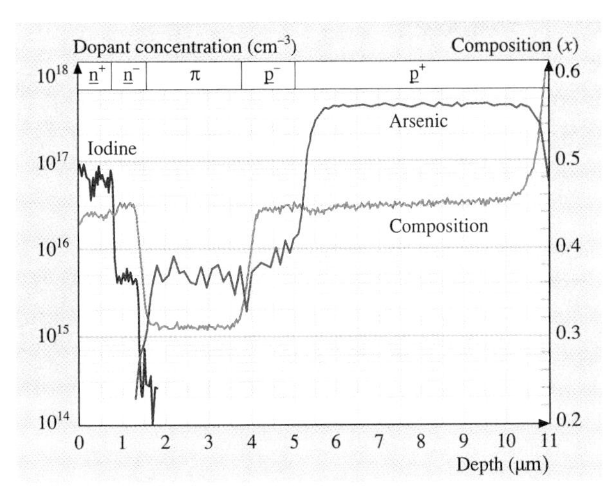

Fig. 15.8 SIMS profile of a $\underline{p}^+\underline{p}^-\pi\underline{n}^-\underline{n}^+$ device structure grown on a 3 inch wafer. (After [15.74])

Both CdTe and GaAs have the zinc blende structure but because the lattice mismatch is so large there is a better match with the orientation relationship (111)||(100). The growth orientation of a CdTe epilayer can be either (100) or (111) according to the nucleation conditions. Tellurium-rich conditions appear to favor the (111) nucleation but temperature and choice of precursors also play an important role. A method for ensuring that the (100) orientation is achieved (thus avoiding twin formation) is to grow a thin nucleation layer of ZnTe that has a lattice parameter between that of CdTe and GaAs.

The majority of the MOVPE growth onto silicon substrates has used a GaAs buffer layer to step the change in lattice parameter between silicon and CdTe. By growing a 12 μm-thick CdTe buffer layer, the X-ray rocking curve width is reduced to 120 arcs, larger than the best values obtained on GaAs substrates [15.82]. X-ray topography reveals the same type of mosaic structure as for CdTe on bulk GaAs substrates. The advantage of thermal expansion match with the silicon multiplexer has been demonstrated with a midwave 256×256 array where the entire structure was grown by MOVPE [15.83]. Hybrid arrays of 480×640 elements have also been made, where MOVPE-grown CdZnTe/GaAs/Si was used as a substrate for LPE growth of the active CMT structure [15.65].

15.3.2 Doping

The earliest FPAs were made in as-grown, undoped MOVPE layers where the p-type nature was due to metal vacancies in the layers. However, for current and future FPAs, the more complex devices such as p/n heterostructures require extrinsic doping and low-temperature annealing in a mercury vapor to remove metal vacancies. Activation of donors is relatively easy using either Group III metals or Group VII halogens. Indium has been used for doping concentrations from $< 10^{15}\,\mathrm{cm}^{-3}$ up to $10^{18}\,\mathrm{cm}^{-3}$. The low concentrations are limited by residual donors and acceptors, with the Hall mobility being a very sensitive measure of compensation. For the p/n structures, the n-type side of the junction is the absorber layer and therefore minority carrier lifetime and diffusion length are also important. An alternative n-type doping approach is to use a halogen such as iodine or chlorine that substitutes on the Group VI sites. The first use of iodine doping [15.84] used a crystalline iodine source, where the vapor was transported via hydrogen flow and a separate injector. The main advantage of iodine over indium is the slower diffusion, which enables more abrupt doped structures to be grown. A liquid source, ethyl iodide (EtI), was used by *Mitra* et al. [15.85] to dope over a range from 10^{15} to $10^{18}\,\mathrm{cm}^{-3}$, with no apparent memory. Current work on doped heterostructures [15.74] uses isobutyl iodide (IBI), as levels over a more useful range of (mid-10^{14} to mid-$10^{17}\,\mathrm{cm}^{-3}$) can be achieved with this source donor.

Acceptor doping, as in LPE processes, proved to be more problematic in MOVPE than donor doping. The first attempts at arsenic doping used AsH_3 that would normally be very stable at growth temperatures of 350 °C. It was proposed by *Capper* et al. [15.86] that the AsH_3 forms an adduct with DMCd in the vapor and causes the deposition of As at a low temperature. It was found that the doping incorporation was much higher in the CdTe layers and that activation of the dopant could be changed by adjusting the VI/II ratio during growth. A second-generation acceptor dopant source (*tris*-dimethyl amino arsine (DMAAs)) is now the preferred way to introduce As into layers. Multilayer, fully doped hetereostructures have been grown [15.74, 87] which demonstrate higher operating temperature performance due to Auger suppression techniques. These heterostructures were grown to investigate the device performance in the 3–5 μm and 8–12 μm IR bands. Figure 15.8 shows a SIMS profile of a 3–5 μm device structure grown on a 3 inch-diameter GaAs substrate. The excellent control over the As concentration allows the active region to be doped at $4–5\times10^{15}\,\mathrm{cm}^{-3}$ in the so-called 'π-layer' of the heterostructure design. The iodine profile shows well-defined transitions that help define the junction depth. Such layers have been fabricated into mesa test and 2-D arrays. These have been indium-bumped onto either test array carriers or Si multiplexers. This material represents the current state-of-the-art in MOVPE.

15.3.3 In Situ Monitoring

The use of optical in situ monitoring probes for MOVPE is helping to both elucidate the kinetic mechanisms and provide monitors suitable for feedback control. Although in situ monitoring is not widely used in CMT growth, there have been a number of notable examples where progress has been made as a result of its application. The choice of monitoring technique depends on the film property that is of interest. For example, laser reflectometry will measure the bulk of the film, measuring film thickness, growth rate and some indication of film composition. Reflection difference spectroscopy (RDS) is sensitive to surface composition and will monitor surface reconstruction and adsorbate composition. Spectroscopic ellipsometry is very sensitive to film composition and is now becoming useful as an in situ monitor with fast multichannel analysis. A number of in situ characterization techniques have been outlined in a review [15.73].

15.4 Molecular Beam Epitaxy (MBE)

An alternative to MOVPE for the production of advanced multilayer heterostructures is growth by molecular beam epitaxy (MBE). Figure 15.9 shows a schematic of a simple molecular beam epitaxial system for the growth of semiconductors. Significant progress was made in CMT MBE technology in the 1990s [15.88–91]. Current CMT MBE technology offers low-temperature growth under an ultrahigh vacuum environment, in situ n-type and p-type doping, and control of composition, doping and interfacial profiles. All of these are essential for the growth of double-layer and multilayer structures for advanced FPAs and other novel devices. A typical system consists of eight interconnected chambers (Fig. 15.10), including two growth chambers, two preparation chambers with integrated ion etching capabilities and entry chambers, an ESCALAB chamber, two buffer chambers and a transport chamber. Each growth chamber is equipped with dual Hg reservoirs, eight effusion cells and dual load locks for effusion cells to enable recharging of the materials without breaking the vacuum. The system is modular in nature and designed to accommodate sample sizes up to 3 inch in diameter.

Normally, growth of CMT is carried out at 180–190 °C on (211) CdZnTe substrates using Hg, CdTe and Te sources. Typically, the CdZnTe substrates are first degreased, then etched in Br-methanol solution, and then rinsed with methanol. The substrates are blown with dry nitrogen and mounted on holders, and then thermally cleaned at 350–400 °C under vacuum to remove oxides prior to growth [15.92]. Although MBE is essentially a nonequilibrium growth process, the vapor pressure of the constituent species over a CMT layer is governed by thermodynamics. It has been shown [15.93] that the vapor pressure of Hg over a CMT layer at 170 °C is $\approx 7\times10^{-4}$ mbar and 8 mbar for the Te-rich and Hg-rich sides of the phase boundary, respectively. The value of the Hg vapor pressure on the Te-rich side of the phase boundary is similar to the beam equivalent pressure of

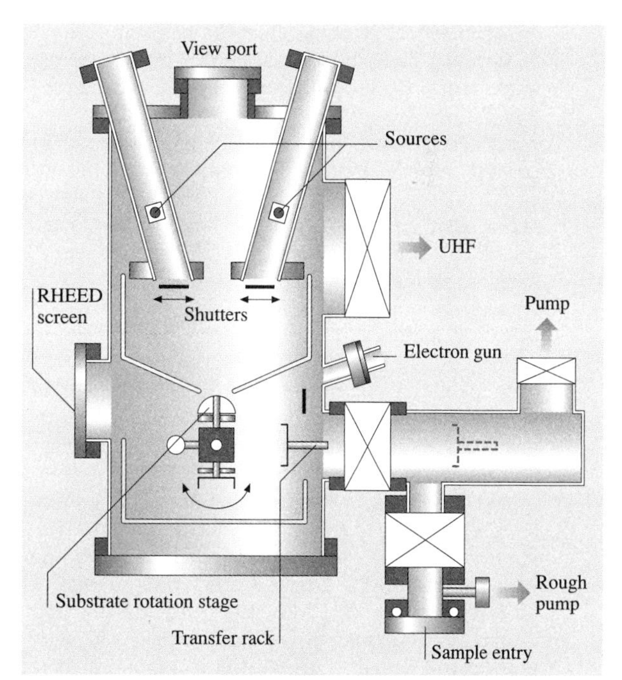

Fig. 15.9 Schematic of a simple molecular beam epitaxial system for the growth of semiconductors

Fig. 15.10 CMT MBE growth facility. (After *O.K. Wu* et al., in [15.4] p. 97)

Hg employed for the growth of CMT by MBE, suggesting that CMT growth occurs on the Te-rich side of the phase boundary, and RHEED studies confirm this. Although Hg is the more mobile species, which is likely to attach at step edges, sufficient mobility of the Cd species is critical for the growth of high-quality films.

The CMT alloy composition can be readily varied by choosing the appropriate beam-flux ratio. Over the range from $x = 0.2 - 0.50$, excellent control of composition can be achieved readily by varying the CdTe source flux with a constant flux of Hg at 3×10^{-4} mbar and Te at 8×10^{-7} mbar during the MBE growth [15.91].

The most widely used n-type dopant for CMT alloys during MBE growth is indium [15.94, 95]. The In concentration can be varied from 2×10^{15} to 5×10^{18} cm^{-3} by adjusting the In cell temperature (450–700 °C) with no evidence of a memory effect. The doping efficiency of In was almost 100%, evident from the Hall measurement and secondary ion mass spectrometry (SIMS) data, for carrier concentrations $< 2 \times 10^{18}$ cm^{-3}. As in several bulk growth techniques, and in LPE and MOVPE processes, donor doping is seen to be much easier than acceptor doping in MBE growth.

A critical issue when growing advanced CMT structures is the ability to grow high-quality p-type materials in situ. As, Sb, N, Ag and Li have all been used as acceptors during MBE growth of CMT, with varying degrees of success [15.96]. Most data available is centered on the use of arsenic, and two approaches have been investigated. The first approach is based on photoassisted MBE to enable high levels of p-type As-doping of CdTe [15.95]. For As-doping during compositionally modulated structure growth, only the CdTe layers in a CdTe-CMT combination are doped, as in MOVPE growth. Since the CdTe does not contain Hg vacancies, and is grown under cation-rich conditions, the As is properly incorporated onto the Te site and its concentration is proportional to the As flux. The structure then interdiffuses after annealing at high temperature to remove residual Hg vacancies, resulting in p-type, homogeneous CMT. The main disadvantage of this approach is that it requires a high-temperature anneal that results in reduced junction and interface control. An alternative approach is to use cadmium arsenide and correct Hg/Te ratios to minimize Hg vacancies during CMT growth [15.94]. As a result, the As is directed to the Group VI sublattice to promote efficient p-type doping. The main growth parameters that determine the properties of As-doped p-type CMT are the growth temperature and Hg/Te flux ratio. A comparison of the net hole concentration and the SIMS measurement indicates that the electrical activity of the As acceptors exceeds 60%.

Lateral compositional and thickness uniformity, evaluated by nine-point FTIR measurements, were performed on a 2.5×2.5 cm^2 sample, and the results showed that the average alloy composition and thickness were $x = 0.219 \pm 0.0006$ and $t = 8.68 \pm 0.064$ μm, respectively [15.91].

The surface morphology of CMT layers is important from a device fabrication point of view. Scanning electron microscopy (SEM) studies indicate that surface morphology of MBE-grown CMT alloys is very smooth for device fabrication, except for occasional small undulations (< 1 μm). The excellent crystal quality of CMT layers grown by MBE is illustrated by X-ray rocking curve data for a LWIR double-layer heterojunction structure. The In-doped n-type (about 8 μm thick) base layer peak has a width of < 25 arcs and is indistinguishable from the CdZnTe substrate. Because the As-doped p-type cap layer is much thinner (about 2 μm) and has a different alloy composition, its peak is broader (45 arcs), but the X-ray FWHM width still indicates high quality [15.91].

Other material properties such as minority carrier lifetime and etch pit density of the material are important for device performance. The lifetime of the photoexcited carriers is among the most important, since it governs the diode leakage current and the quantum efficiency of a detector. In the case of In-doped n-type layers ($x = 0.2 - 0.3$), results show that the lifetime ranges from 0.5–3 μs depending upon the x value and carrier concentration. Measured lifetimes at 77 K approach the Auger recombination limit [15.97].

Extended defects including dislocations, pinholes, particulates, inclusions, microtwins, precipitates and inhomogeneities occur over lengths of several micrometers and are often observable by selective etching and microscopy [15.59]. Optical microscopy studies of etched MBE-grown CMT base layers indicate that the EPD of device-quality alloys is on the order of $2\times10^5\,cm^{-2}$, which is in good agreement with the lifetime data and comparable to the best LPE materials for device applications.

15.4.1 Double-Layer Heterojunction Structures

Two approaches have been used to achieve these structures, namely ion implantation and in situ doping, in CMT MBE [15.96]. The advantage of the in situ doping approach is that it is a simple layer-by-layer growth process and so it is relatively easy to grow multilayer structures. Reproducible p-type doping with low defect density is difficult to achieve, and it requires very stringent passivation for mesa structures. However, high-performance CMT 64×64 and 128×128 FPAs for MWIR and LWIR infrared detection using CMT double-layer heterojunction structures (DLHJ) grown by MBE have been demonstrated [15.98]. The performances of detectors fabricated from MWIR and LWIR layers compare very well to those from the LPE-based production process.

15.4.2 Multilayer Heterojunction Structures

In recent years, as infrared technology has continued to advance, the demand for multispectral detectors has grown. The original growth process developed for the DLHJ structures was applied and extended for the growth of n-p-p-n multilayer heterojunction (MLHJ) structures for two-color detector applications [15.99]. A SIMS profile of an MLHJ structure confirms that the in situ doping process is able to incorporate $> 2\times10^{18}\,cm^{-3}$ As and control the n-type doping at $1{-}2\times10^{15}\,cm^{-3}$ for both absorbing layers [15.100]. X-ray rocking curve measurement indicates that the FWHM is about 50 arcs for this structure, suggesting high-quality material. This dual-band detector is basically a four-layer structure with two p–n junctions, one for each spectral band. The top and bottom layers are the IR absorbers while the middle two layers are transparent. The spectral response of this MW/LW two-color detector is shown in Fig. 15.11. This preliminary demonstration indicated that CMT MBE in situ doping technology is capable of growing multilayer structures for advanced IRFPAs.

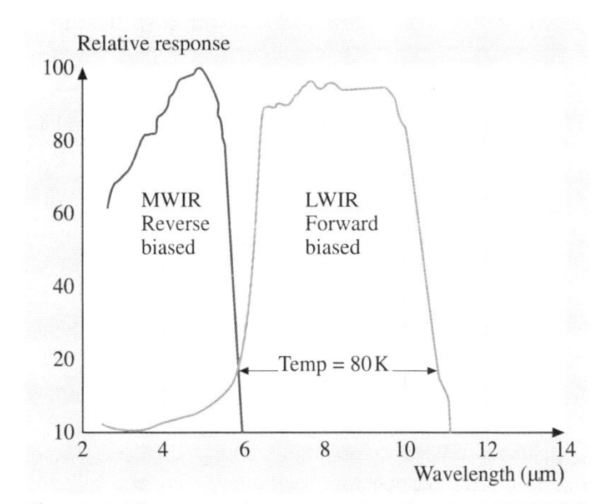

Fig. 15.11 The spectral response data for the MW-LWIR two-color detectors fabricated from a *n-p-p-n* multilayer heterojunction structure. The data indicate that the MWIR is detected in negative bias mode and LWIR is detected under forward bias conditions, as expected. (After *O.K. Wu* et al., in [15.4] p. 97)

15.4.3 CMT and CdZnTe Growth on Silicon

The move to using Si substrates for the epitaxial growth of CMT infrared IRFPAs is motivated by several important technological factors [15.91]. Primarily there is the need to establish a thermal expansion match between the Si readout electronics chip and the array of CMT infrared detectors. As the size of such CMT hybrid IRFPAs increases, long-term thermal cycle reliability can be compromised by the thermal expansion mismatch between Si and bulk CdZnTe substrates. One approach to solving this problem requires the fabrication of the CMT array on a Si substrate, rather than bulk CdZnTe, so that the array's thermal expansion is constrained to match that of the Si readout chip. Considerable research activity [15.101–104] has been directed towards achieving epitaxial deposition of high crystalline quality CMT films on Si substrates. Beyond their thermal expansion match to the readout electronics ship, Si substrates for IRFPAs offer additional advantages when compared with CdZnTe substrates; for example, they are more readily available in larger sizes with superior mechanical strength and at substantially lower cost. The use of Si substrates also avoids potential problems associated with outdiffusion of fast-diffusing impurities, such as Cu, that has been identified [15.105] as a recurring problem with CdZnTe substrates. Finally, development of the technology for epitaxial growth of CMT on Si will ultimately be a requisite technology should monolithic integration of IR detector and readout electronics on a single Si chip become a goal of future IRFPA development.

However, the most serious technical challenge faced when fabricating device-quality epitaxial layers of CMT on Si is the reduction in the density of threading dislocations that results from the accommodation of the 19% lattice constant mismatch and the large difference in thermal expansion coefficients between Si and CMT. Dislocation density is known to have a direct effect on IR detector performance [15.58], particularly at low temperature. All efforts to fabricate CMT IR detectors on Si substrates have relied upon the prior growth of CdZnTe buffer layers on Si. Growth of $\approx 5\,\mu m$ of CdZnTe is required to allow dislocation annihilation processes to decrease the dislocation density to low $10^6\,cm^{-2}$ [15.104].

As an additional step, initiation layers of ZnTe have been used to facilitate parallel MBE deposition of CdTe(001) on Si(001) [15.98, 102]; ZnTe nucleation layers are also commonly used for the same purpose for the growth of CdZnTe on GaAs/Si substrates by other vapor-phase techniques [15.106]. CdTe(001) films with rocking curves as narrow as 78 arcsec and EPD of $1-2\times10^6\,cm^{-2}$ have been demonstrated with this technique. Both (111)- and (001)-oriented MBE CdTe/Si substrates have been used as the basis for demonstrating LPE-grown CMT detectors [15.98, 101]. The (001)-oriented CdTe/Si films have been used in demonstrations of 256×256 CMT hybrid arrays on Si [15.98].

Current state-of-the-art MBE material on four-inch-diameter Si substrates has been discussed by *Varesi* et al. [15.107, 108]. Dry etching is used to produce array sizes of 128×128 and 1024×1024 with performances equivalent to LPE material. Similar material, grown on CdZnTe this time, by another group [15.109] in the MW and SW regions is used in astronomical applications (see [15.110]).

Other recent applications of MBE-grown CMT include very long wavelength arrays (onto (211)B CdZnTe substrates) by *Philips* et al. [15.111], two-color (MW $4.5\,\mu m$/SW $2.5\,\mu m$) arrays of 128×128 diodes [15.69, 112], gas detectors in the $2-6\,\mu m$ region [15.113] and $1.55\,\mu m$ avalanche photodiodes using Si substrates [15.66]. All of these applications demonstrate the versatility of the CMT MBE growth technique.

One final point to note about the current devices being researched in MOVPE and MBE (and to a lesser extent LPE) processes is that growth is no longer of single layers from which the detector is made; instead the materials growers are actually producing the device structures within the grown layer. This is particularly true of the fully doped heterostructures grown by MOVPE and MBE shown in Figs. 15.8 and 15.11.

15.5 Alternatives to CMT

Rogalski [15.1, 114] has provided details about several Hg-based alternatives to CMT for infrared detection. He concludes that only HgZnTe and HgMnTe are serious candidates from the range of possibilities. Theoretical considerations of *Sher* et al. [15.115] showed that the Hg–Te bond is stabilized by the addition of ZnTe, unlike the destabilization that occurs when CdTe is added, as in CMT.

The pseudobinary phase diagram of HgZnTe shows even more separation of solidus and liquidus than the equivalent for CMT, see Fig. 15.12. This leads to large segregation effects, large composition variations for small temperature changes, and the high Hg vapor pressure presents the usual problems of containment. HgTe and MnTe are not completely miscible over the entire range; the single-phase region is limited to $\lessapprox x \lessapprox 0.35$. The solidus–liquidus separation in the HgTe-MnTe pseudobinary is approximately half that in CMT, so for equivalent wavelength uniformity requirements any crystals of the former must be much more uniform than CMT crystals.

Three methods: Bridgman, SSR and THM are the most popular ones for the bulk growth of HgZnTe and HgMnTe. The best quality crystals have been produced by THM [15.116], with uniformities of ±0.01 in both the axial and radial directions for HgZnTe. For HgMnTe, *Bodnaruk* et al. [15.117] produced crystals of $0.04 < x < 0.2$ with uniformities of ±0.01 and ±0.005 in the axial and radial directions, while *Gille* et al. [15.118] grew $x = 0.10$ crystals with ±0.003 along

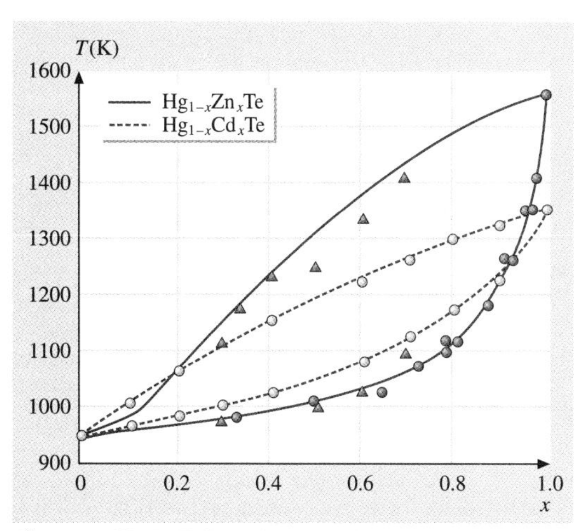

Fig. 15.12 HgTe-ZnTe and HgTe-CdTe pseudobinary phase diagrams. (After [15.120])

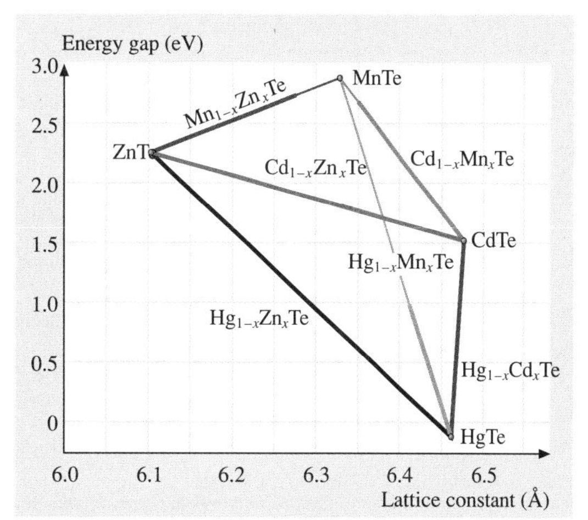

Fig. 15.13 Energy gap versus lattice parameter in Hg-based ternary alloy systems at room temperature. The *bold lines* indicate ranges of x for which homogeneous crystal phases form. (After *A. Rogalski*, in [15.6] p. 377)

a 16 mm-diameter slice. *Becla* et al. [15.119] achieved decreased radial variations by applying a magnetic field.

These bulk crystals are suitable for some device architectures, but higher performance is obtained in more sophisticated structures that can only be made by epitaxial techniques. The advantages are the usual ones of lower growth temperature, lower Hg vapor pressure, shorter growth times and less precipitation with improved uniformity of composition. These features have been discussed in detail and descriptions of the various epitaxial processes given by *Rogalski* [15.114, 120, 121]. LPE growth of HgCdZnTe and HgCdMnTe by *Uchino* and *Takita* [15.122] showed that incorporation of Zn or Mn improved the uniformity of composition. HgMnTe films have been grown by MOCVD using the IMP technique, with both n- and p-type layers being produced. As in all epitaxial growth techniques, there is a need for suitable (nearly lattice-matched) large-area single-crystal substrates. Figure 15.13 shows the lattice constants of the Hg-based alloy systems at room temperature. There are larger differences between the lattice parameters of HgTe and ZnTe and HgTe and MnTe than between HgTe and CdTe. At first sight it would appear that CdMnTe would be a suitable substrate system for these ternary alloys, but CdMnTe crystals grown by the Bridgman process are highly twinned and therefore unsuitable for use as substrates.

The device aspects of these alternative ternary systems are discussed in the chapter by Baker in this Handbook.

References

15.1 A. Rogalski: *IR Detectors and Emitters: Materials and Devices*, ed. by P. Capper, C. T. Elliott (Kluwer, Boston 2000)

15.2 M. Reine: *Encyclopedia of Modern Optics* (Academic, London 2002) p. 392

15.3 I. M. Baker: *Handbook of Infrared Detection Technologies*, ed. by M. Henini, M. Razeghi (Elsevier, Oxford 2003) Chap. 8

15.4 P. Capper (Ed.): *Narrow-Gap II–VI Compounds for Optoelectronic and Electromagnetic Applications* (Chapman & Hall, London 1997)

15.5 P. Capper (Ed.): *Properties of Narrow Gap Cadmium-Based Compounds*, EMIS Datarev. Ser. (IEE, London 1994)

15.6 P. Capper, C. T. Elliott (Eds.): *IR Detectors and Emitters: Materials and Devices* (Kluwer, Boston 2000)

15.7 P. Capper (Ed.): *Bulk Crystal Growth of Electronic, Optical and Optoelectronic Materials* (Wiley, Chichester 2005)

15.8 W. F. H. Micklethwaite: Semicond. Semimet. **18**, 48 (1981) Chap. 3

15.9 P. W. Kruse: Semicond. Semimet. **18**, 1 (1981) Chap.1
15.10 R. Triboulet, T. Nguyen Duy, A. Durand: J. Vac. Sci. Technol. A **3**, 95 (1985)
15.11 W. E. Tennant, C. Cockrum, J. Gilpin, M. A. Kinch, M. B. Reine, R. P. Ruth: J. Vac. Sci. Technol. B **10**, 1359 (1992)
15.12 T. C. Harman: J. Electron. Mater. **1**, 230 (1972)
15.13 A. W. Vere, B. W. Straughan, D. J. Williams: J. Cryst. Growth **59**, 121 (1982)
15.14 L. Colombo, A. J. Syllaios, R. W. Perlaky, M. J. Brau: J. Vac. Sci. Technol. A **3**, 100 (1985)
15.15 R. K. Sharma, V. K. Singh, N. K. Mayyar, S. R. Gupta, B. B. Sharma: J. Cryst. Growth **131**, 565 (1987)
15.16 L. Colombo, R. Chang, C. Chang, B. Baird: J. Vac. Sci. Technol. A **6**, 2795 (1988)
15.17 J. Ziegler: US Patent 4,591,410 (1986)
15.18 W. M. Higgins, G. N. Pultz, R. G. Roy, R. A. Lancaster: J. Vac. Sci. Technol. A **7**, 271 (1989)
15.19 J. H. Tregilgas: Prog. Cryst. Growth Charact. **28**, 57 (1994)
15.20 P. Capper, J. Harris, D. Nicholson, D. Cole: J. Cryst. Growth **46**, 575 (1979)
15.21 P. Capper: Prog. Cryst. Growth Charact. **28**, 1 (1994)
15.22 A. Yeckel, and J.J. Derby: Paper given at 2002 US Workshop on Physics and Chemistry of II–VI Materials, San Diego, USA (2002)
15.23 P. Capper, and J.J.G. Gosney: U.K. Patent 8115911 (1981)
15.24 P. Capper, C. Maxey, C. Butler, M. Grist, J. Price: Mater. Electron. Mater. Sci. **15**, 721 (2004)
15.25 R. Triboulet: Prog. Cryst. Growth Charact. **28**, 85 (1994)
15.26 Y. Nguyen Duy, A. Durand, J. Lyot: Mater. Res. Soc. Symp. Proc **90**, 81 (1987)
15.27 A. Durand, J. L. Dessus, T. Nguyen Duy, J. Barbot: Proc. SPIE **659**, 131 (1986)
15.28 P. Gille, F. M. Kiessling, M. Burkert: J. Cryst. Growth **114**, 77 (1991)
15.29 P. Gille, M. Pesia, R. Bloedner, N. Puhlman: J. Cryst. Growth **130**, 188 (1993)
15.30 M. Royer, B. Jean, A. Durand, R. Triboulet: French Patent No. 8804370 (1/4/1988)
15.31 R. U. Bloedner, P. Gille: J. Cryst. Growth **130**, 181 (1993)
15.32 B. Chen, J. Shen, S. Din: J. Electron Mater. **13**, 47 (1984)
15.33 D. A. Nelson, W. M. Higgins, R. A. Lancaster: Proc. SPIE **225**, 48 (1980)
15.34 C.-H. Su, G. Perry, F. Szofran, S. L. Lehoczky: J. Cryst. Growth **91**, 20 (1988)
15.35 R. R. Galazka: J. Cryst. Growth **53**, 397 (1981)
15.36 B. Bartlett, P. Capper, J. Harris, M. Quelch: J. Cryst. Growth **47**, 341 (1979)
15.37 A. Durand, J. L. Dessus, T. Nguyen Duy: Proc. SPIE **587**, 68 (1985)
15.38 P. Capper, C. D. Maxey, C. L. Jones, J. E. Gower, E. S. O'Keefe, D. Shaw: J. Electron Mater. **28**, 637 (1999)
15.39 M. A. Kinch: Mater. Res. Soc. Symp. Proc. **90**, 15 (1987)
15.40 R. Pratt, J. Hewett, P. Capper, C. Jones, N. Judd: J. Appl. Phys. **60**, 2377 (1986)
15.41 R. Pratt, J. Hewett, P. Capper, C. L. Jones, M. J. T. Quelch: J. Appl. Phys. **54**, 5152 (1983)
15.42 F. Grainger, I. Gale, P. Capper, C. Maxey, P. Mackett, E. O'Keefe, J. Gosney: Adv. Mater. Opt. Electron. **5**, 71 (1995)
15.43 A. W. Vere: Proc. SPIE **659**, 10 (1986)
15.44 D. J. Williams, A. W. Vere: J. Vac. Sci. Technol. A **4**, 2184 (1986)
15.45 J. H. Tregilgas, J. D. Beck, B. E. Gnade: J. Vac. Sci. Technol. A **3**, 150 (1985)
15.46 C. Genzel, P. Gille, I. Hahnert, F. M. Kiessling, P. Rudolph: J. Cryst. Growth **101**, 232 (1990)
15.47 P. Capper, C. Maxey, C. Butler, M. Grist, J. Price: J. Cryst. Growth **275**, 259 (2005)
15.48 T. Tung: J. Cryst. Growth **86**, 161 (1988)
15.49 T.-C. Yu, R. F. Brebrick: *Properties of Narrow Gap Cadmium-Based Compounds*, EMIS Datarev. Ser., ed. by P. Capper (IEE, London 1994) p. 55
15.50 M. G. Astles: *Properties of Narrow Gap Cadmium-based Compounds*, EMIS Datarev. Ser., ed. by P. Capper (IEE, London 1994) p. 1
15.51 T. Tung, L. V. DeArmond, R. F. Herald: Proc. SPIE **1735**, 109 (1992)
15.52 P. W. Norton, P. LoVecchio, G. N. Pultz: Proc. SPIE **2228**, 73 (1994)
15.53 G. H. Westphal, L. Colombo, J. Anderson: Proc. SPIE **2228**, 342 (1994)
15.54 D. W. Shaw: J. Cryst. Growth **62**, 247 (1983)
15.55 L. Colombo, G. H. Westphal, P. K. Liao, M. C. Chen, H. F. Schaake: Proc. SPIE **1683**, 33 (1992)
15.56 I. B. Baker, G. J. Crimes, J. Parsons, E. O'Keefe: Proc. SPIE **2269**, 636 (1994)
15.57 P. Capper, E. S. O'Keefe, C. D. Maxey, D. Dutton, P. Mackett, C. Butler, I. Gale: J. Cryst. Growth **161**, 104 (1996)
15.58 R. S. List: J. Electron. Mater. **22**, 1017 (1993)
15.59 S. Johnson, D. Rhiger, J. Rosbeck: J. Vac. Sci. Technol. B **10**, 1499 (1992)
15.60 M. Yoshikawa: J. Appl. Phys. **63**, 1533 (1988)
15.61 P. Capper: J. Vac. Sci. Technol. B **9**, 1667 (1991)
15.62 C. A. Cockrum: Proc. SPIE. **2685**, 2 (1996)
15.63 W. A. Radford, R. E. Kvaas, S. M. Johnson: *Proc. IRIS Specialty Group on Infrared Materials* (IRIS, Menlo Park 1986)
15.64 K. J. Riley, A. H. Lockwood: Proc. SPIE **217**, 206 (1980)
15.65 S. M. Johnson, J. A. Vigil, J. B. James: J. Electron. Mater. **22**, 835 (1993)
15.66 T. DeLyon, A. Hunter, J. Jensen, M. Jack, V. Randall, G. Chapman, S. Bailey, K. Kosai: Paper given at 2002 US Workshop on Physics and Chemistry of II–VI Materials, San Diego, USA (2002)
15.67 S. Johnson, J. James, W. Ahlgren: Mater. Res. Soc. Symp. Proc. **216**, 141 (1991)
15.68 P. R. Norton: Proc. SPIE **2274**, 82 (1994)

15.69 G. Destefanis, A. Astier, J. Baylet, P. Castelein, J.P. Chamonal, E. De Borniol, O. Gravand, F. Marion, J.L. Martin, A. Million, P. Rambaud, F. Rothan, J.P. Zanatta: J. Electron. Mater. **32**, 592 (2003)

15.70 T.F. Kuech, J.O. McCaldin: J. Electrochem. Soc. **128**, 1142 (1981)

15.71 S.J.C. Irvine, J.B. Mullin: J. Cryst. Growth **55**, 107 (1981)

15.72 A.C. Jones: J. Cryst. Growth **129**, 728 (1993)

15.73 S.J.C. Irvine, J. Bajaj: Semicond. Sci. Technol. **8**, 860 (1993)

15.74 C.D. Maxey, J. Camplin, I.T. Guilfoy, J. Gardner, R.A. Lockett, C.L. Jones, P. Capper: J. Electron. Mater. **32**, 656 (2003)

15.75 J. Tunnicliffe, S. Irvine, O. Dosser, J. Mullin: J. Cryst. Growth **68**, 245 (1984)

15.76 D.D. Edwall: J. Electron. Mater. **22**, 847 (1993)

15.77 S. Murakami: J. Vac. Sci. Technol B **10**, 1380 (1992)

15.78 S.J.C. Irvine, D. Edwall, L. Bubulac, R.V. Gil, E.R. Gertner: J. Vac. Sci. Technol. B **10**, 1392 (1992)

15.79 D.W. Snyder, S. Mahajan, M. Brazil: Appl. Phys. Lett. **58**, 848 (1991)

15.80 P. Mitra, Y.L. Tyan, F.C. Case: J. Electron. Mater. **25**, 1328 (1996)

15.81 A.M. Kier, A. Graham, S.J. Barnett: J. Cryst. Growth **101**, 572 (1990)

15.82 S.J.C. Irvine, J. Bajaj, R.V. Gil, H. Glass: J. Electron. Mater. **24**, 457 (1995)

15.83 S.J.C. Irvine, E. Gertner, L. Bubulac, R.V. Gil, D.D. Edwall: Semicond. Sci. Technol. **6**, C15 (1991)

15.84 C.D. Maxey, P. Whiffin, B.C. Easton: Semicond. Sci. Technol. **6**, C26 (1991)

15.85 P. Mitra, Y.L. Tyan, T.R. Schimert, F.C. Case: Appl. Phys. Lett. **65**, 195 (1994)

15.86 P. Capper, C. Maxey, P. Whiffin, B. Easton: J. Cryst. Growth **97**, 833 (1989)

15.87 C.D. Maxey, C.J. Jones: . Proc. SPIE **3122**, 453 (1996)

15.88 R.D. Rajavel, D. Jamba, O.K. Wu, J.A. Roth, P.D. Brewer, J.E. Jensen, C.A. Cockrum, G.M. Venzor, S.M. Johnson: J. Electron. Mater. **25**, 1411 (1996)

15.89 J. Bajaj, J.M. Arias, M. Zandian, D.D. Edwall, J.G. Pasko, L.O. Bubulac, L.J. Kozlowski: J. Electron. Mater. **25**, 1394 (1996)

15.90 J.P. Faurie, L.A. Almeida: Proc. SPIE **2685**, 28 (1996)

15.91 O.K. Wu, T.J. deLyon, R.D. Rajavel, J.E. Jensen: *Narrow-Gap II–VI Compounds for Optoelectronic and Electromagnetic Applications*, ed. by P. Capper (Chapman & Hall, London 1997) p. 97

15.92 O.K. Wu, D.R. Rhiger: *Characterization in Compound Semiconductor Processing*, ed. by Y. Strausser, G.E. McGuire (Butterworth-Heinemann, London 1995) p. 83

15.93 T. Tung, L. Golonka, R.F. Brebrick: J. Electrochem. Soc. **128**, 451 (1981)

15.94 O. Wu, D. Jamba, G. Kamath: J. Cryst. Growth **127**, 365 (1993)

15.95 J. Arias, S. Shin, D. Copper: J. Vac. Sci. Technol. A **8**, 1025 (1990)

15.96 O.K. Wu: Mater. Res. Soc. Symp. Proc. **340**, 565 (1994)

15.97 V. Lopes, A.J. Syllaios, M.C. Chen: Semicond. Sci. Technol. **8**, 824 (1993)

15.98 S.M. Johnson, T.J. de Lyon, C. Cockrum: J. Electron. Mater. **24**, 467 (1995)

15.99 G. Kamath, and O. Wu: US Patent Number 5,028,561, July 1, 1991

15.100 O.K. Wu, R.D. Rajavel, T.J. deLyon: Proc. SPIE **2685**, 16 (1996)

15.101 F.T. Smith, P.W. Norton, P. Lo Vecchio: J. Electron. Mater. **24**, 1287 (1995)

15.102 J.M. Arias, M. Zandian, S.H. Shin: J. Vac. Sci. Technol. B **9**, 1646 (1991)

15.103 R. Sporken, Y. Chen, S. Sivananthan: J. Vac. Sci. Technol. B **10**, 1405 (1992)

15.104 T.J. DeLyon, D. Rajavel, O.K. Wu: Proc. SPIE. **2554**, 25 (1995)

15.105 J.P. Tower, S.P. Tobin, M. Kestigian: J. Electron. Mater. **24**, 497 (1995)

15.106 N. Karam, R. Sudharsanan: J. Electron. Mater. **24**, 483 (1995)

15.107 J.B. Varesi, A.A. Buell, R.E. Bornfreund, W.A. Radford, J.M. Peterson, K.D. Maranowski, S.M. Johnson, D.F. King: J. Electron. Mater. **31**, 815 (2002)

15.108 J.B. Varesi, A.A. Buell, J.M. Peterson, R.E. Bornfreund, M.F. Vilela, W.A. Radford, S.M. Johnson: J. Electron. Mater. **32**, 661 (2003)

15.109 M. Zandian, J.D. Garnett, R.E. DeWames, M. Carmody, J.G. Pasko, M. Farris, C.A. Cabelli, D.E. Cooper, G. Hildebrandt, J. Chow, J.M. Arias, K. Vural, D.N.B. Hall: J. Electron. Mater. **32**, 803 (2003)

15.110 I.S. McLean: Paper given at 2002 US Workshop on Physics and Chemistry of II–VI Materials, San Diego, USA (2002)

15.111 J.D. Philips, D.D. Edwall, D.L. Lee: J. Electron. Mater. **31**, 664 (2002)

15.112 L.A. Almeida, M. Thomas, W. Larsen, K. Spariosu, D.D. Edwall, J.D. Benson, W. Mason, A.J. Stolz, J.H. Dinan: J. Electron. Mater. **31**, 669 (2002)

15.113 J.P. Zanatta, F. Noel, P. Ballet, N. Hdadach, A. Million, G. Destefanis, E. Mottin, E. Picard, E. Hadji: J. Electron. Mater. **32**, 602 (2003)

15.114 A. Rogalski: *New Ternary Alloy Systems for Infrared Detectors* (SPIE Optical Engineering, Bellingham 1994)

15.115 A. Sher, A.B. Chen, W.E. Spicer, C.K. Shih: J. Vac. Sci. Technol. A **3**, 105 (1985)

15.116 R. Triboulet: J. Cryst. Growth **86**, 79 (1988)

15.117 O.A. Bodnaruk, I.N. Gorbatiuk, V.I. Kalenik: Neorg. Mater. **28**, 335 (1992)

15.118 P. Gille, U. Rössner, N. Puhlmann: Semicond. Sci. Technol. **10**, 353 (1995)

15.119 P. Becla, J-C. Han, S. Matakef: J. Cryst. Growth **121**, 394 (1992)

15.120 A. Rogalski: Prog. Quantum Electron. **13**, 299 (1989)
15.121 A. Rogalski: Infrared Phys. **31**, 117 (1991)
15.122 T. Uchino, K. Takita: J. Vac. Sci. Technol. A **14**, 2871 (1996)

16. Wide-Bandgap II–VI Semiconductors: Growth and Properties

Wide-bandgap II–VI compounds are been applied to optoelectronic devices, especially light-emitting devices in the short-wavelength region of visible light, because of their direct gap and suitable bandgap energies. Many methods have been extensively applied to grow high-quality films and bulk single crystals from the vapor and liquid phases.

This chapter firstly discusses the basic properties and phase diagrams of wide-bandgap II–VI compounds such as ZnS, ZnO, ZnSe, ZnTe, CdSe and CdTe. Then the growth methods and recent progress in films and bulk crystal growth are reviewed. In the epitaxial growth methods, the focus is on liquid-phase epitaxy (LPE), vapor-phase epitaxy (VPE) containing conventional VPE, hot-wall epitaxy (HWE), metalorganic chemical vapor deposition (MOCVD) or metalorganic phase epitaxy (MOVPE), molecular-beam epitaxy (MBE) and atomic-layer epitaxy (ALE). In bulk crystal growth, two typical growth methods, chemical/physical vapor transport (CVT/PVT) and Bridgman techniques, are introduced.

Wide-bandgap II–VI compounds are expected to be one of the most vital materials for high-performance optoelectronics devices such as light-emitting diodes (LEDs) and laser diodes (LDs) operating in the blue or ultraviolet spectral range. Additionally, the high ionicity of these compounds makes them good candidates for high electro-optical and electromechanical coupling. The basic promises of wide-bandgap materials can be found in Fig. 16.1.

Thin films were commonly grown using the conventional vapor-phase epitaxy (VPE) method for 60 s. With the development of science and technology, new and higher requirements arose for material preparation. For this reason, novel epitaxial growth techniques were developed, including hot-wall epitaxy (HWE) [16.1], metalorganic chemical vapor deposition (MOCVD) [16.2], molecular-beam epitaxy (MBE) [16.3], metalorganic molecular-beam epitaxy (MOMBE) [16.4] and atomic-layer epitaxy (ALE) [16.5]. Using these growth methods, film thickness can be controlled, and quality can be improved.

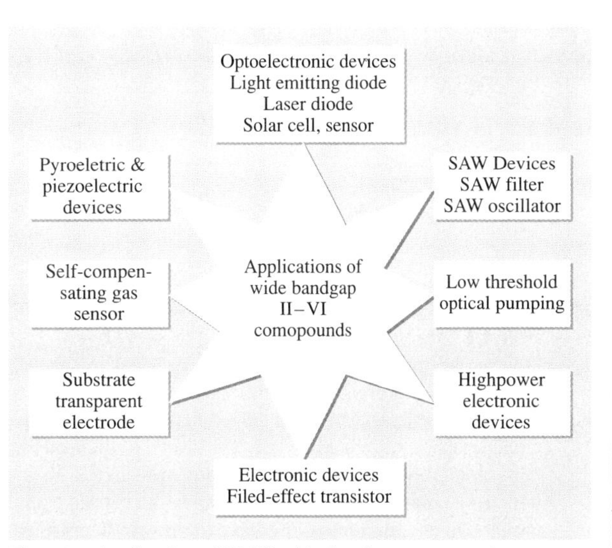

Fig. 16.1 Application of II–VI wide-bandgap compounds

On the other hand, basic research work into growing bulk crystals of wide-bandgap II–VI compounds has been carried out. Focus was put on high-purity, high-quality, large single crystals [16.6–10]. Since the electrical and optical properties of semiconductor compounds are drastically affected by impurities and native defects, purity and quality are very important for fundamental research and engineering application where they are used as substrates. Bulk single crystals of these wide-bandgap II–VI compounds have been grown from the vapor, liquid and solid phases. Vapor-phase growth includes chemical vapor transport (CVT) and physical vapor transport (PVT) methods; liquid-phase methods includes growth from the melt or solvent. Among these growth methods, melt growth is most suitable to produce sizable bulk crystals for relatively short growth duration. Growth methods for films and bulk crystals of wide-bandgap II–VI compounds are summarized in Fig. 16.2 and the details can be found in Sects. 16.2 and 16.3.

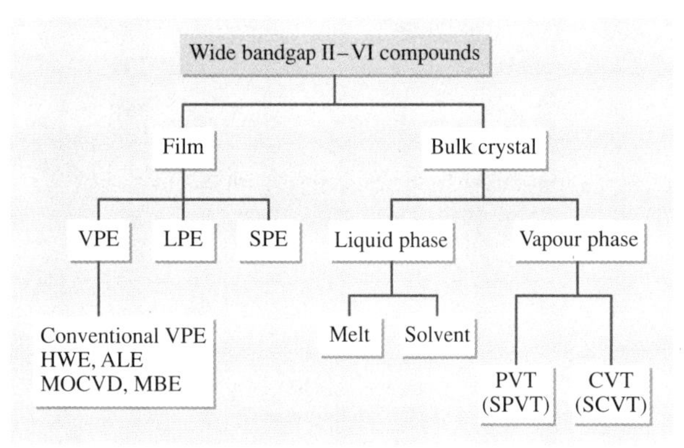

Fig. 16.2 Film and bulk-crystal growth techniques for II–VI wide-bandgap compounds

This chapter firstly describes the physical and chemical properties of these wide-bandgap II–VI compounds, then reviews the growth techniques and introduces the main results in preparing film and bulk single crystals of ZnS, ZnO, ZnSe, ZnTe, CdTe, and so on.

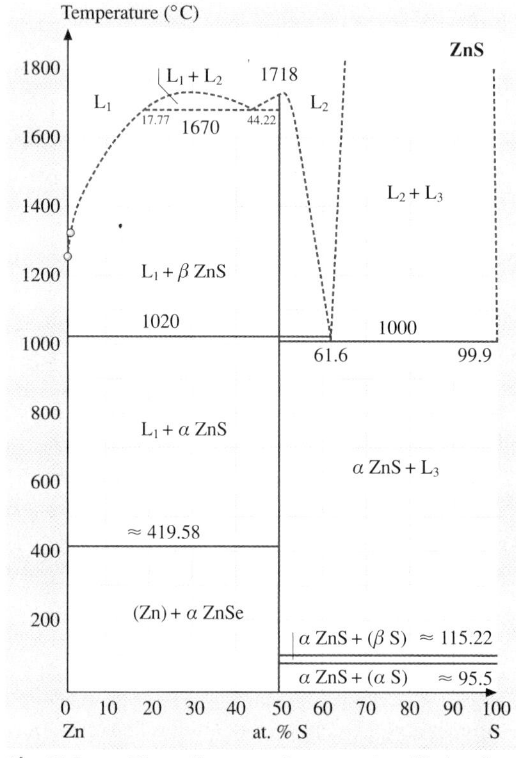

Fig. 16.3a–g Phase diagrams of some main wide-bandgap II–VI compounds. (**a**) ZnS, (**b**) ZnSe, (**c**) ZnSe phase diagram near the congruent point; (**d**) ZnTe; (**e**) CdSe; (**f**) CdTe; (**g**) CdTe phase diagram near the congruent point

16.1 Crystal Properties

16.1.1 Basic Properties

Wide-bandgap compound semiconductors have higher melting points. Due to their high ionicity, the overheating phenomenon occurs when they are heated to their melting point. Owing to the higher vapor pressures at their melting points, it is difficult to grow bulk crystals from melt. On the other hand, it is easy to grow bulk crystals as well as their films from the vapor phase. Therefore, before introducing film and bulk crystal growth, it is necessary to review their physical and chemical properties. Table 16.1 shows the properties of some of the main II–VI compound semiconductor materials [16.11–24].

16.1.2 Phase Diagram

It is necessary to understand the phase diagram to grow high-quality film and bulk single crystals. Figure 16.3

Table 16.1 Properties of some wide-bandgap II–VI compound semiconductors

Material Property	ZnS	ZnO	ZnSe	ZnTe	CdS	CdSe	CdTe
Melting point (K)	2038 (WZ, 150 atm)	2248	1797	1513	2023 (WZ, 100 atm)	1623	1370 (ZB)
Energy gap E_g at 300 K (eV)(ZB*/WZ*)	3.68/3.911	–/3.4	2.71/–	2.394	2.50/2.50	–/1.751	1.475
$dE_g/dT(\times10^{-4}$ eV/K) ZB/WZ	4.6/8.5	–/9.5	4.0/–	5.5/–	–/5.2	–/4.6	5.4/–
Structure	ZB/WZ	WZ	ZB/WZ	ZB	WZ	WZ	ZB
Bond length (μm)	2.342 (WZ)	1.977 (WZ)	2.454 (ZB)	2.636 (ZB)	2.530 (ZB)	2.630 (ZB)	2.806 (ZB)
Lattice constant (ZB) a_0 at 300 K (nm)	0.541	–	0.567	0.610	0.582	0.608	0.648
ZB nearest-neighbor dist. at 300 K (nm)	0.234	–	0.246	0.264	0.252	0.263	0.281
ZB density at 300 K (g/cm^3)	4.11	–	5.26	5.65	4.87	5.655	5.86
Lattice constant (WZ) at 300 K (nm)							
$a_0 = b_0$	0.3811	0.32495	0.398	0.427	0.4135	0.430	–
c_0	0.6234	0.52069	0.653	0.699	0.6749	0.702	–
c_0/a_0	1.636	1.602	1.641	1.637	1.632	1.633	–
WZ density at 300 K (g/cm^3)	3.98	5.606	–	–	4.82	5.81	–
Symmetry ZB/WZ	C6me/F43m	–/C6me	–/F43m	–/F43m	C6me/F43m	C6me/F43m	–/–
Electron affinity χ (eV)			4.09	3.53	4.79	4.95	4.28
Stable phase(s) at 300 K	ZB & WZ	WZ	ZB	ZB	ZB & WZ	ZB & WZ	ZB
Solid–solid phase transition temperature (K)	1293	–	1698	–	–	403	1273(?)
Heat of crystallization ΔH_{LS} (kJ/mol)	44	62	52	56	58	45	57
Heat capacity C_P (cal/mol K)	11.0	9.6	12.4	11.9	13.2	11.8	–
Ionicity (%)	62	62	63	61	69	70	72
Equilibrium pressure at c.m.p. (atm)	3.7	–	1.0	1.9	3.8	1.0	0.7
Minimum pressure at m.p. (atm)	2.8	7.82	0.53	0.64	2.2	0.4–0.5	0.23
Specific heat capacity (J/gK)	0.469	–	0.339	0.16	0.47	0.49	0.21
Thermal conductivity (W cm^{-1}K^{-1})	0.27	0.6	0.19	0.18	0.2	0.09	0.01
Thermo-optical cofficient $(dn/dT)(\lambda = 10.6\,\mu m)$	4.7	–	6.1	–	–	–	11.0
Electrooptical coefficient r_{41} (m/V) $(\lambda = 10.6\,\mu m)$	2×10^{-12}	–	2.2×10^{-12}	4.0×10^{-12} ($r_{41} = r_{52} = r_{63}$)	–	–	6.8×10^{-12}

m.p. – melting point; c.m.p. – congruent melting point; ZB – zinc blende; WZ – wurtzite

Table 16.1 (continued)

Material Property	ZnS	ZnO	ZnSe	ZnTe	CdS	CdSe	CdTe
Linear expansion coefficient ($10^{-6}\,K^{-1}$) ZB/WZ	–/6.9	2.9/7.2	7.6/–	8.0/–	3.0/4.5	3.0/7.3	5.1/–
Poisson ratio	0.27		0.28				0.41
Dielectric constant $\varepsilon_0/\varepsilon_\infty$	8.6/5.2	8.65/4.0	9.2/5.8	9.3/6.9	8.6/5.3	9.5/6.2	2.27/–
Refractive index ZB/WZ	2.368/2.378	–/2.029	2.5/–	2.72/–	–/2.529	2.5/–	2.72/–
Absorption coeff. (including two surfaces) ($\lambda = 10.6\,\mu m$)(cm^{-1})	≤ 0.15	–	$1-2\times10^{-3}$	–	≤ 0.007	≤ 0.0015	≤ 0.003
Electron effective mass (m^*/m_0)	–0.40	–0.27	0.21	0.2	0.21	0.13	0.11
Hole effective mass m^*_{dos}/m_0	–	–	0.6	circa 0.2	0.8	0.45	0.35
Electron Hall mobility (300) K for n = lowish (cm^2/Vs)	165	125	500	340	340	650	1050
Hole Hall mobility at 300 K for p = lowish (cm^2/Vs)	5	–	30	100	340	–	100
Exciton binding energy (meV)	36	60	21	10	30.5	15	12
Average phonon energy (meV) ZB/WZ	16.1/17.1	–	15.1/–	10.8/–	–/13.9	18.9/25.4	5.8/–
Elastic constant ($10^{10}\,N/m^2$)							
C_{11}	1.01±0.05	–	8.10±0.52	0.72±0.01	–	–	5.57
C_{12}	0.64±0.05	–	4.88±0.49	0.48±0.002	–	–	3.84
C_{44}	0.42±0.04	–	4.41±0.13	0.31±0.002	–	–	2.095
Knoop hardness (N/cm^2)	0.18	0.5	0.15	0.13	–	–	0.10
Young's modulus	10.8 Mpsi	–	10.2 Mpsi	–	45 GPa	5×10^{11} dyne/cm^2	3.7×10^{11} dyne/cm^2

m.p. – melting point; c.m.p. – congruent melting point; ZB – zinc blende; WZ – wurtzite

shows the phase diagrams reported for ZnS [16.25], ZnSe [16.25, 26], ZnTe [16.25], CdSe [16.27] and CdTe [16.28, 29]. Although much work has been done, there some exact thermodynamic data are still lacking, especially details close to the congruent point. Unfortunately, the phase diagram of ZnO is not available in spite of its growing importance in applications.

16.2 Epitaxial Growth

Epitaxial growth of wide-bandgap II–VI compounds was mainly carried out using liquid-phase epitaxy (LPE), or VPE. VPE includes several techniques, such as conventional VPE, hot-wall epitaxy (HWE), metalorganic chemical vapor deposition (MOCVD) or metalorganic phase epitaxy (MOVPE), molecular-beam epitaxy (MBE), metalorganic molecular-beam epitaxy (MOMBE) and atomic-layer epitaxy (ALE), etc. Each

Table 16.2 Strengths and weaknesses of several epitaxial growth techniques

LPE	Thermodynamic equilibrium growth Easy-to-use materials Low-temperature growth High purity Multiple layers Thickness control not very precise Poor surface/interface morphology	**MOCVD**	Gaseous reaction for deposition Precise composition Patterned/localized growth Potentially easier large-area multiple-wafer scale-up Low-temperature growth High-vapor-pressure materials growth allowed About 1 ML/s deposition rate Expensive equipment Safety precautions needed
HWE	Easy-to-use materials Low cost Thermodynamic equilibrium Hard to grow thick layers Thickness control not very precise		
VPE	Easy to operate Economic Thinner layers High growth rates Easier composition control High temperature (800–1000 °C)	**MBE and MOMBE**	Physical vapor deposition Ultra-high-vacuum environment About 1 ML/s deposition rate In situ growth-front monitoring Precise composition Low growth rate Sophisticated equipment Limit for high-vapor-pressure materials growth (MBE)
ALE	Gaseous reaction for deposition Low-temperature growth Precise composition Low growth rate Safety precautions needed		

of these methods has its advantages and disadvantages. They are summarized in Table 16.2.

In the case of hetero-epitaxy, the mismatch between substrate material and epitaxial layer affects the growing structure and quality of the epitaxial layer. The mismatch should be made as small as possible when choosing the pair of materials (substrate and epitaxial material). Furthermore, the difference between the thermal expansion coefficients of the pair of materials has to be considered to obtain high-quality epitaxial layer [16.30].

16.2.1 The LPE Technique

LPE growth occurs at near-thermodynamic-equilibrium conditions. There are two growth methods. The first is called equilibrium cooling, in which the saturated solution is in contact with the substrate and the temperature is lowered slowly, the solution becomes supersaturated; meanwhile a slow epitaxial growth on the substrate is initiated. The second is the step-cooling process ,in which the saturated solution is cooled down a few degrees (5–20 K) to obtain a supersaturated solution. The substrate is inserted into the solution, which is kept at this cooled temperature. Growth occurs first due to the supersaturation, and will slow down and stop finally. For both techniques, if the substrate is dipped in sequence into several different melt sources, multiple layer structures can be grown. LPE can successfully and inexpensively grow homo- and heterostructures. As the growth is carried out under thermal equilibrium, an epilayer with a very low native defect density can be obtained.

The LPE method can be used to grow high-quality epilayers, such as ZnS [16.31], ZnSe [16.31, 32], ZnSSe [16.33], ZnTe [16.34], etc. *Werkhoven* et al. [16.32] grew ZnSe epilayers by LPE on ZnSe substrates in a low-contamination-level environment. In their study, the width of bound exciton lines in low-temperature photoluminescence spectra was used

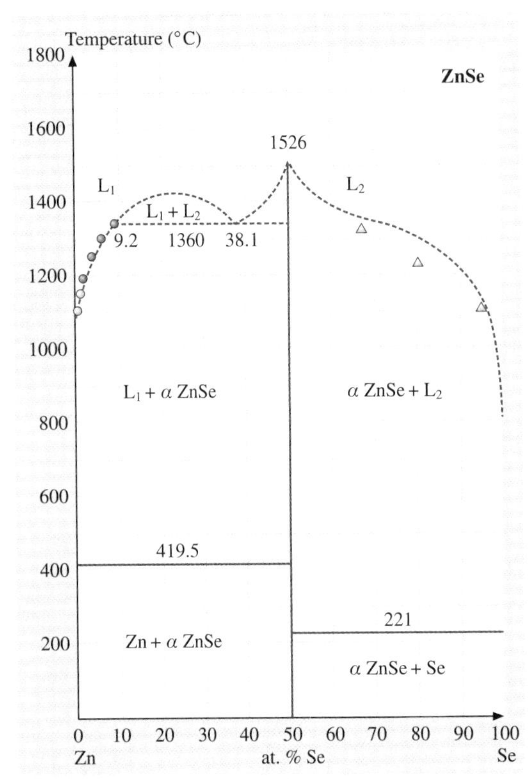

Fig. 16.4 A typical VPE growth system

to define the quality of the material, and the energy of the lines was used to identify trace impurities. The photoluminescence (PL) results showed that the ZnSe epitaxial layer has the high quality. The sharpest spectra occurred in layers grown rapidly on a previously grown buffer layer, indicating the importance of impurity outdiffusion from the substrate into the growing layer. The sharpness of these bound exciton lines indicates that the total concentration of electrically active impurities $(N_A + N_D)$ was below $10^{17}\ cm^{-3}$.

16.2.2 Vapor-Phase Epitaxy Techniques

Conventional VPE

A typical VPE growth system is shown in Fig. 16.4. In VPE growth, thin films are formed by the deposition of atoms from the vapor phase. There are two types of transport mechanisms for the source materials, physical vapor deposition (PVD) without any chemical reaction, and chemical vapor deposition (CVD), where the formation of the deposited film is the result of a chemical reaction of the precursors on the substrate. In VPE growth, there are several important parameters, such as the source temperature, the substrate temperature, the flow rate of the carrier gas, the growth pressure, and so on. These determine the growth rate, composition, and crystallinity of the epitaxial layers.

The VPE technique is the most popular in semiconductor epitaxial growth. Since the vapor pressures of all wide-bandgap II–VI materials are high, their epitaxial layers can be grown by this method. As an example, high-quality ZnS single-crystal films have been grown on a Si substrate using hydrogen as a carrier gas [16.35, 36]. Furthermore, the Iida group [16.37] doped N and P into a ZnS epilayer and studied their behavior in details. The N and P were expected to compensate the native donor state and to result in an insulating material. The results showed that the doped acceptors N and P reduced the donor density and an insulating material was obtained. Later, this group was successful in preparing a p-type ZnS epilayer using NH_3 as an acceptor dopant [16.38]. Many efforts have been made to grow ZnSe epilayers on different substrates, especially on GaAs in the past two decades [16.39, 40]. p-type ZnSe was also obtained by this technique [16.41]. In addition, other compounds, ZnTe [16.42], CdS [16.43], CdSe [16.44], and CdTe [16.45], were also studied using this technique.

The HWE Technique

Hot-wall epitaxy [16.1] has proved to be a very successful growth method for II–VI compound epitaxial layers. Its principal characteristic is the growth of thin films under conditions near thermodynamic equilibrium. Compared with other VPE methods, the HWE technique has the advantages of low cost, simplicity, convenience, and relatively high growth rate. In particular, it can control deviation from stoichiometry during growth of an epilayer.

A schematic diagram of the improved HWE system is showed in Fig. 16.5; it consists of four independent furnaces. The source material placed at the middle is transported to the substrate. The region of the growth reactor between the source and substrate, called the hot wall, guarantees a nearly uniform and isotropic flux of molecules onto the substrate surface. To control the deviation from stoichiometry, the reservoir part is placed at the bottom with a constituent element.

HWE has been applied to growing II–VI compound epilayers, such as CdTe [16.46], CdS [16.47], CdSe [16.48], ZnTe [16.49], and also to producing heterostructures for laser and photovoltaic detector fab-

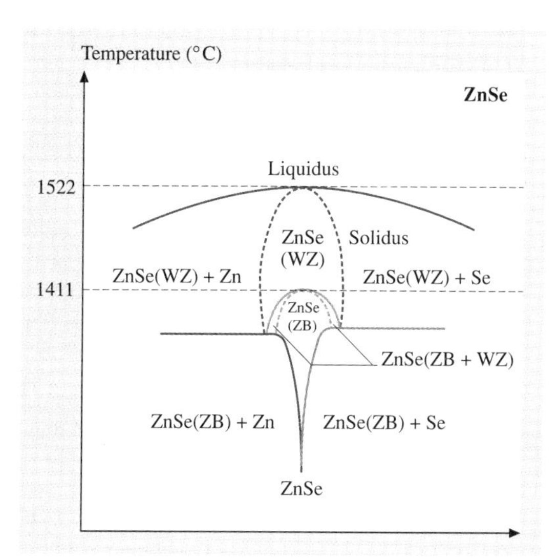

Fig. 16.5 Diagram of a typical HWE system growth chamber

rication [16.50]. Most research using HWE technique has focused on CdTe growth. *Wang* et al. [16.46] optimized growth conditions and grew high-quality CdTe epitaxial films using the HWE apparatus shown in Fig. 16.5. All the CdTe epilayers show mirror-like surfaces. Results from PL and X-ray diffraction (XRD) show that CdTe epilayers on GaAs suffer from a biaxial compressive stress, that this stress is rapidly relaxed within a thickness of about 5 μm, and that it remains in the epilayer up to a film thickness of 15 μm. Although this heterosystem has a 14.6% lattice mismatch and −26% thermal expansion mismatch at 300 K, high-quality CdTe epilayers, with a full-width half-maximum (FWHM) of 0.26 meV for bound-exciton emission lines in 4.2-K PL, about 90 arcs for (400) diffraction in four-crystal XRD spectrum, were prepared by selecting suitable growth conditions and epilayer thickness.

Recently, significant results have been achieved for CdTe/Si (111) epilayer growth by HWE. *Lalev* et al. [16.51] reported that high-quality CdTe (111) epilayers with *A* polarity were directly grown on hydrogen-terminated Si (111) without any preheating treatment. Through the originally designed two-step growth regime, the crystal quality of CdTe film was significantly improved, and the best FWHM value of 118 arcs from four-crystal rocking curves was obtained for a 5-μm-thick epilayer.

The MOCVD Technique

MOCVD or MOVPE is an improvement over conventional VPE. Since its introduction in 1968 [16.52], this technique has been established as one of the techniques for epitaxial growth of compound semiconductors both for research and production. The factors that have allowed MOCVD to reach this popularity are the purity and abruptness of the grown layers together with the flexibility of the technique, which makes the growth of almost all compound semiconductors possible. This abrupt transition in the composition of the epitaxial structure is necessary for the fabrication of digital or analog alloy system.

The development history of MOCVD technique is equivalent to that of source precursors. Since ZnSe epilayers were grown by MOCVD [16.52], many source precursors of II–VI elements have been developed. $Zn(CH_3)_2$ dimethylzinc (DMZn) and $Zn(C_2H_5)_2$ diethylzinc (DEZ) were used at the beginning of MOCVD growth [16.53]. ZnSe and ZnS films were grown using these metalorganic sources and inorganic H_2Se or H_2S. Unfortunately, the quality of these films was very poor. For this reason, $Se(CH_3)_2$ dimethylselenide (DMSe) and $Se(C_2H_5)_2$ diethylselenium (DESe) were developed [16.54]. The quality of epilayers was greatly improved. From then, II and VI elemental gas precursors were proposed one after another. *Wright* et al. grew a ZnSe film using (DMZn-$(NEt_3)_2$) triethylamine adduct of dimethylzinc [16.55]. *Hirata* et al. [16.56] and *Nishimura* et al. [16.57] proposed methylselenol (MSeH) and tertiarybutylselenol (*t*-BuSeH) as Se sources, respectively. Methylallylselenide (MASe) [16.58], diallyl-selenide (DASe) [16.59], t-butylallylselenide (*t*-BuASe) [16.60], tertiarybutyl-selenide (*Dt*-BuSe) [16.61], were also used as Se sources. *Fujita* et al. found methylmercaptan (MSH) as an S source [16.53]. Besides these, many other source precursors of II–VI elements, such as $Cd(CH_3)_2$ dimethylcadmium (DMCd) [16.62], $Te(CH_3)_2$ dimethyltelluride (DMTe) [16.62], $Te(C_2H_5)_2$ diethyltelluride (DETe), $Te(C_3H_7)_2$ diisopropyltelluride (DIPTe) [16.63], $S(C_2H_5)_2$ diethylsulfide (DES), $S(C_4H_9)_2$ ditertiarybutylsulfide (DTBS), $(C_4H_9)SH$ tertiarybutylthiol (tBuSH), have been used.

The great advantage of using metalorganics is that they are volatile at moderately low temperatures. Since all constituents are in the vapor phase, precise electronic control of gas flow rates and partial pressures is possible. This, combined with pyrolysis reactions that are relatively insensitive to temperature, allows efficient and reproducible deposition.

The substrate wafer is placed on a graphite susceptor inside a reaction vessel and heated by a radio-frequency (RF) induction heater. The growth temperature depends on the type of compounds grown. Growth is carried out in a hydrogen atmosphere at a pressure of 100–700 torr. The growth precursors decompose on contact with the hot substrate to form epitaxial layers. Each layer is formed by switching the source gases to yield the desired structure.

The films of almost all wide-bandgap II–VI compounds have been grown by MOCVD technique. Most work has been done on p-ZnSe epilayers in the past two decades [16.64–66]. The highest hole concentration of $8.8\times10^{17}\ cm^{-3}$ was reported with a NH_3 doping source [16.67]. Recently, quantum wells (QW) and quantum dots (QD) of these wide-bandgap compounds have become the focus. Successful pulsed laser operation at 77 K in ZnCdSe/ZnSe/ZnMgSSe QW-structure separated-confinement heterostructures has been realized [16.68].

MBE and MOMBE

MBE was developed at the beginning of the 1970s to grow high-purity high-quality compound semiconductor epitaxial layers on some substrates [16.69, 70]. To date, it has become a very important technique for growing almost all semiconductor epilayers. An MBE system is basically a vacuum evaporation apparatus. The pressure in the chamber is commonly kept below $\approx 10^{-11}$ torr. Any MBE process is dependent on the relation between the equilibrium vapor pressure of the constituent elements and that of the compound [16.71]. There are a number of features of MBE that are generally considered advantageous for growing semiconducting films: the growth temperature is relatively low, which minimizes any undesirable thermally activated processes such as diffusion; the epilayer thickness can be controlled precisely; and the introduction of different vapor species to modify the alloy composition and to control the dopant concentration can be conveniently achieved by adding different beam cells with proper shutters. These features become particularly important in making structures involving junctions.

Metalorganic molecular-beam epitaxy growth (MOMBE) is one of the variations of the MBE system [16.72, 73]. The difference is that metalorganic gaseous sources are used as the source materials. Therefore, this growth technique has the merits of MOCVD and MBE.

MBE or MOMBE techniques have been used to grow epilayers of almost all wide-bandgap II–VI semiconductors [16.74,75]. Due to its features, it is very successful in growing super-thin layers, such as single quantum wells (SQW), multiple quantum wells (MQW) [16.76,77] and nanostructures [16.78].

In nanostructures, quantum dot (QD) structures have attracted a lot of attention in recent years. This field represents one of the most rapidly developing areas of current semiconductor. They present the utmost challenge to semiconductor technology, rendering possible fascinating novel devices. QD are nanometer-size semiconductor structures where charge carriers are confined in all three spatial dimensions. They are neither atomic nor bulk semiconductor, but may best be described as artificial atoms.

In the case of heteroepitaxial growth there are three different growth modes [16.79]: (a) Frank–van der Merwe (FM) or layer-by-layer growth, (b) Volmer–Weber (VW) or island growth, and (c) Stranski–Krastanov (SK) or layer-plus-island growth. Which growth mode will be adopted in a given system depends on the surface free energy of the substrate, (σ_s), that of the film, (σ_f), and the interfacial energy (σ_i). Layer-by-layer growth mode occurs when $\Delta\sigma = \sigma_f + \sigma_i - \sigma_s = 0$. The condition for FM-mode growth is rigorously fulfilled only for homoepitaxy, where $\sigma_s = \sigma_f$ and $\sigma_i = 0$. If the FM-mode growth condition is not fulfilled, then three-dimensional crystals form immediately on the substrate (VW mode). For a system with $\Delta\sigma = 0$ but with a large lattice mismatch between the substrate and the film, initial growth is layer-by-layer. However, the film is strained. As the film grows, the stored strain energy increases. This strained epilayer system can lower its total energy by forming isolated thick islands in which the strain is relaxed by interfacial misfit dislocations, which leads to SK growth in these strained systems. The SK growth mode occurs when there is a lattice mismatch between the substrate and the epilayer, causing the epilayer to be strained, which results in the growth of dot-like self-assembled islands. Wire-like islands can grow from dot-like islands via a shape transition which helps strain relaxation.

For nanostructure fabrication, a thin epilayer is usually grown on a substrate. This two-dimensional (2-D) layer is used to fabricate lower-dimensional structures such as wires (1-D) or dots (0-D) by lithographic techniques. However, structures smaller than the limits of conventional lithography techniques can only be obtained by self-assembled growth utilizing the principles of SK or VW growth. For appropriate growth conditions, self-assembled epitaxial islands can be grown in reasonably well-controlled sizes [16.80].

Because wide-bandgap II–VI materials typically have stronger exciton–phonon interactions than III–V materials, their nanostructures are expected to be very useful in fabricating optoelectronics devices and in exploring the exciton nature in low-dimensional structure. Self-assembled semiconductor nanostructures of different system, such as CdSe/ZnSe [16.81], ZnSe/ZnS [16.82], CdTe/ZnTe [16.83], CdS/ZnSe [16.84], are thought to be advantageous for future application. MBE/MOMBE [16.81, 84], MOCVD [16.82], HWE [16.85] are the main growth techniques used to obtain such structures. MBE is the most advanced technique for the growth of controlled epitaxial layers. With the advancement of nanoscience and nanotechnology, lower-dimensional nanostructures are being fabricated by lithographic techniques from two-dimensional epitaxial layers. Alternately, self-assembled, lower-dimensional nanostructures can be fabricated directly by self-assembly during MBE growth.

Atomic-Layer Epitaxy

ALE is a chemical vapor deposition technique [16.5] where the precise control of the system parameters (pressure and temperature) causes the reaction of adsorption of the precursors to be self-limiting and to stop with the completion of a single atomic layer. The precursors are usually metalorganic molecules. The special feature of ALE is that the layer thickness per cycle is independent of subtle variations of the growth parameters. The growth rate is only dependent on the number of growth cycles and the lattice constant of the deposited material. The conditions for thickness uniformity are fulfilled when material flux on each surface unit is sufficient for monolayer saturation. In an ALE reactor, this means freedom in designing the precursor transport and its interaction with the substrates.

The advantages obtainable with ALE depend on the material to be processed and the type of application. In single-crystal epitaxy, ALE may be a way to obtain a lower epitaxial crystal-growth temperature. It is also a method for making precise interfaces and material layers needed in superlattice structures and super-alloys. In thin-film applications, ALE allows excellent thickness uniformity over large areas. The process has primarily been developed for processing of compound materials. ALE is not only used to grow conventional thin films of II–VI wide-bandgap compounds [16.5,86,87], but is also a powerful method for the preparation of monolayers (ML) [16.88].

16.3 Bulk Crystal Growth

Bulk crystal is the most important subject studied in recent decades. The quality of bulk crystals is the most important aspect of electronic device design. To date, many growth methods have been developed to grow high-quality crystals. Significant improvements have been made in bulk crystal growth with regard to uniformity, reproducibility, thermal stability, diameter control, and impurity and dopant control. According to the phase balance, crystals can be grown from vapor phase, liquid (melt) phase, and solid phase.

16.3.1 The CVT and PVT Techniques

Crystal growth from the the vapor phase is the most basic method. It has advantages that growth can be performed at lower temperatures. This can prevent from phase transition and undesirable contamination. Therefore, this method has commonly been used to grow II–VI compound semiconductors.

Crystal growth techniques from the vapor phase can be divided into chemical vapor transport (CVT) and physical vapor transport (PVT). CVT is based on chemical transport reactions that occur in a closed ampoule having two different temperature zones. Figure 16.6 shows a typical schematic diagram of the CVT technique. In the high-temperature region, the source AB reacts with the transport agent X:

$$2AB + 2X_2 \leftrightarrow 2AX_2(gas) + B_2(gas) \qquad (16.1)$$

In the low-temperature region, the reverse reaction takes place. The whole process continues by back-diffusion of the X_2 generated in the lower-temperature region. The transport agent X usually employed is hydrogen (H_2), a halogen (I_2, Br_2, Cl_2), a halide (HCl, HBr), and so on. For example, I_2 has been used as a transport agent for ZnS, ZnSe, ZnTe and CdS [16.89]; HCl, H_2, Cl_2, NH_3 [16.90], and C and CH_4 [16.91] have been used as the transport agents for ZnO. According to [16.89]: the typical growth temperature for ZnS is 1073–1173 K, for ZnSe 1023–1073 K, for ZnTe 973–1073 K; ΔT is 5–50 K; the concentration of the transport agent is

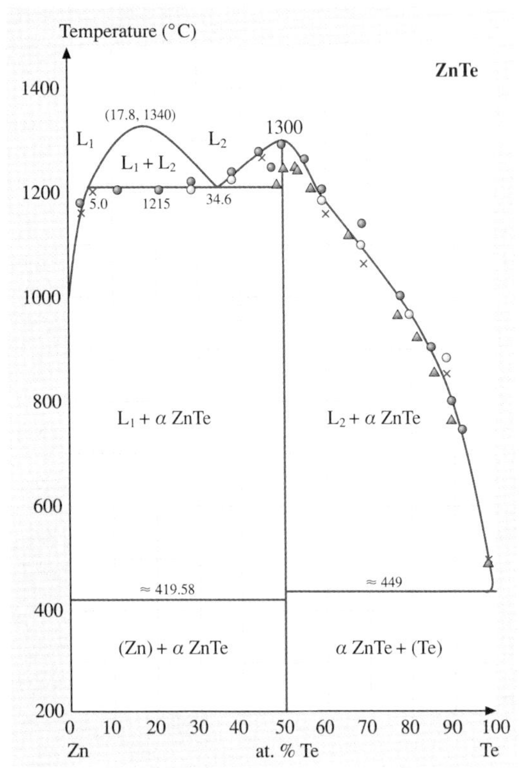

Fig. 16.6 Diagram of a conventional chemical vapor transport system

0.5–5 mg/cm^3 of the ampoules vapor space; the aspect ratios are 5–17 at ampoule diameters of 10–20 mm. According to [16.91]: the growth temperature for ZnO is 1228–1273 K, and ΔT is 5–10 K. The transport rate does not strongly depend on the initial amount of carbon when the concentration of the transport agent is over 0.3 mg/cm^3.

The PVT method is similar to CVT, but the transport agent is not used. This technique is based on the dissociative sublimation of compounds. Initially, the Piper–Polich method was developed, in 1961 [16.92]. *Prior* [16.93] improved the PVT method using a reservoir to control the deviation from stoichiometry; the experimental arrangement is shown in Fig. 16.7. The constituent element is placed in the reservoir. The reservoir temperature can be calculated according to the solid–vapor equilibrium

$$2\mathrm{AB} \leftrightarrow 2\mathrm{A(gas)} + \mathrm{B_2(gas)}\,. \tag{16.2}$$

The total pressure (p) in ampoule is given by

$$\begin{aligned} p &= p_\mathrm{A} + p_{\mathrm{B}_2} \\ &= p_\mathrm{A} + K \cdot p_\mathrm{A}^{-2} \\ &= (K/p_{\mathrm{B}_2})^{1/2} + p_{\mathrm{B}_2}\,, \end{aligned} \tag{16.3}$$

where p_A and p_{B_2} are the partial pressures of the group II and VI elements respectively and $K = p_\mathrm{A}^{-2} \cdot p_{\mathrm{B}_2}$ is the equilibrium constant of (16.2). At any temperature, there is minimum total pressure (p_min), which corresponds to the condition,

$$p_\mathrm{A} = 2p_{\mathrm{B}_2} = 2^{1/3}K^{2/3}\,. \tag{16.4}$$

Under this condition, the vapor-phase composition is stoichiometric and growth rate is maximum [16.94]. After modifying this method to use a closed ampoule, it was applied to grow high-purity and high-quality crystals of II–VI compounds.

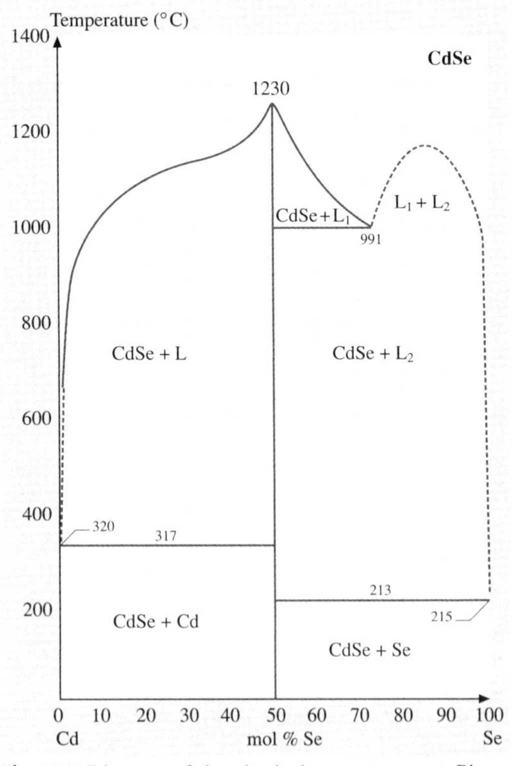

Fig. 16.7 Diagram of the physical vapor transport Piper–Polich method

The PVT of II–VI compounds takes advantage of the volatility of both components of the compound semiconductor. This same volatility, coupled with typically high melting points, makes melt growth of these materials difficult. In the PVT process, an ampoule containing a polycrystalline source of the desired II–VI compound is heated to a temperature that causes the compound to sublime at a rate conducive to crystal growth. The ampoule is typically placed in a furnace having a temperature gradient over the length of the ampoule, so that the polycrystalline source materials sublime at the end with the higher temperature. The end of the ampoule where the crystal is to be grown is then maintained at a lower temperature. This temperature difference causes supersaturation, and vaporized molecules from source materials eventually deposit at the cooler end. In order to control the deviation from stoichiometry, a reservoir is often used (Fig. 16.7). One of the constituent elements is placed in it. By selecting the proper growth conditions, the rate of deposition can be set to a value leading to growth of high-quality crystals. Typically, PVT growth of II–VI compounds is carried out at temperatures much lower than their melting points; this gives benefits in terms of reduced defects, which are related to the melt growth of II–VI compounds such as voids and/or inclusions of excess components of the compound, and also helps to reduce the contamination of the growing crystal from the ampoule. Other effects, such as the reduction of point defects, are also typically found when crystals grown by PVT are compared to crystals grown by melt techniques. Although claims have been made that the lower temperatures of physical vapor transport crystal growth should also reduce the twinning found in most of the cubic II–VI compound crystals, the reduction is not usually realized in practice. The assumption that the twinning is a result of cubic/hexagonal phase transitions is not found to be the determining factor in twin formation.

Ohno et al. [16.95] grew cubic ZnS single crystals by the iodine transport method without a seed. By means of Zn-dip treatment, this low-resistivity crystal was used for homoepitaxial MOCVD growth, and a metal–insulator–semiconductor(MIS)-structured blue LED, which yielded an external quantum efficiency as high as 0.05%. They found that crystal quality was significantly improved by prebaking the ZnS powder in H_2S gas prior to growth. The growth rate also increased by three times.

Isshiki et al. [16.96] purified zinc by a process consisting of vacuum distillation and overlap zone melting in pure argon. Using refined zinc and commercial high-purity Se, high-quality ZnSe single crystals were grown by the same method, as reported by *Huang* and *Igaki* [16.97]. The emission intensities of donor-bound exciton (I_2) are remarkably small. The emission intensities of the radiative recombinations of free excitons (E_X) are very strong [16.98]. These intensities indicated the crystal had a very high purity and a very low donor concentration, and they suggest that the purity of the grown crystal strongly depends on the purity of the starting materials. This method is suitable for preparing high-purity crystals, since a purification effect is expected during growth. Impurities with a higher vapor pressure will condense at the reservoir portion and those with a lower vapor pressure will remain in the source crystal. This effect was confirmed by the PL results [16.99]. As for these crystals, photoexcited cyclotron resonance measurements have been attempted and cyclotron resonance signals due to electrons [16.100] and heavy holes [16.101] have been detected for the first time. The cyclotron mobility of electrons under $B = 7\,\mathrm{T}$ is $2.3 \times 10^5\,\mathrm{cm^2/Vs}$. This indicates that the quality of the grown crystals is very high. Furthermore, the donor concentration in the crystal is estimated to be $4 \times 10^{14}\,\mathrm{cm^3}$ by analyzing the temperature dependence of the cyclotron mobility [16.99].

The crystals are grown in a self-seeded approach by the CVT or PVT techniques introduced above. This limits single-crystal volume to several $\mathrm{cm^3}$. Meanwhile, grain boundaries and twins are easy to form during growth. In order to solve these problems, seeded chemical vapor transport (SCVT) and seeded physical vapor transport (SPVT), the so-called modified Lely method, have been developed [16.102]. The difference between SCVT/SPVT and CVT/PVT is that a seed is set in the crystal growth space before growth starts. The most successful method of eliminating twin formation has usually been by using a polycrystal or single-crystal seed. Even this seeding cannot assure complete elimination of twinning unless seeding is done carefully. The usual method of using small seeds and increasing the diameter of the growing crystal are dependent on the preparation and condition of the walls of the ampoule and the furnace profiles required to eliminate spurious nucleation from the walls. Since the use of a seed crystal provides better control over the nucleation process, high-quality single crystals can be grown [16.103, 104]. Using this technique, sizable single crystals of II–VI wide-bandgap compounds has been commercialized.

Fujita et al. [16.105] grew ZnS single crystals as large as $24\,\mathrm{mm} \times 14\,\mathrm{mm} \times 14\,\mathrm{mm}$ by the SCVT method using iodine as a transport agent. The average linear growth rate was about 10^{-6} cm/s. The crystal size depended strongly on the ampoule geometry and the temperature difference between the seed and solvent. The study of the electrical properties showed that the annealed crystal was n-type.

16.3.2 Hydrothermal Growth

The hydrothermal technique is a method for growing crystal from aqueous solvent [16.106]. Figure 16.8 shows a diagram of hydrothermal techniques. The hydrothermal method of crystal growth has several advantages: (1) due to the use of a closed system, it is easier to control oxidization or maintain conditions that allow the synthesis of phases that are difficult to attain by other methods, such as compounds of elements in oxidation states, especially for transition-metal compounds; (2) crystal grow occurs under lower thermal strain, and thus may contain a lower dislocation density than when the crystal is grown from a melt, where large thermal gradients exist; (3) the method has proven to be very useful for the synthesis of the so-called low-temperature phases; (4) it can be employed for large-scale synthesis of piezoelectric, magnetic, optic, ceramic, and many other special materials; (5) hydrothermal synthesis results in rapid convection and very efficient solute transfer, which results in comparatively rapid growth of larger, purer, and dislocation-free crystals. The most successful example of obtaining II–VI compounds is growing single ZnO crystals.

ZnO crystals are considered extrinsic n-type piezoelectric semiconductors. Undoped crystals have a typical resistivity of 0.1–100 Ω cm and a drift mobility of 10–125 cm^2/Vs. Low carrier concentrations can be approached by special growth and annealing methods. ZnO crystal is quite transparent in the range 0.4–6.0 μm. Slight absorption is sometimes found around 2.2–2.3 μm and an additional slight absorption is found at 3.42 μm. Since 1953, *Walker* [16.107] and

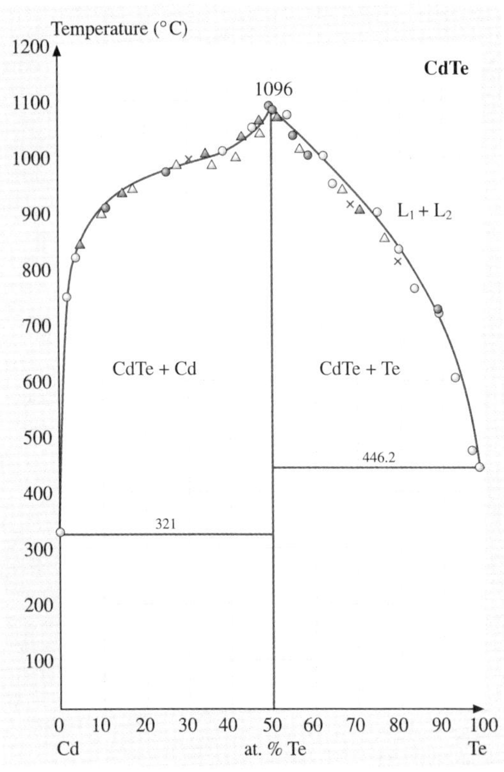

Fig. 16.8 Diagram of a typical hydrothermal technique for growing ZnO single crystals

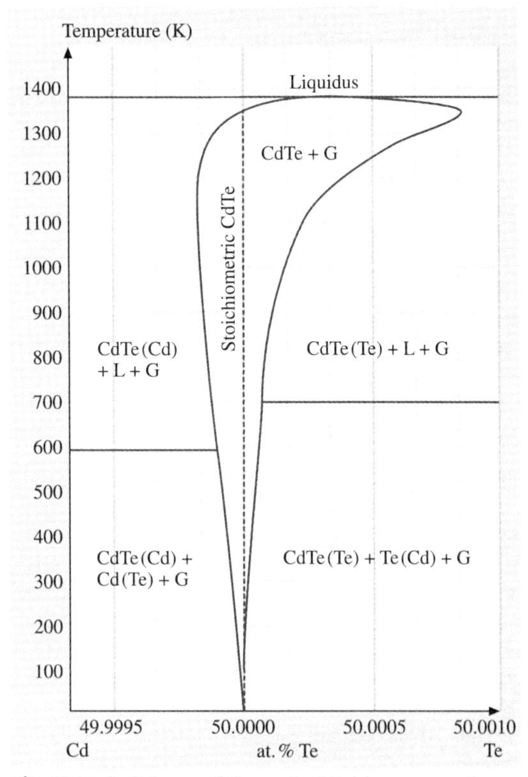

Fig. 16.9a,b Scheme of the vertical Bridgman growth system (a), and its temperature profiles (b)

many other researchers [16.108–111] have grown large ZnO single crystals with hydrothermal techniques and other methods.

The seed, suspended by a Pt wire, and sintered ZnO strings as a source material (nutrient), together with a KOH (3M) and LiOH (1M) aqueous solution, were put into a Pt crucible [16.112] (the hydrothermal conditions are different in different papers). The seed crystals and the source material were separated by a Pt baffle. The crucible was sealed by welding and put into an autoclave. This hydrothermal autoclave is made of high-strength steel. Then, the autoclave was put into a vertical furnace. The temperature of the autoclave was raised to about 673 K, which produced 0.1 GPa of pressure. the growth temperature was monitored by a thermocouple inserted in the autoclave. Seed crystals grew to about 10 mm or bigger after two weeks. The crystal habit of hydrothermal ZnO crystals grown on basal plane seeds shows that growth direction in [0001] is faster 3 times than [000$\bar{1}$] [16.113].

16.3.3 Bridgman and Gradient Freezing (GF) Method

From the viewpoint of industrial production, melt growth is the most useful for obtaining large single crystals. VPE growth has limitations with regard to crystal size and productivity. The Bridgman technique is a typical crystal growth method from melt. Bridgman growth can be simply understood in terms of a molten charge that passes through a temperature gradient at a slow speed and solidifies when the temperature is below the melting point of this material. If the ampoule and furnace are stationary and the temperature is gradually reduced by keeping the temperature gradient at the interface constant, this growth process is called the gradient freezing (GF) method [16.114]. In Bridgman or GF growth, single crystals can be grown using either seeded or unseeded ampoules or crucibles.

The Bridgman method has been most extensively used to grow II–VI wide-bandgap compounds such as CdTe, ZnTe and ZnSe crystals, because of the simplicity of the growth apparatus, the high growth rate and the availability of crystals of appropriate size and quality. There are two Bridgman method techniques: the high-pressure technique [16.115] and the closed technique [16.116]. In the former, it is inevitable that a compositional deviation from stoichiometry occurs during melting. Since the properties and structural perfection of these compound crystals are correlated very strongly with this nonstoichiometry [16.117], compositional deviation must be controlled during melt growth. *Omino* et al. [16.116] and *Wang* et al. [16.118, 119] have adopted a closed double crucible to prevent deviation of the melt from stoichiometric composition during Bridgman growth. Figure 16.9 shows a closed vertical Bridgman growth furnace and its temperature profiles.

In Bridgman growth, the temperature gradient (G) and growth rate (R) are very important parameters since they determine the shape of the solid–liquid interface. For this reason, the relationship between temperature gradient and growth rate is investigated. The experi-

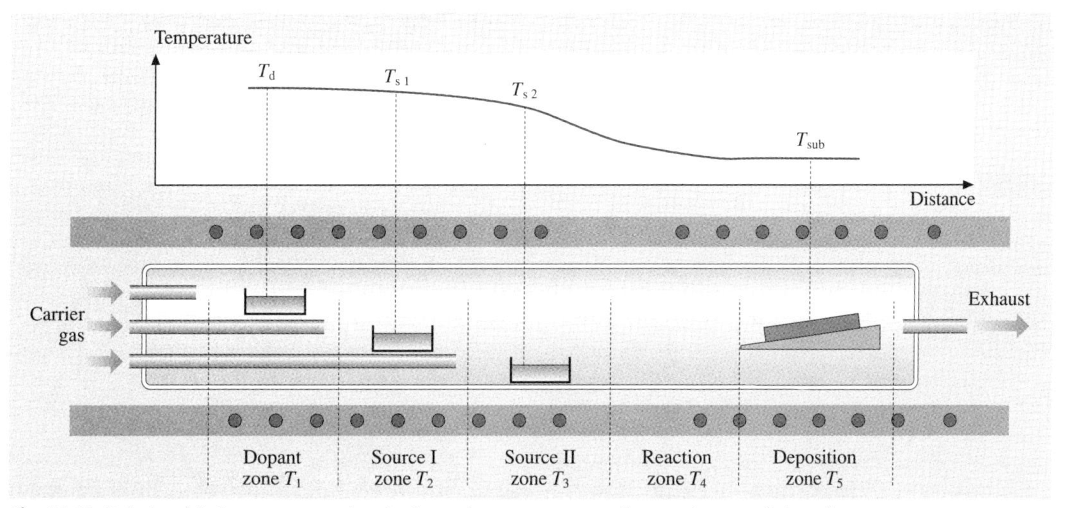

Fig. 16.10 Relationship between growth velocity and temperature gradient at the growth interfaces

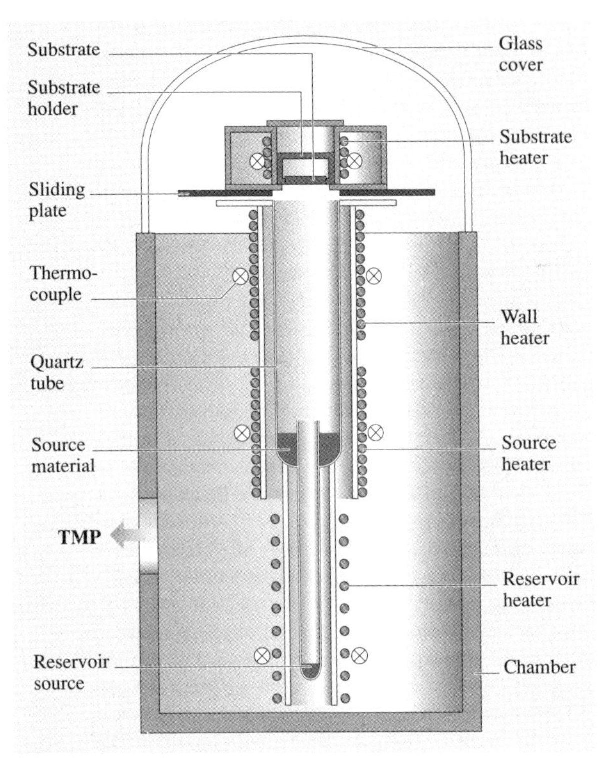

Fig. 16.11 Photograph of twin-free ZnSe single crystal grown by the vertical Bridgman technique

mental results are summarized in Fig. 16.10 [16.120, 121]. The experimental results show that, to grow a ZnSe single crystal, it is necessary that the G/R [(temperature) gradient/(growth) rate] value should be limited to 57–175 K h/cm^2. The most suitable value of G/R, assessed from the determined optimum temperature gradient and growth rate, is 83 K h/cm^2. *Wang* et al. [16.118, 119] found the optimum conditions to include: a special temperature program for removing the gas bubbles generated in melt, an overheating temperature of 76 K from the melting point of 1797 K, a temperature gradient of 30 K/cm and a growth rate of 3.6 mm/h as marked by the open squares in Fig. 16.10. Under these growth conditions, twin-free high-quality ZnSe single crystals (Fig. 16.11) were grown using a polycrystalline seed. Chemical etching on the cleaved (110) plane revealed that the average value of the etch pit density (EPD) is about 2×10^5 cm^{-2}. The rocking curves of four-crystal XRD showed a full-width at half-maximum (FWHM) value of 19 arcs. The resolved intensive free-exciton, bound-exciton emission lines and the weak donor-acceptor-pair (DAP) emission bands are observed in the PL spectra at 4.2 K. The FWHM of the I_1^d emission was smaller than 0.5 meV. On the other hand, the deep-level emission bands were almost not observed. All these results suggest that the ZnSe single crystals grown by this method are of very high quality.

Asahi et al. [16.122] successfully grew ZnTe single crystals with a diameter of 80 mm and a length of 50 mm by the vertical gradient freezing (VGF) method. In this method, a high-pressure furnace was used and the melt was encapsulated by B_2O_3 during crystal growth. The growth direction was nearly $\langle 111\rangle$ or $\langle 110\rangle$. When long ZnTe crystals were grown, polycrystals were found at the tail. It seems to be difficult to grow an ingot longer than 50 mm. The researchers believed that this is because the shape of the solid–liquid interface easily becomes concave against the liquid at the tail owing to the low thermal conductivity of ZnTe. Evaluation of the crystals showed that the FWHMs of the rocking curve measured by XRD were about 20 arcs, and the EPDs were 5×10^3–1×10^4 cm^{-2}.

16.3.4 The Traveling Heater Method (THM)

The traveling heater method (THM) [16.123] is a solution growth process whereby polycrystalline feed material with an average constant composition is progressively dissolved under the influence of a temperature gradient, followed by deposition in single-crystal form onto a seed with the same composition. The growth proceeds by the relative translation of the heater and charge. THM is particularly useful for the growth of binary and ternary semiconductor alloys, such as CdTe and CdZnTe [16.124]. In such materials, the wide separation between the solidus and liquidus in the pseudo-binary (CdTe–ZnTe) phase diagram imposes a monotonic variation in composition of the solid in the melt growth processes. The THM process ensures a constant macroscale composition in the crystal grown. Since the process takes place at a temperature below the melting point, contamination from the container is also reduced. The reduced operating temperature also leads to a lower ambient pressure within the growth environment, and a reduced risk of ampoule fracture.

16.3.5 Other Methods

Besides the growth methods described, there are many other techniques for growing single crystals of II–VI

Table 16.3 Growth methods for films and bulk crystals of wide-bandgap II–VI compounds

	Epitaxial growth						Bulk growth			
	LPE	VPE	HWE	ALE	MOCVD	MBE	CVT	PVT	Hydrothermal	Bridgman
ZnS	•	•		•	•	•	•	•		
ZnO					•	•	•		•	
ZnSe	•	•	•	•	•	•	•	•		•
ZnTe	•	•	•	•	•	•	•	•		•
CdS		•	•	•	•	•	•	•		•
CdSe		•	•	•	•	•	•	•		•
CdTe		•	•	•	•	•	•	•		•

wide-bandgap compounds. Zone melting [16.125] and solid-state recrystallization (SSR) [16.126] are often used to grow bulk crystals of ZnSe, ZnS and CdTe.

Recently, *Asahi* et al. [16.127] proved that B_2O_3 is a suitable encapsulant for ZnTe melt growth. Furthermore, B_2O_3 and a total weight of 6N Zn and Te was charged into a pBN crucible. Then this crucible was put into a high-pressure furnace with five heaters. A ZnTe seed was used to pull the ZnTe crystal. Before growth started, the starting materials were heated to 1573 K and kept at this temperature for several hours. The pressure in the growth furnace was kept at 1.5–2 MPa using Ar gas during growth. The temperature gradient on the solid–liquid surface was about 10–20 K/cm. The growth rate was 2–4 mm/h. Under these growth conditions, ZnTe single crystals with a diameter of 80 mm and a height of 40 mm were successfully grown using a combined GF/Kyropoulos method [16.127] (Fig. 16.12).

Some growth methods for film and bulk crystals are summarized in Table 16.3.

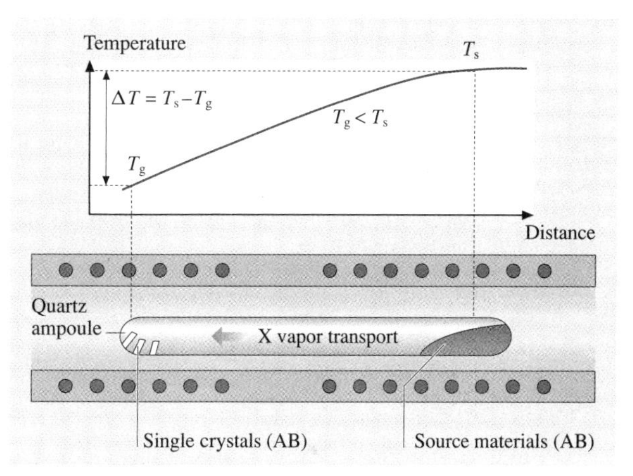

Fig. 16.12 ZnTe single crystal grown by a combination of the GF and Kyropoulos methods

16.4 Conclusions

It was expected that II–VI wide-bandgap compounds would become applicable to optoelectronic devices, especially LEDs and LDs in the short-wavelength visible-light region. However, as they are very strongly bonded with a nearly equal balance of covalent and ionic bonding, it is challenging to grow high-purity high-quality single crystals. Another problem with their application to devices is the difficulty of controlling the conductive type. This is because native defects commonly occur in these semiconductors. These native defects can have either a donor or acceptor character, or even be amphoteric, and they act as compensating centers. Furthermore, these defects react with dopant impurities to form complexes. This makes it difficult to reverse the conductive type. For example, in the case of ZnSe, it is still difficult to obtain low-resistivity p-type crystals or epilayers that can be used to fabricate devices. Therefore, a significant improvement in the understanding of the fundamental physical and chemical properties needs to be achieved. In particular, the specifications for many applications are very demanding, and considerable progress needs to be made in growth, particularly in the areas of reproducibility, convenient shape, conductivity, and structural perfection.

References

16.1 A. Lopez-Otero: Thin Solid Films **49**, 1 (1978)
16.2 H. M. Manasevit, W. I. Simpson: J. Electrochem. Soc. **118**, 644 (1971)
16.3 L. L. Chang, R. Ludeke: *Epitaxial Growth, Part A*, ed. by J. W. Matthews (Academic, New York 1975) p. 37
16.4 E. Veuhoff, W. Pletschen, P. Balk, H. Luth: J. Cryst. Growth **55**, 30 (1981)
16.5 T. Suntola: Mater. Sci. Rep. **4**, 261 (1989)
16.6 M. M. Faktor, R. Heckingbottom, I. Garrett: J. Cryst. Growth **9**, 3 (1971)
16.7 I. Kikuma, M. Furukoshi: J. Cryst. Growth **41**, 103 (1977)
16.8 Y. V. Korostelin, V. J. Kozlovskij, A. S. Nasibov, P. V. Shapkin: J. Cryst. Growth **159**, 181 (1996)
16.9 J. F. Wang, A. Omino, M. Isshiki: Mater. Sci. Eng. **83**, 185 (2001)
16.10 S. H. Song, J. F. Wang, G. M. Lalev, L. He, M. Isshiki: J. Cryst. Growth **252**, 102 (2003)
16.11 H. Harmann, R. Mach, B. Sell: In: *Current Topics Mater. Sci.*, Vol. 9, ed. by E. Kaldis (North-Holland, Amsterdam 1982) pp. 1–414
16.12 P. Rudolph, N. Schäfer, T. Fukuda: Mater. Sci. Eng. **15**, 85 (1995)
16.13 R. Shetty, R. Balasubramanian, W. R. Wilcox: J. Cryst. Growth **100**, 51 (1990)
16.14 K. W. Böer: *Survey of Semiconductor Physics, Vol. 1: Electrons and Other Particales in Bulk Semiconductors* (Van Nostrand, New York 1990)
16.15 C. M. Wolf, N. Holonyak, G. E. Stillman: *Physical Properties of Semiconductors* (Prentice Hall, New York 1989)
16.16 L. Smart, E. Moore: *Solid State Chemistry*, 2nd edn. (Chapman Hall, New York 1995)
16.17 E. Lide(Ed.): *Handbook of Chemistry and Physics*, 2nd edn. (CRC, Boca Raton 1973)
16.18 J. Singh: *Physics of Semiconductors and Their Heterostructures* (McGraw–Hill, New York 1993)
16.19 N. Yamamoto, H. Horinaka, T. Miyauchi: Jpn. J. Appl. Phys. **18**, 225 (1997)
16.20 H. Neumann: Kristall Technik **15**, 849 (1980)
16.21 J. Camassel, D. Auvergne, H. Mathieu: J. Phys. Colloq. **35**, C3–67 (1974)
16.22 W. Shan, J. J. Song, H. Luo, J. K. Furdyna: Phys. Rev. **50**, 8012 (1994)
16.23 K. A. Dmitrenko, S. G. Shevel, L. V. Taranenko, A. V. Marintchenko: Phys. Status Solidi B **134**, 605 (1986)
16.24 S. Logothetidis, M. Cardona, P. Lautenschlager, M. Garriga: Phys. Rev. B **34**, 2458 (1986)
16.25 R. C. Sharma, Y. A. Chang: J. Cryst. Growth **88**, 192 (1988)
16.26 H. Okada, T. Kawanaka, S. Ohmoto: J. Cryst. Growth **165**, 31 (1996)
16.27 N. Kh. Abrikosov, V. F. Bankina, L. B. Poretzkaya, E. V. Skudnova, S. N. Chichevskaya: *Poluprovodnikovye chalkogenidy i splavy na ikh osnovje* (Nauka, Moscow 1975) (in Russian)
16.28 R. F. Brebrick: J. Cryst. Growth **86**, 39 (1988)
16.29 M. R. Lorenz: *Physics and Chemistry of II–VI Compounds*, ed. by M. Aven, J. S. Prener (North Holland, Amsterdam 1967) pp. 210–211
16.30 T. Yao: Optoelectron. Dev. Technol. **6**, 37 (1991)
16.31 H. Nakamura, M. Aoki: Jpn. J. Appl. Phys. **20**, 11 (1981)
16.32 C. Werkhoven, B. J. Fitzpatrik, S. P. Herko, R. N. Bhargave, P. J. Dean: Appl. Phys. Lett. **38**, 540 (1981)
16.33 H. Nakamura, S. Kojima, M. Wasgiyama, M. Aoki: Jpn. J. Appl. Phys. **23**, L617 (1984)
16.34 V. M. Skobeeva, V. V. Serdyuk, L. N. Semenyuk, N. V. Malishin: J. Appl. Spectrosc. **44**, 164 (1986)
16.35 P. Lilley, P. L. Jones, C. N. W. Litting: J. Mater. Sci. **5**, 891 (1970)
16.36 T. Matsumoto, T. Morita, T. Ishida: J. Cryst. Growth **53**, 225 (1987)
16.37 S. Zhang, H. Kinto, T. Yatabe, S. Iida: J. Cryst. Growth **86**, 372 (1988)
16.38 S. Iida, T. Yatabe, H. Kinto: Jpn. J. Appl. Phys. **28**, L535 (1989)
16.39 P. Besomi, B. W. Wessels: J. Cryst. Growth **55**, 477 (1981)
16.40 T. Kyotani, M. Isshiki, K. Masumoto: J. Electrochem. Soc. **136**, 2376 (1989)
16.41 N. Stucheli, E. Bucher: J. Electron. Mater. **18**, 105 (1989)
16.42 M. Nishio, Y. Nakamura, H. Ogawa: Jpn. J. Appl. Phys. **22**, 1101 (1983)
16.43 N. Lovergine, R. Cingolani, A. M. Mancini, M. Ferrara: J. Cryst. Growth **118**, 304 (1992)
16.44 O. De. Melo, E. Sánchez, S. De. Roux, F. Rábago-Bernal: Mater. Chem. Phys., **59**, 120 (1999)
16.45 M. Kasuga, H. Futami, Y. Iba: J. Cryst. Growth **115**, 711 (1991)
16.46 J. F. Wang, K. Kikuchi, B. H. Koo, Y. Ishikawa, W. Uchida, M. Isshiki: J. Cryst. Growth **187**, 373 (1998)
16.47 J. Humenberger, G. Linnet, K. Lischka: Thin Solid Films **121**, 75 (1984)
16.48 F. Sasaki, T. Mishina, Y. Masumoto: J. Cryst. Growth **117**, 768 (1992)
16.49 B. J. Kim, J. F. Wang, Y. Ishikawa, S. Sato, M. Isshiki: Phys. Stat. Sol. (a) **191**, 161 (2002)
16.50 A. Rogalski, J. Piotrowski: Prog. Quantum Electron. **12**, 87 (1988)
16.51 G. M. Lalev, J. Wang, S. Abe, K. Masumoto, M. Isshiki: J. Crystal Growth **256**, 20 (2003)
16.52 H. M. Manasevit: Appl. Phys. Lett. **12**, 1530 (1968)

16.53 Sg. Fujita, M. Isemura, T. Sakamoto, N. Yoshimura: J. Cryst. Growth **86**, 263 (1988)
16.54 H. Mitsuhashi, I. Mitsuishi, H. Kukimoto: J. Cryst. Growth **77**, 219 (1986)
16.55 P.J. Wright, P.J. Parbrook, B. Cockayne, A.C. Jones, E.D. Orrell, K.P. O'Donnell, B. Henderson: J. Cryst. Growth **94**, 441 (1989)
16.56 S. Hirata, M. Isemura, Sz. Fujita, Sg. Fujita: J. Cryst. Growth **104**, 521 (1990)
16.57 S. Nishimura, N. Iwasa, M. Senoh, T. Mukai: Jpn. J. Appl. Phys. **32**, L425 (1993)
16.58 K.P. Giapis, K.F. Jensen, J.E. Potts, S.J. Pachuta: Appl. Phys. Lett. **55**, 463 (1989)
16.59 S.J. Pachuta, K.F. Jensen, S.P. Giapis: J. Cryst. Growth **107**, 390 (1991)
16.60 M. Danek, J.S. Huh, L. Foley, K.F. Jenson: J. Cryst. Growth **145**, 530 (1994)
16.61 W. Kuhn, A. Naumov, H. Stanzl, S. Bauer, K. Wolf, H.P. Wagner, W. Gebhardt, U.W. Pohl, A. Krost, W. Richter, U. Dümichen, K.H. Thiele: J. Cryst. Growth **123**, 605 (1992)
16.62 J.K. Menno, J.W. Kerri, F.H. Robert: J. Phys. Chem. B **101**, 4882 (1997)
16.63 H.P. Wagner, W. Kuhn, W. Gebhardt: J. Cryst. Growth **101**, 199 (1990)
16.64 N.R. Taskar, B.A. Khan, D.R. Dorman, K. Shahzad: Appl. Phys. Lett. **62**, 270 (1993)
16.65 Y. Fujita, T. Terada, T. Suzuki: Jpn. J. Appl. Phys. **34**, L1034 (1995)
16.66 J. Wang, T. Miki, A. Omino, K.S. Park, M. Isshiki: J. Cryst. Growth **221**, 393 (2000)
16.67 M.K. Lee, M.Y. Yeh, S.J. Guo, H.D. Huang: J. Appl. Phys. **75**, 7821 (1994)
16.68 A. Toda, T. Margalith, D. Imanishi, K. Yanashima, A. Ishibashi: Electron. Lett. **31**, 1921 (1995)
16.69 A. Cho: J. Vac. Sci. Tech. **8**, S31 (1971)
16.70 C.T. Foxon: J. Cryst. Growth **251**, 130 (2003)
16.71 T. Yao: *The Technology and Physics of Molecular Beam Epitaxy*, ed. by E.H.C. Parker (Plenum, New York 1985) Chap. 10, p. 313
16.72 E. Veuhoff, W. Pletschen, P. Balk, H. Luth: J. Cryst. Growth **55**, 30 (1981)
16.73 M.B. Panish, S. Sumski: J. Appl. Phys. **55**, 3571 (1984)
16.74 Y.P. Chen, G. Brill, N.K. Dhar: J. Cryst. Growth **252**, 270 (2003)
16.75 H. Kato, M. Sano, K. Miyamoto, T. Yao: J. Cryst. Growth **237–239**, 538 (2002)
16.76 M. Imaizumi, M. Adachi, Y. Fujii, Y. Hayashi, T. Soga, T. Jimbo, M. Umeno: J. Cryst. Growth **221**, 688 (2000)
16.77 W. Xie, D.C. Grillo, M. Kobayashi, R.L. Gunshor, G.C. Hua, N. Otsuka, H. Jeon, J. Ding, A.V. Nurmikko: Appl. Phys. Lett. **60**, 463 (1992)
16.78 S. Guha, A. Madhukar, K.C. Rajkumar: Appl. Phys. Lett. **57**, 2110 (1990)
16.79 E. Bauer, J.H. van der Merwe: Phys. Rev. B **33**, 3657 (1986)
16.80 J. Drucker, S. Chapparro: Appl. Phys. Lett. **71**, 614 (1997)
16.81 S.H. Xin, P.D. Wang, A. Yin, C. Kim, M. Dobrowolska, J.L. Merz, J.K. Furdyna: Appl. Phys. Lett. **69**, 3884 (1996)
16.82 M.C. Harris Liao, Y.H. Chang, Y.H. Chen, J.W. Hsu, J.M. Lin, W.C. Chou: Appl. Phys. Lett. **70**, 2256 (1997)
16.83 Y. Terai, S. Kuroda, K. Takita, T. Okuno, Y. Masumoto: Appl. Phys. Lett. **73**, 3757 (1998)
16.84 M. Kobayashi, S. Nakamura, K. Wakao, A. Yoshikawa, K. Takahashi: J. Vac. Sci. Technol. B **16**, 1316 (1998)
16.85 S.O. Ferreira, E.C. Paiva, G.N. Fontes, B.R.A. Neves: J. Appl. Phys. **93**, 1195 (2003)
16.86 M.A. Herman, J.T. Sadowski: Cryst. Res. Technol. **34**, 153 (1999)
16.87 M. Ahonen, M. Pessa, T. Suntola: Thin Solid Films **65**, 301 (1980)
16.88 M. Ritala, M. Leskelä: Nanotechnology **10**, 19 (1999)
16.89 H. Hartmann: J. Cryst. Growth **42**, 144 (1977)
16.90 M. Shiloh, J. Gutman: J. Cryst. Growth **11**, 105 (1971)
16.91 S. Hassani, A. Tromson-Carli, A. Lusson, G. Didier, R. Triboulet: Phys. Stat. Sol. (b) **229**, 835 (2002)
16.92 W.W. Piper, S.J. Polich: J. Appl. Phys. **32**, 1278 (1961)
16.93 A.C. Prior: J. Electrochem. Soc. **108**, 106 (1961)
16.94 T. Kiyosawa, K. Igaki, N. Ohashi: Trans. Jpn. Inst. Metala **13**, 248 (1972)
16.95 T. Ohno, K. Kurisu, T. Taguchi: J. Cryst. Growth **99**, 737 (1990)
16.96 M. Isshiki, T. Tomizono, T. Yoshita, T. Ohkawa, K. Igaki: J. Jpn. Inst. Metals **48**, 1176 (1984)
16.97 X.M. Huang, K. Igaki: J. Cryst. Growth **78**, 24 (1986)
16.98 M. Isshiki, T. Yoshita, K. Igaki, W. Uchida, S. Suto: J. Cryst. Growth **72**, 162 (1985)
16.99 M. Isshiki: J. Cryst. Growth **86**, 615 (1988)
16.100 T. Ohyama, E. Otsuka, T. Yoshita, M. Isshiki, K. Igaki: Jpn. J Appl. Phys. **23**, L382 (1984)
16.101 T. Ohyama, K. Sakakibara, E. Otsuka, M. Isshiki, K. Igaki: Phys. Rev. B **37**, 6153 (1988)
16.102 Y.M. Tairov, V.F. Tsvetkov: J. Cryst. Growth **43**, 209 (1978)
16.103 G. Cantwell, W.C. Harsch, H.L. Cotal, B.G. Markey, S.W.S. McKeever, J.E. Thomas: J. Appl. Phys. **71**, 2931 (1992)
16.104 Yu.V. Korostelin, V.I. Kozlovsky, A.S. Nasibov, P.V. Shapkin: J. Cryst. Growth **161**, 51 (1996)
16.105 S. Fujita, H. Mimoto, H. Takebe, T. Noguchi: J. Cryst. Growth **47**, 326 (1979)
16.106 K. Byrappa: *Hydrothermal Growth of Crystal*, ed. by K. Byrappa (Pergamon, Oxford 1991)
16.107 A.C. Walker: J. Am. Ceram. Soc. **36**, 250 (1953)
16.108 R.A. Laudice, E.D. Kolg, A.J. Caporaso: J. Am. Ceram. Soc. **47**, 9 (1964)
16.109 M. Suscavage, M. Harris, D. Bliss, P. Yip, S.-Q. Wang, D. Schwall, L. Bouthillette, J. Bailey, M. Callahan, D.C. Look, D.C. Reynolds, R.L. Jones, C.W. Litton: MRS Internet J. Nitride Semicond. Res **4S1**, G3.40 (1999)

16.110 L. N. Demianets, D. V. Kostomarov: Ann. Chim. Sci. Mater. **26**, 193 (2001)
16.111 N. Ohashi, T. Ohgaki, T. Nakata, T. Tsurumi, T. Sekiguchi, H. Haneda, J. Tanaka: J. Kor. Phys. Soc. **35**, S287 (1999)
16.112 D. C. Look, D. C. Reynolds, J. R. Sizelove, R. L. Jones, C. W. Litton, G. Gantwell, W. C. Harsch: Solid State Commun. **105**, 399 (1988)
16.113 T. Sekiguchi, S. Miyashita, K. Obara, T. Shishido, N. Sakagami: J. Cryst. Growth **214/215**, 72 (2000)
16.114 P. Höschl, Yu. M. Ivanov, E. Belas, J. Franc, R. Grill, D. Hlidek, P. Moravec, M. Zvara, H. Sitter, A. Toth: J. Cryst. Growth **184/185**, 1039 (1998)
16.115 T. Fukuda, K. Umetsu, P. Rudolph, H. J. Koh, S. Iida, H. Uchiki, N. Tsuboi: J. Cryst. Growth **161**, 45 (1996)
16.116 A. Omino, T. Suzuki: J. Cryst. Growth **117**, 80 (1992)
16.117 I. Kikuma, M. Furukoshi: J. Cryst. Growth **71**, 136 (1985)
16.118 J. F. Wang, A. Omino, M. Isshiki: J. Cryst. Growth **214/215**, 875 (2000)
16.119 J. Wang, A. Omino, M. Isshiki: J. Cryst. Growth **229**, 69 (2001)
16.120 J. F. Wang, A. Omino, M. Isshiki: Mater. Sci. Eng. B **83**, 185 (2001)
16.121 P. Rudolph, N. Schäfer, T. Fukuda: Mater. Sci. Eng. R **15**, 85 (1995)
16.122 T. Asahi, A. Arakawa, K. Sato: J. Cryst. Growth **229**, 74 (2001)
16.123 M. Ohmori, Y. Iwase, R. Ohno: Mater. Sci. Eng. B **16**, 283 (1999)
16.124 R. Triboulet: Prog. Cryst. Growth Char. Mater. **128**, 85 (1994)
16.125 H. H. Woodbury, R. S. Lewandowski: J. Cryst. Growth **10**, 6 (1971)
16.126 R. Triboulet: Cryst. Res. Technol. **38**, 215 (2003)
16.127 T. Asahi, T. Yabe, K. Sato: The Japan Society of Applied Physics and Related Societies, Extended Abstracts, The 50th Spring Meeting, (2003) p. 332

17. Structural Characterization

The aim of this chapter is to convey the basic principles of X-ray and electron diffraction, as used in the structural characterization of semiconductor heterostructures. A number of key concepts associated with radiation–material and particle–material interactions are introduced, with emphasis placed on the nature of the signal used for sample interrogation. Various modes of imaging and electron diffraction are then described, followed by a brief appraisal of the main techniques used to prepare electron-transparent membranes for TEM analysis. A number of case studies on electronic and photonic material systems are then presented in the context of a growth or device development program; these emphasize the need to use complementary techniques when characterizing a given heterostructure.

The functional properties of semiconductors emanate from their atomic structures; indeed, the interrelationship between materials processing, microstructure and functional properties lies at the heart of semiconductor science and technology. Therefore, if we are to elucidate how the functional properties of a semiconductor depend on the processing history (the growth or device fabrication procedures used), then we must study the development of the microstructure of the semiconductor by applying an appropriate combination of analytical techniques to the given bulk crystal, heterostructure or integrated device structure.

The main aim of this chapter is to provide a general introduction to the techniques used to characterize the structures of semiconductors. Thus, we consider techniques such as X-ray diffraction (XRD) and electron diffraction, combined with diffraction contrast imaging, alongside related techniques used for chemical microanalysis, since modern instruments such as analytical electron microscopes (AEMs) provide a variety of operational modes that allow both structure and chemistry to be investigated, in addition to functional activity. For example, chemical microanalyses of the fine-scale structures of materials can be performed within a scanning electron microscope (SEM) and/or a transmission electron microscope (TEM), using the techniques of energy dispersive X-ray (EDX) analysis, wavelength dispersive X-ray (WDX) analysis or electron energy loss spectrometry (EELS). In addition, electrical and optical properties of semiconductors can also be investigated in situ using the techniques of electron beam induced conductivity (EBIC) or cathodoluminescence (CL), respectively. Techniques such as X-ray photoelectron spectrometry (XPS; also known as electron spectroscopy for chemical analysis, ESCA), secondary ion mass spectrometry (SIMS) or Rutherford backscattering spectrometry (RBS), can also be used to study semiconductor chemistry. We should also mention reflection high-energy electron diffraction (RHEED), which can be used for the rapid structural assessment of the near surface of bulk or thin film samples.

A far from exhaustive list of acronyms one might come across when assessing a given sample is incorporated into the general list of abbreviations at the start of this book. We can organize these techniques used to characterize materials into groups based on a number of viewpoints: for example, with respect to the material property being investigated; whether they are destructive or nondestructive; bulk or near-surface assessment techniques; based on radiation–material or particle–material interactions, or based on elastic or inelastic scattering processes, or whether they are diffraction-, imaging- or spectroscopic-based techniques.

In this broad introduction to the structural characterization of semiconductors, we focus on the interaction of a material with radiation and/or particles. Thus, we start by considering the interactions of photons, electrons or ions with a sample and the nature of the signals used for structural or chemical microanalysis. We then briefly focus on the techniques of X-ray and electron diffraction, and issues regarding the formation of TEM images. The aim is simply to convey an appreciation of the underlying principles, the applicability of these characterization techniques and the information provided by them. We also briefly consider sample preparation, and the chapter closes with a variety of TEM-based case studies of semiconductor heterostructures, which are included to illustrate some of the approaches used to characterize fine-scale microstructure, as well as to emphasise the need for complementary analysis when assessing the interrelationships between processing, structure and functional properties. Much literature already exists in this area, and a selection of references is provided at the end of the chapter that tackle many of the topics we cover here in more detail [17.1–15].

17.1 Radiation–Material Interactions

Each part of the electromagnetic spectrum has quantum energies that can be used to elicit certain forms of excitation at the atomic or molecular level. Different parts of the electromagnetic spectrum will interact with matter in different ways, according to the energy states within the material, allowing absorption or ionization effects to occur. The salient features of these various radiation–material interaction processes are summarized in the schematics shown in Fig. 17.1.

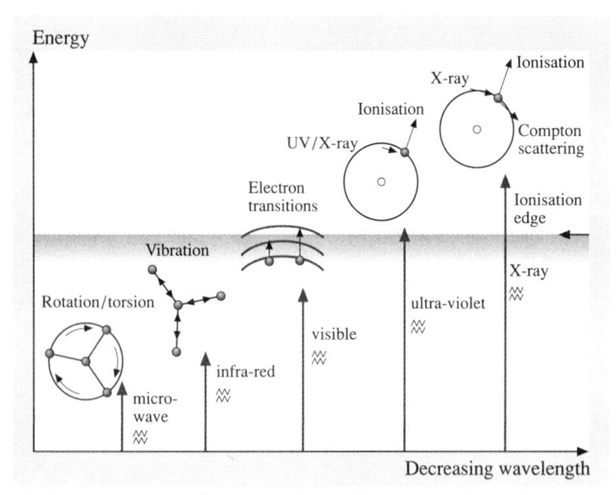

Fig. 17.1 Schematic illustrating the various interactions of energetic radiation with a molecule or atom

As the quantum energy increases from radio waves, through microwaves, to infrared and visible light, absorption increases, whilst specific quantized ionization effects come into play upon moving further into the ultraviolet and X-ray parts of the spectrum. Microwave and infrared radiation, for example, can interact with the quantum states of molecular rotation and torsion, leading to the generation of heat for example. Strong absorption also occurs within metallic conductors, leading to the induction of electric currents. Visible and ultraviolet light can elevate electrons to higher energy levels in what is known as the photoelectric effect, which is essentially the liberation of electrons from matter by short-wavelength electromagnetic radiation when all of the incident radiation energy is transferred to an electron. This process can be explained in terms of the absorption of discrete photon energies, with electrons being emitted when the photon energy exceeds the material's work function for the case of weakly bound electrons or the binding energy for more strongly bound inner shell electrons (Fig. 17.2a).

X-ray and γ-ray quantum energies are generally too high to be completely absorbed in direct electron

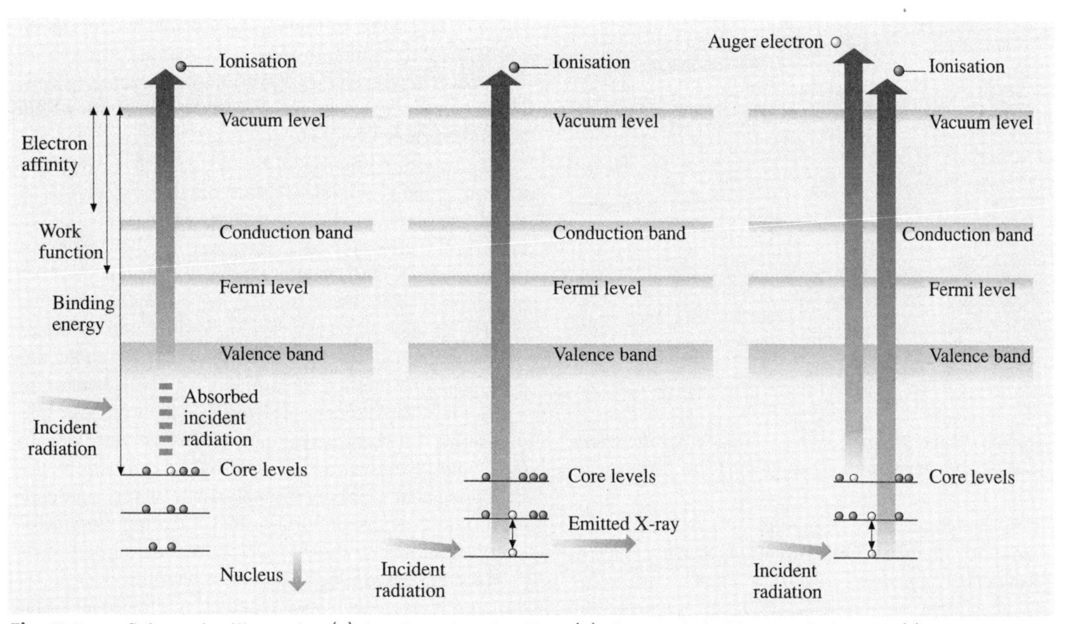

Fig. 17.2a–c Schematics illustrating (**a**) the photoelectric effect; (**b**) characteristic X-ray emission and (**c**) Auger electron emission

transitions, but they can induce ionization, with the displacement of electrons from atoms to form ions. The relaxation of a high-energy electron to the vacated state can lead to the emission of a characteristic X-ray (Fig. 17.2b) or the emission of an Auger electron (Fig. 17.2c). During the process of ionization, some of the incident photon energy is transferred to the ejected electron in the form of kinetic energy, and the scattering of a lower energy photon (longer wavelength X-ray) occurs, termed Compton scattering. X-rays can also be scattered elastically by shell electrons, without the loss of energy, through a process of absorption and re-emission.

17.2 Particle–Material Interactions

The interaction of an energetic particle with the surface of a material is most commonly associated with the process of sputtering – the non-thermal removal of atoms from a surface under ion bombardment. Transfer of momentum to the surface atoms is followed by a chain of collision events leading to the ejection of matrix atoms. Figure 17.3 illustrates the various signals produced when a particle interacts with a material, depending on the energy available. In broad terms, processes in the range 10^4–10^5 eV are associated with ion implantation, the 10–10^3 eV energy range is associated with sputtering, whilst the creation of activated point defects occurs in the 1–10^3 eV range. At lower 0.1–100 eV energies, desorption of surface impurity atoms occurs, whilst energies in the range 0.01–1 eV are associated with the enhanced mobility of surface condensing particles, such as those required during growth. In practical terms, the process of sputtering is most efficient when the masses of the incident and ejected particles are similar, whilst it is also dependent on the sputtering gas pressure, the energy spread of the particles, the bias conditions and the sample geometry.

In this context, we should briefly mention the technique of SIMS, which uses heavy ions, typically in the range 2–30 keV, with an ion current density of $\approx 1\ \mathrm{mA/cm^2}$. The sputtered atoms consist of a mixture of neutrals and ions, and the latter may be mass-spectrally analyzed in order to perform elemental depth profiling. The sensitivity of the technique is very high – for example on the scale of dopant concentrations

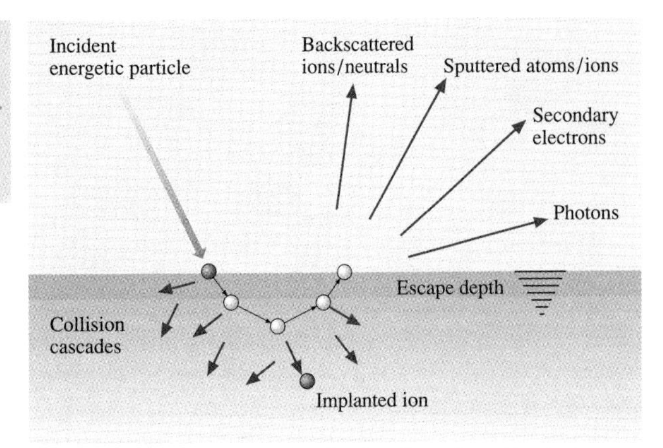

Fig. 17.3 Schematic illustrating the signals produced by the interaction of an energetic ion beam with a sample

within semiconductors – but standards are required for quantitative analysis. One variant of SIMS makes use of lower energy primary ions (0.5–2 keV, $\approx 1\,\text{nA/cm}^2$) with an almost negligible sputter rate, thus enabling surface chemical analysis. Conversely, the technique of RBS makes use of a very high energy (2–3 MeV) beam of light ions bombarding a sample normal to its surface. An ion such as helium is chosen to avoid the effects of sputtering, whilst high energy is required to overcome the problem of ion neutralization and the screening interaction potential between the ion and the nucleus associated with techniques such as low-energy ion scattering (LEIS). As an energetic positive ion penetrates the sample it loses energy, mainly due to collisions with electrons, but occasionally (and more significantly) with nuclei. The energy of the backscattered ion depends on the depth of the collision and the mass of the target atom. By measuring the energy spectrum of the recoiling ions, information on elemental composition and depth within a sample can be obtained.

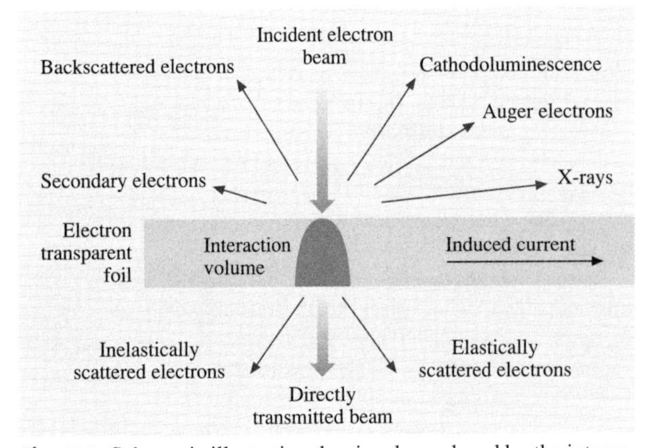

Fig. 17.4 Schematic illustrating the signals produced by the interaction of an energetic electron beam with a thin sample foil

By way of comparison, Fig. 17.4 illustrates the variety of signals produced from the interaction of an energetic beam of electrons with a thin semiconductor sample. In order to make sense of the origins of the many different signals, we must consider the phenomenon of electron scattering, which underpins them all. The process of elastic (Rutherford) scattering arises from an electrostatic (Coulomb) interaction with the nucleus and surrounding electrons of an atom, leading to a change in the incident electron direction without loss of energy. Elastically scattered electrons contribute to the formation of diffraction patterns and diffraction contrast images in TEM. Conversely, inelastically scattered electrons have, by definition, lost a certain amount of energy. In this context, core-shell interaction processes produce scattered electrons whose energies depend on the atomic number of the scatterer, and an analysis of the loss of electron energy (up to $\approx 1000\,\text{eV}$) is the basis of the technique of electron energy loss spectroscopy (EELS). In addition, plasmon scattering (5–30 eV loss) can occur, due to the interactions of incident electrons with waves in the conduction band of a metal, or the bonding electrons of non-metals. The signatures from plasmon-scattered electrons can dominate the low-energy regimes of EELS spectra, providing information on sample thickness. Phonon scattering ($\approx 1\,\text{eV}$ loss) can also occur, which is the interaction of incident electrons with quantized atomic vibrations within a sample, leading to the production of heat. In an interaction of an electron beam with a bulk sample, nearly all of the incident energy ends up being dissipated through such phonon interactions. The probability of each type of scattering interaction is commonly expressed either as a cross-section, representing the apparent area the scattering process presents to the electron, or as a mean free path, which is the average distance the electron travels before being scattered.

A variety of secondary events also occur as a direct consequence of these primary electron scattering processes. For example, following interactions with a high-energy beam of electrons, excited atoms may subsequently relax in a number of different ways (in a similar fashion to the processes induced by incident high-energy X-ray photons). If core-level (inner shell) electrons are displaced, the relaxation of electrons from higher energy shells to the lower energy core states can

Table 17.1 Overview of characterization techniques

Technique	Primary beam	Energy	Signals detected	Assessment	Spatial resolution	Elements detected	Detection limit
AES	Electron	0.5–10 keV	Auger electron	Surface composition	lateral ≈ 200 nm (LaB_6 source) lateral ≈ 20 nm (FE source) depth ≈ 2–20 nm	Li–U	≈ 0.1–1 at % (sub-monolayer) accuracy ≈ 30%
RBS	Ion (He atoms)	> 1 MeV	Ion (He atoms)	Depth composition & thickness	lateral ≈ 1 mm depth ≈ 5–20 nm	B–U	≈ 0.001–10 at %
SEM / EDS	Electron	0.3–30 keV	Electron (SE, BSE) X-ray (characteristic)	Surface morphology & composition	≈ 1–5 nm (SE) < 1 µm (BSE) lateral > 0.3 µm (EDS) depth ≈ 0.5–3 µm (EDS)	B–U	≈ 0.1–1 at % accuracy ≈ 20% (depends on matrix)
SIMS	Ion	1–30 keV	Ion (secondary)	Depth trace composition	lateral ≈ 60 µm (Dynamic SIMS) lateral ≈ 1 µm (Static SIMS) depth ≈ 2–20 nm	H–U	≈ 10^{-10} – 10^{-5} at. %
TEM /EDS / EELS	Electron	100–400 keV	Electron (elastic, inelastic) X-ray (characteristic)	Structure and chemistry of thin sections (high resolution)	≈ 0.1–0.3 nm lateral > 2 nm (EDS) lateral ≈ 1 nm (EELS) energy resolution ≈ 1 eV (EELS)	up to U	≈ 0.1–1 at % accuracy ≈ 20% (depends on matrix)
XPS	X-ray	1–10 keV	Photoelectron	Surface composition (chemical bonding)	lateral ≈ 10 µm–2 mm depth ≈ 1–10 nm	Li–U	≈ 0.1 – 1 at % (sub-monolayer) accuracy ≈ 30%
XRD	X-ray	1–10 keV	X-ray	Structure	lateral ≈ 10 µm depth ≈ 0.1–10 µm	Low Z may be difficult to detect	≈ 3 at % in a two-phase mixture (≈ 0.1 at % for synchrotron) accuracy ≈ 10%
XRF	X-ray	30 kV / 20 mA	X-ray (fluorescent)	Composition	lateral ≈ 0.1–10 mm depth ≈ 10 nm	Na–U	≈ ppb - ppm, accuracy ≈ 10%

lead to the discrete emission of X-ray photons characteristic of the atomic number of the element concerned. This is referred to as the K, L, M, N series, and Moseley's law states that the square root of the frequency of the characteristic X-rays of this series, for certain elements, is linearly related to the atomic number. Discrimination of these characteristic X-rays, as a function of energy or wavelength, provides the basis for the EDX or the WDX techniques, respectively. Alternatively, a secondary process of Auger electron emission may again occur, particularly for low atomic number materials, whereby outer electrons are ejected with a characteristic kinetic energy. It should also be noted that characteristic X-rays may also induce the emission of lower energy X-rays within a sample, and this is the basis of X-ray fluorescence (XRF) and the origin of background scintillation during EDX analysis.

In addition, if an outer (valence) electron state is vacant, relaxation across the band-gap of a semiconductor may occur with the emission of light, and this constitutes the basis of the CL technique. Alternatively, a current may be induced within a sample and non-radiative recombination pathways in the presence of structural defects (and a collection junction) provide the contrast mechanism for the EBIC technique that profiles the electrical activity within a crystalline semiconductor. Also, incident electrons that interact with atomic nuclei may become backscattered with energies comparable to the incident energy, and used to image a sample surface with a contrast that is dependent on the average local composition. Secondary (low-energy, $< 50\,\text{eV}$) electrons (SEs), produced by a variety of mechanisms, may also be emitted and escape from the near-surface of a sample. SEs can be used to obtain topographic images of irregular surfaces since they are easily absorbed.

Table 17.1 lists the most commonly used characterization techniques based on X-ray, ion or electron beam interactions with a semiconductor, and broadly indicates their limits of applicability.

17.3 X-Ray Diffraction

The basic concepts behind radiation–material interactions and scattering link into the concept of diffraction, whereby the spatial distribution and intensity of scattered X-rays or electrons provide information on the arrangement of atoms in a periodic sample. The theory of wave–particle duality proposes that an electron may be considered to be a wave rather than a particle when discussing diffraction. Electrons are scattered by electric fields within a crystal whilst X-rays are scattered by shell electrons. Nevertheless, the geometry of diffraction is very similar in both cases, being governed by Bragg's law.

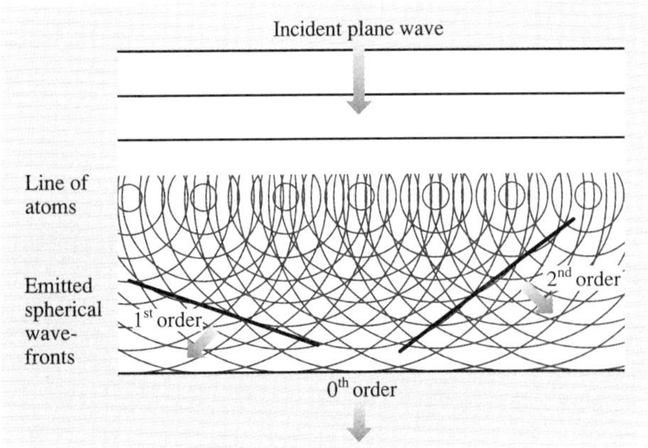

The basic principles associated with diffraction (in reflection, transmission or glancing angle geometry) are generally introduced with reference to X-ray scattering and interference. X-rays are a form of energetic electromagnetic radiation of wavelength $\approx 10^{-10}$–10^{-11} m, of comparable size to the spacing of atoms within a solid. A crystal lattice comprises a regular array of atoms; the electron clouds around these act as point sources for spherical X-ray wavelets, through a process of absorption and re-emission, when interaction with an incident beam of X-ray occurs. Constructive interference of the scattered wavelets, termed the Huygens's principle (Fig. 17.5), occurs in preferred directions, depending on the Bravais lattice of the crystal and the X-ray wavelength. The positions of the resultant maxima in scattering intensity may be used to deduce crystal plane spacings and hence the structure of an unknown sample. Geometrical considerations show that the scattering angles corresponding to diffracted intensity maxima can be described by the Bragg equation $n\lambda = 2d \sin\theta$ (sometimes expressed as $\lambda = 2d_{\text{hkl}} \sin\theta_{\text{hkl}}$). Formally,

Fig. 17.5 Schematic illustrating Huygens's principle, with a reconstruction of spherically emitted X-ray wavefronts providing diffracted intensity in specific directions

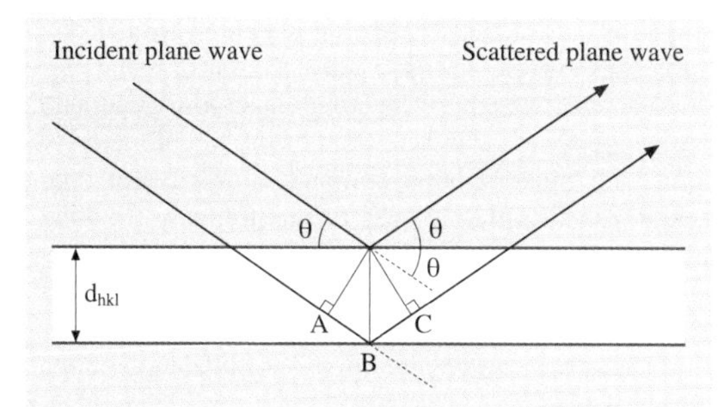

Fig. 17.6 Geometric illustration of Bragg diffraction. For constructive interference, $n\lambda = \text{AB} + \text{BC} = 2d \sin\theta$

this equation describes the minimum condition for the coherent diffraction of a monochromatic X-ray beam from a set of planes of a primitive lattice. Figure 17.6 illustrates the geometrical conditions associated with Bragg diffraction from a set of {hkl} planes spaced d_{hkl} apart, with X-rays incident at a Bragg angle θ being diffracted through an angle 2θ. The path difference between the X-rays 'reflected' from successive planes must be equivalent to an integer number of wavelengths n for constructive interference to occur.

In order to interpret the information contained in the *intensity* of the diffracted beam, which is measured in practice, it is necessary to consider the amplitude of the elastically scattered waves. (The phase of the scattered beam is much more difficult to measure.) A useful concept to introduce at this stage is that of the atomic scattering factor, f, which is a measure of the amplitude of the wave scattered by an atom, which depends on the number of electrons in the atom. Formally, the atomic scattering factor is the ratio of the amplitude of the wave scattered by an atom to the amplitude scattered in the same direction by a free classical electron. (For the purpose of such discussions, an electron orbiting an atom is considered to be a free classical electron.) The scattered amplitude (and hence intensity) varies with direction, with higher angles θ having lower amplitudes.

When X-rays interact with a periodic crystal lattice, it is considered that each atom scatters with an amplitude f into an hkl reflection, while there is a summation of all of the scattering amplitudes from different atoms, with phase differences that depend on hkl and the relative positions of the atoms. This leads to the concept of the structure factor, F_{hkl}, which is the total scattering amplitude from all of the atoms in one unit cell of a lattice. Formally, the structure factor is the ratio of the amplitude scattered by a unit cell into an hkl reflection to that scattered in the same direction by a free classical electron. The phase difference between waves scattered from two different atoms depends on the Miller indices of the reflection being considered and the fractional coordinates of the atoms within the unit cell. In general terms, for a reflection from a set of {hkl} planes, the phase difference ϕ between the wave scattered by an atom at the origin and that scattered by an atom with fractional coordinates x, y, z is given by $\phi = 2\pi(\text{h}x + \text{k}y + \text{l}z)$. In theoretical terms, the resultant intensity of the scattered beam is denoted $I_{hkl} \propto |F_{hkl}|^2$, where $F_{hkl} = \Sigma f_j(\theta)\exp(\text{i}\phi_j)$, which is a summation of the individual scattered sinusoidal waves, performed over both phase and amplitude. In a practical diffraction experiment, however, the combination of photoionization and Compton scattering can act to diminish the scattered beam intensity. Consideration also needs to be given to the effects of absorption, along with the effect of multiplicity, which arises from the number of symmetrical variants of a unit cell, and the geometrical and polarization factors specific to a given experimental arrangement.

The crystallographic structure of an unknown material can, nevertheless, be analyzed via the diffraction of X-rays of known wavelength. For example, for a crystal system with orthogonal axes, the general formula which relates plane spacing, d_{hkl}, to the plane index {hkl} and the lattice parameters a, b, c is $1/(d_{hkl})^2 = \text{h}^2/a^2 + \text{k}^2/b^2 + \text{l}^2/c^2$. For a cubic system, this simplifies to $d_{hkl} = a/\sqrt{(\text{h}^2+\text{k}^2+\text{l}^2)} = a/\sqrt{N}$. Thus, $\lambda = 2a\sin\theta/\sqrt{N}$, from which $N = (4a^2/\lambda^2)\sin^2\theta$ and hence $N \propto \sin^2\theta$. Accordingly, the 2θ angles of scattering arising from the process of X-ray diffraction may be used to identify values for N, against which hkl indices can be assigned and the lattice identified. Information is contained within the intensities and widths of the diffracted X-ray peaks, in addition to their positions. The process of formal identification can be automated by electronically referencing crystallographic databases containing the positions and relative magnitudes of the strongest diffraction peaks from known compounds. Systematic or partial absences may also occur depending on the lattice symmetry, and the intensities of some peaks may be weak due to the motif of the crystal lattice – the number of atoms sited on each lattice point. For example, sphalerite or wurtzite lattices may show weak reflections that are absent for face-centered cubic or hexagonal close-packed structures, respectively.

Partial differentiation of the Bragg equation gives $2(\sin\theta)\delta d + 2d(\cos\theta)\delta\theta = 0$, from which $\delta d/\delta\theta = -d\cot\theta$. Thus, for a fixed error in θ, the error in d_{hkl} will minimize as $\cot\theta$ tends to zero (as θ tends

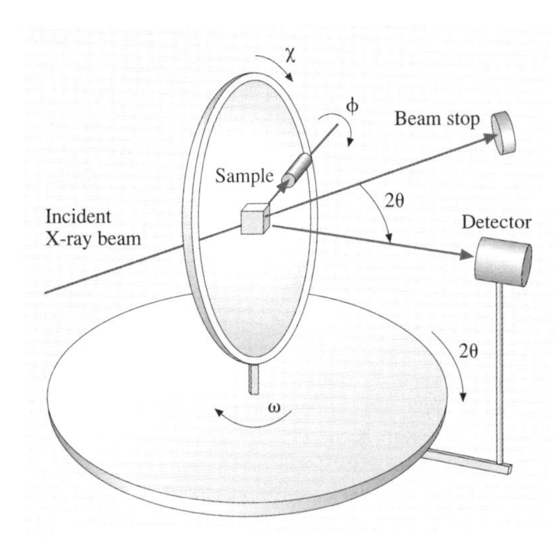

Fig. 17.7 X-ray four-circle diffractometer

to 90°). Diffraction techniques are therefore more accurate when measurements are made at high 2θ angles. Similarly, it can be shown that the sensitivities and resolutions of diffraction measurements improve at high angles. The sensitivity of the X-ray diffraction technique, for example, is sufficiently high to enable subtle stress measurements to be made, with elastic changes in plane spacing leading to small shifts in diffracted peak positions. Accordingly, diffraction techniques may be used to investigate temperature-dependent order–disorder transitions within alloys and preferred orientation effects. The high accuracy of lattice parameter measurement similarly enables good compositional analysis of an alloy, for example by assuming Vegard's law, that the alloy exhibits a linear dependence of lattice parameter on composition between the extremes of composition. Brief reference should be made here to the related technique of neutron diffraction, which is able to sample over a much larger range of d_{hkl} spacings, which makes it particularly useful for the ab initio determination of very complex structures. The very rigorous approach of Rietveld refinement can also be used to model a complete diffraction profile, deducing the structure from first principles using atom positions as fitted parameters.

The principle behind the generation of X-rays from characteristic transitions, used when performing chemical microanalyses of unknown samples by EDX or WDX, also provides the route to defining an X-ray source from a known target sample for the purpose of XRD. In practical terms, the X-ray tube of a diffractometer is an evacuated vessel in which electrons from a hot filament are focused onto a cooled metal target, such as Cu or Ni. The X-ray spectrum generated consists of X-rays characteristic of the target material, along with a background emission of X-rays of continuous wavelengths, termed Bremsstrahlung (braking radiation), which are caused by the acceleration of electrons in the vicinity of nuclei. The X-rays emerging from the tube, through a window made from a material of low atomic number, can be filtered and collimated to define a beam of specific wavelength.

The precise geometry used in a diffraction experiment will depend on the form of the sample and the information content required. For example, crystalline samples such as semiconductor heterostructures can be rotated within a cylindrical geometrical framework that enables successive sets of crystal planes to be brought into play for detection. Figure 17.7 illustrates the basic geometrical arrangements for a four-circle diffractometer that allows diffraction spectra to be acquired. Computer-controlled rotation of the sample around ω, χ and ϕ axes, as the detector is rotated about the 2θ axis, enables the positions and intensities of hkl reflections to be recorded. By way of example, Fig. 17.8 shows a $2\theta/\omega$ plot recorded from a heteroepitaxial GaN/GaAs(001) sample, with reflections attributable to both epilayer and substrate, whilst the full width at half maximum (FWHM) values for the diffraction peaks provide a measure of the mosaic spread of the subgrains within the sample.

Since amorphous materials do not exhibit long-range order, their diffraction profiles show diffuse intensities rather than well-defined maxima. Partially crystalline materials may show broad diffraction peaks, from which

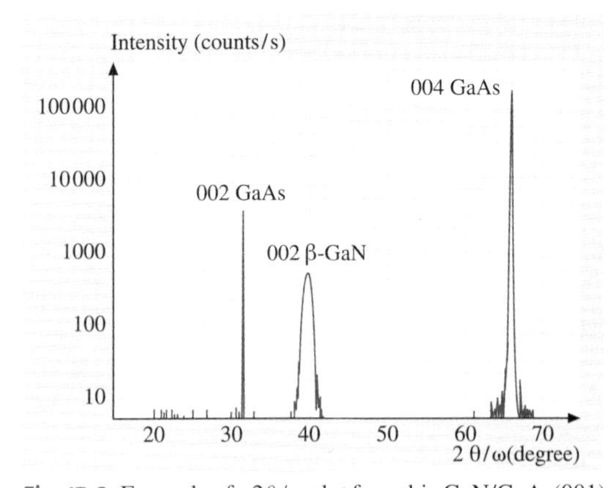

Fig. 17.8 Example of a $2\theta/\omega$ plot for cubic GaN/GaAs(001)

it is possible to approximate the crystal size from the λ–peak-width relationship. Other X-ray diffraction techniques include Laue back reflection, that can be used conveniently to orient bulk single crystals, for example for sectioning prior to use as substrates for heteroepitaxial growth. Alternatively, the Debye–Scherrer method can be used for powder samples, since a significant number of crystal grains will always be in an orientation that satisfies the Bragg equation for each set of {hkl} planes. In this scattering arrangement, the diffracted rays form cones coaxial with the incident X-ray beam, with each cone of diffracted rays corresponding to a Bragg reflection from a specific set of lattice planes in the sample. A cylindrical strip of photographic film can be used to detect the diffracted intensity.

To reiterate, it is the combination of Bragg's law and the structure factor equation that enables the directions and intensities of beams scattered from a crystal to be predicted. In this context, it is instructive to briefly compare XRD with electron diffraction. Electrons are scattered by the periodic potential – the electric field – within a crystal lattice, whilst X-rays are scattered by shell electrons. Since X-rays and electrons exhibit comparable and comparatively small wavelengths, respectively, on the scale of the plane spacings of a crystal lattice, this equates to large and small angles of scattering, respectively, for the diffracted beams. Accordingly, in principle XRD techniques offer greater accuracy than electron diffraction for the measurement of lattice parameters. It should also be noted that XRD is essentially a kinematic process based on single scattering events, whilst electron diffraction is potentially more complex due to the possibility of dynamic (or plural) scattering processes which can affect the generated intensities. Also, electrons are more strongly absorbed than X-rays, so there is need for very thin sample foils, typically $< 1\,\mu m$, for the purposes of transmission electron diffraction (TED) experiments. However, electrons are more easily scattered by a crystal lattice than X-rays, albeit through small angles, so an electron-transparent sample foil is capable of producing intense diffracted beams. X-rays require a much greater interaction volume to achieve a considerable diffraction intensity. The effectiveness of the technique of electron diffraction becomes most apparent when combined with TEM-based chemical microanalysis imaging techniques. This enables features such as small grains and embedded phases, or linear or planar defect structures such as dislocations and domain boundaries, to be investigated in detail.

Before describing some variants of the electron diffraction technique, a few concepts related to imaging and modes of operation of the TEM need to be introduced.

17.4 Optics, Imaging and Electron Diffraction

The aim of a microscope-based system is to image an object at high magnification, with optimum resolution and without distortion. The concepts of magnification and resolution associated with imaging in electron microscopy are usually introduced via light ray diagrams for optical microscopy. The constraints on achieving optimum resolution in TEM are generally considered to be lens aberration and astigmatism. The concepts of depth of field and depth of focus must also be considered.

If we consider the objective lens shown in Fig. 17.9, a single lens is characterized by a focal length f and a magnification M. The expression $1/f = 1/u + 1/v$ relates the focal length to the object distance u and the image distance v for a thin convex lens. The magnification of this lens is then given by $M = v/u = f/(u-f) = (v-f)/f$, from which it is apparent that $u-f$ must be small and positive for a large magnification to be obtained. In practice, a series of lenses are used to achieve a high magnification overall whilst minimizing distortion effects. For the combined projection microscope system shown in Fig. 17.9, magnification scales as $M' = (v-f)(v'-f')/ff'$.

Resolution is defined as being the smallest separation of two points on an object that can be reproduced distinctly within an image. The resolution of an optical lens system is diffraction-limited since light must pass through a series of apertures, and so a point source is imaged as a set of Airy rings. Formally, the minimum resolvable separation of two point sources, imaged as two overlapping sets of Airy rings, is given by the Rayleigh criterion, whereby the center of one set of Airy rings overlaps the first minimum of the second set of Airy rings. The defining equation for resolution is given by $r = 0.61\lambda/n \sin\alpha$, where λ is the wavelength of the imaging radiation, n is the refractive index of the lens, and α is the semiangle subtended at the lens. The combined term $n \sin\alpha$ is the 'numerical aperture' of the lens. Thus, resolution can be improved by decreasing λ

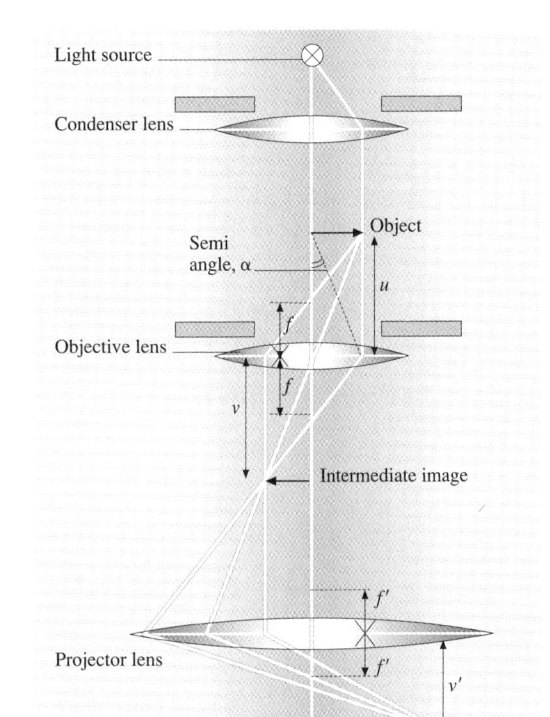

Fig. 17.9 Schematic ray diagram for a transmission projection optical microscope

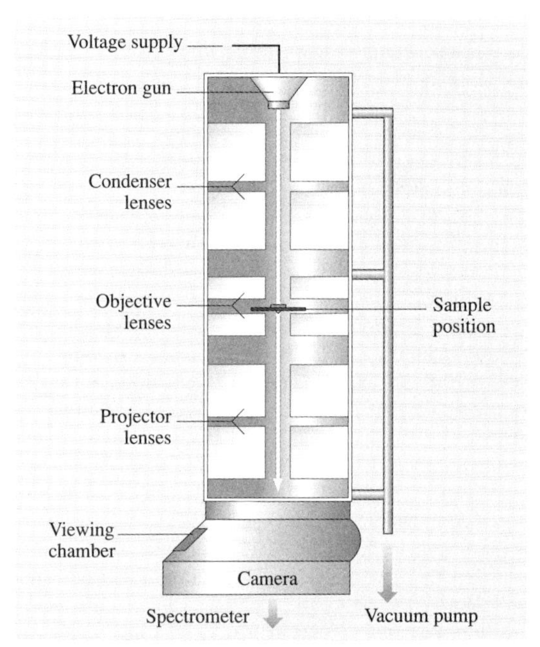

Fig. 17.10 Simplified lens configuration of a TEM

or by increasing n or α. It is noted that magnification beyond the resolution limit of the system is possible, but there is no benefit to achieving this, and so it is termed 'empty magnification.'

The performance of a TEM can be described in analogous terms, comprising an electron gun and a series of electromagnetic lenses for sample illumination, image formation and magnification: the condenser, objective and projector systems, respectively (Fig. 17.10). The electron source can be either a hot filament (e.g. W or LaB_6) for thermionic emission, or a hot or cold cathode (e.g. ZrO_2 coated W) for field emission. The gun and lenses are traditionally assembled in a column with a viewing screen and camera at the bottom to record images or diffraction patterns from an electron-transparent sample placed within the objective pole piece. A range of differently sized apertures are located in the condenser system to help collimate the electron probe, in the projector system to select a region of sample from which a diffraction pattern may be formed, and in the objective lens just below the sample to select the transmitted beam(s) used to form an image. The path of travel of the electron beam through the entire electron-optic column must be under conditions of high vacuum, considering the ease of absorption of electrons in air. The electrons are accelerated by high voltage, typically 100 or 200 kV, although there are TEMs that operate at MV conditions, depending on the intended application. An accelerated, high-energy electron acquires significant kinetic energy and momentum. It can also be represented by a wavelength (corrected to take into account relativity) that can be approximated in nanometers by $\lambda = [1.5/(V + 10^{-6}\,V^2)]^{0.5}$. By way of example, an electron beam within a conventional TEM operating at 200 kV has a relativistically corrected wavelength of 0.0025 nm, as compared with the range of $\approx$ 400–700 nm for visible light. Since resolution is wavelength-limited, the technique of TEM is, in principle, able to provide a vast improvement in resolution over conventional optical systems. The equation for resolution becomes approximated by $r \approx 0.61\lambda/\alpha$ since $n = 1$ for electromagnetic lenses and $\sin\alpha \approx \alpha$ in view of the small angles associated with electron diffraction. A value of $\alpha \approx 150$ mrad would suggest a resolution approaching $\approx$ 0.01 nm. However, this level of resolution for TEM is generally not achievable in practice due to the effect of lens aberration, which de-

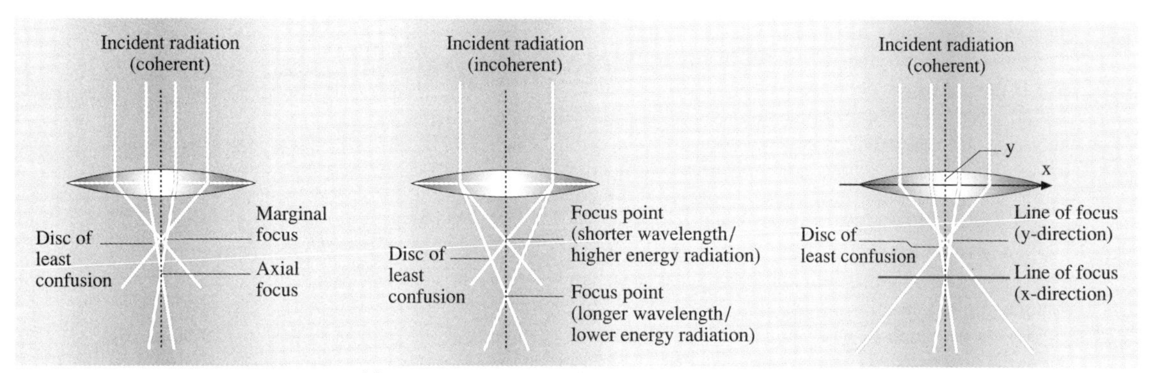

Fig. 17.11a–c Schematic illustration of (a) spherical aberration; (b) chromatic aberration and (c) astigmatism

grades the resolution of an image to a 'disc of least confusion' above the theoretical wavelength-limited resolution of the electron-optic system. Resolution values of $\approx 0.2\,\text{nm} - 0.3\,\text{nm}$ are associated with the best conventional instruments. (When charting the historical development of TEM, improvements in resolution have depended on the construction of microscopes operating at higher voltages, in order to capitalize on the benefits from the reduced electron wavelength. However, it should be noted that recent improvements in techniques of compensating for spherical aberration have now enabled a resolution of $\approx 0.1\,\text{nm}$ to be achieved for intermediate-voltage, field emission gun instruments.)

The variation in focal length of a lens as a function of the distance of a beam from the center of the lens results in rays traveling further from the optic axis being brought to focus closer to the lens (Fig. 17.11a). This spread in path lengths of rays traveling from an object to the image plane is termed spherical aberration. In this case, the radius of the disc of least confusion is given by $r_s = C_s\alpha^3$, when referred back to the object, where C_s is termed the spherical aberration coefficient and α is again the semiangle, in radians, subtended at the lens. Spherical aberration can be limited by reducing α; in other words by 'stopping down' the lens by using smaller apertures (in this case, the term α can be represented by a term β corresponding to the 'aperture collection angle'.). However, this conflicts with the large value of α needed to optimize resolution, so a balance is required for optimum resolution, which is achieved when $\alpha_{opt} \sim (\lambda/C_s)^{1/4}$, corresponding to an effective resolution of $r_{opt} \sim \lambda^{3/4}C_s^{1/4}$. Lenses also exhibit different focal lengths for electrons with different wavelengths; such a spread of wavelengths can arise from slight fluctuations in the accelerating voltage or from inelastic scattering (energy loss) processes within the specimen, and this effect is termed chromatic aberration (Fig. 17.11b). The radius of the disc of least confusion in this instance is given by $r_c = C_c\alpha(\Delta E/E_0)$, where C_c is the chromatic aberration coefficient, E_0 the accelerating voltage and ΔE the spread in electron energy.

An additional aberration, termed astigmatism, arises from the asymmetry of a lens about the optic axis. The different focal lengths of a lens for different orientations leads to a loss of sharpness of the image at focus.

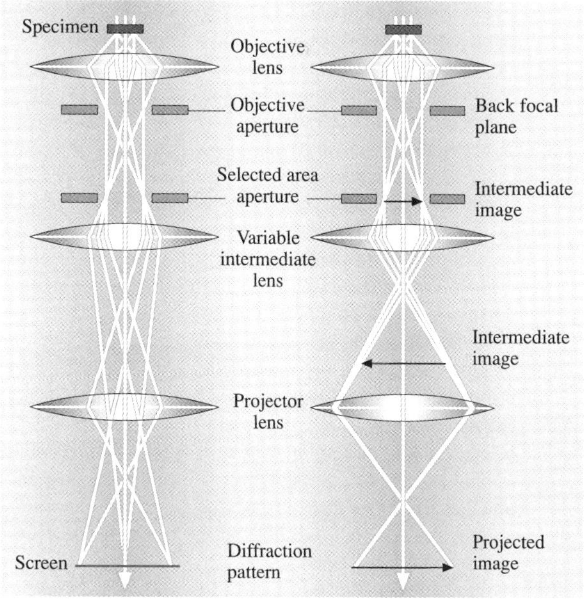

Fig. 17.12a,b Schematic ray diagrams showing the projection of a (a) diffraction pattern and (b) an image onto a TEM viewing screen

One measure of the astigmatism is the distance between the two foci formed at right angles along the optic axis (Fig. 17.11c), in contrast to a single point of focus. Astigmatism may be compensated for by using electromagnetic stigmators that generate a compensating field to bring the rays back into common focus. Distortion of an image can also occur due to slight variations in magnification with radial distance from the optic axis, leading to so-called 'pin-cushion' or 'barrel' distortion. This effect can become particularly noticeable at very low magnification.

The term 'depth of field' is defined as the distance along the optic axis that an object can be moved without noticeably reducing the resolution. This effect is again dependent on the radius of the disc of least confusion that can be tolerated and α. The depth of field approximates to λ/α^2, and a typical value of a few tens of nm for electron microscopy means that every point within the thickness of a typical electron-transparent foil can be imaged at focus. Conversely, the 'depth of focus' corresponds to the maximum permissible spacing between the imaging screen and the photographic plate or CCD used to record an image. The depth of focus approximates to $\lambda M^2/\alpha^2$, and since this works out at many meters, the viewing screen and recording system of an electron microscope need not coincide.

Following on from these general considerations affecting the process of imaging, we now move on to briefly describe the conventional modes of operation of a TEM. The accelerating stack and condenser system of the microscope defines the high-energy electron probe incident at a thin sample, within which many complex interactions occur and various signals are produced, as detailed earlier. A diffraction pattern is initially formed in the back focal plane of the objective lens and the recombination of diffracted beams allows the reconstruction of an inverted image in the first image plane. Changing the strengths of the intermediate lenses allows the back focal plane (corresponding to a projection of reciprocal space) or image plane (corresponding to a projection of real space) to be observed on the viewing screen (Fig. 17.12a,b). Placing an aperture with a selected area located in an intermediate image plane around an area of interest within a projected image enables us to ensure that only beams from that particular area of the sample contribute to the diffraction pattern viewed on the screen.

It is sometimes convenient to think of the intersection of an Ewald sphere with a reciprocal lattice when describing the construction and projection of a diffraction pattern. A reciprocal lattice is constructed from

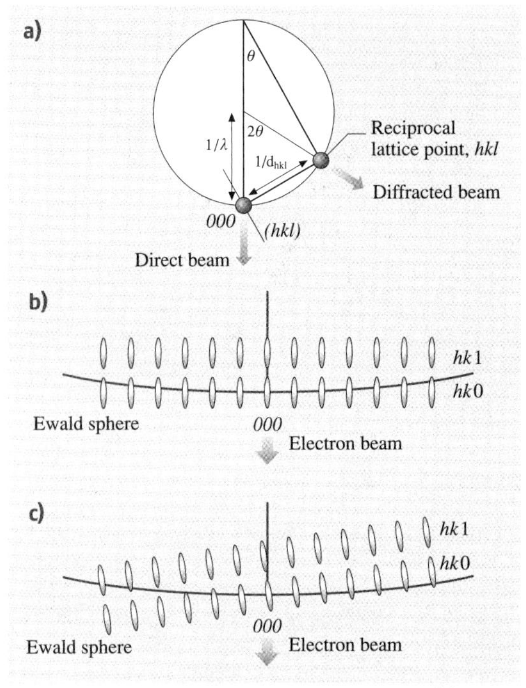

Fig. 17.13a–c Ewald sphere construction for **(a)** X-ray diffraction (large angle scattering); **(b)** electron diffraction (zone axis projection) and **(c)** electron diffraction (tilted projection)

a crystal lattice such that any vector from the origin to a diffracted spot is normal to a particular plane in the real lattice, with a reciprocal length given by the plane spacing. Thus, a three-dimensional crystal lattice can be transformed into a three-dimensional reciprocal lattice. The Ewald sphere can be thought of as a geometrical construction of radius $1/\lambda$ intersecting the reciprocal lattice. Figure 17.13a illustrates the construction of an Ewald sphere for the case of large-angle scattering (X-ray diffraction). The direction of the incident beam at the sample corresponds to the direction of the radius of the Ewald sphere, and the point of emergence of this vector at the sphere surface coincident with a reciprocal lattice point is taken as the origin of the spectrum or diffraction pattern. The same construction applies for a beam of electrons, but the sphere surface has a very shallow curvature relative to the reciprocal lattice spacing, due to the very small value of λ relative to d_{hkl} (Fig. 17.13b). Also, diffraction from a thin crystal is associated with a lengthening of the reciprocal lattice spots into rods in a direction parallel to the electron beam.

Consequently, if the electron beam is incident along a low index zone axis, the Ewald sphere approximates to a plane and intersects a layer of reciprocal lattice points, and a two-dimensional array of diffraction spots is projected. In very general terms, it is considered that a set of {hkl} planes is at the Bragg condition when the reciprocal lattice point corresponding to hkl falls on the surface of the Ewald sphere. For the case of large angles of diffraction, an outer ring of diffraction spots may be observed, termed a high-order Laue zone (HOLZ), since the Ewald sphere has sufficient curvature to intersect with a neighboring layer of points within the reciprocal lattice (Fig. 17.13c).

We should also mention the Kikuchi lines that arise in diffraction patterns due to the elastic scattering of incoherently scattered electrons. The intensities of these Kikuchi lines increase with increasing thickness of the sample foil, and the line spacings are the same as the spacings of the diffraction spots from the associated crystal planes. Kikuchi lines move as the sample is tilted and hence can be used to establish very precise crystal orientations for the purpose of image contrast or convergent beam electron diffraction (CBED) experiments.

Referring back to Fig. 17.12, a diffracted beam may be selected using an aperture inserted into the back focal plane of the objective lens and used to form an image. If the undeviated, transmitted beam is used, then a bright field image is formed where the areas that diffract strongly appear dark. Image contrast also arises from a mass-thickness effect, whereby thicker or high-density regions of material scatter more strongly and hence appear dark. Alternatively, centered dark field images can be created by aligning a diffracted beam, termed a $\mathbf{g}_{hkl}$ reflection, down the optic axis of the microscope (in contrast to moving the aperture over the diffraction spot, which would increase aberrations and degrade resolution). Areas where the {hkl} planes diffract strongly appear bright in such cases. Weak beam images can also be produced under dark field imaging conditions, with the sample tilted slightly away from a strong Bragg condition. The trade-off, in this instance, is reduced image contrast for improved resolution, which allows fine detailed features, such as partial dislocations bounding dissociated dislocations, to be delineated.

By way of example, Fig. 17.14 shows the different contrasts obtained from a plan-view $Si/Si_{0.96}Ge_{0.04}/Si(001)$ sample foil imaged under bright-field and weak-beam diffraction conditions, respectively. The sample foil contains orthogonal arrays

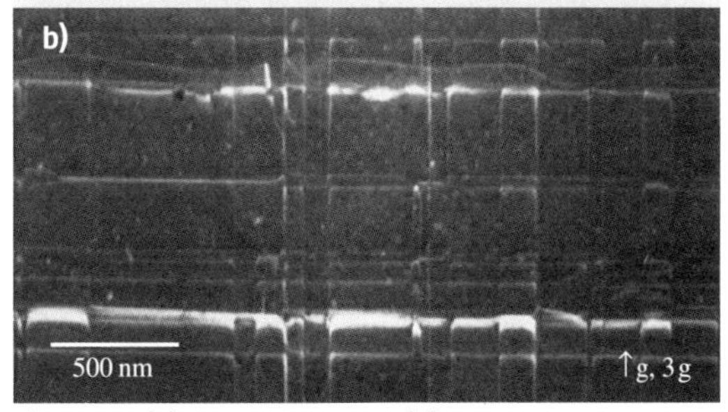

Fig. 17.14 (**a**) Bright-field and (**b**) weak-beam plan-view TEM images of Si/SiGe/Si(001)

of interfacial misfit dislocations that are generated as a consequence of the relaxation of the strain arising from the different lattice parameters of Si and $Si_{0.96}Ge_{0.04}$ [17.16]. The strain fields around the dislocations lead to a local deviation away from the exact Bragg condition, which can be used to delineate the (approximate) positions of the dislocation cores. The weak beam technique allows the complex dislocation tangle to be resolved more clearly. Imaging such sample foils using different diffraction vectors produces different contrasts that allow the nature of the dislocations to be precisely determined. Hence, the dislocation reaction mechanisms responsible for microstructure development can, in principle, be identified.

Some examples are now presented that illustrate how electron diffraction can contribute to the structural characterization of semiconductors. The use of conventional TED combined with image contrast analysis to assess the defect microstructure of a semiconductor heterostructure is initially considered. The application of the CBED technique, sometimes termed microdiffraction, in order to determine the polarities of noncentrosymmetric crystals is then described. We also focus on the RHEED technique, which may be used for rapidly assessing the near-surface microstructures of semiconductor thin films.

17.4.1 Electron Diffraction and Image Contrast Analysis

Figure 17.15 shows a TED pattern corresponding to a highly symmetric, low-index, zone axis projection, acquired from an electron-transparent foil of epitaxial GaN grown on the {0001} basal plane of sapphire, viewed in cross-section under spread beam conditions, in other words near-parallel illumination.

It is generally instructive to view such diffraction patterns alongside other images of the sample. Figure 17.16 compares high-resolution electron microscopy (HREM) and conventional TEM (CTEM) images of heteroepitaxial GaN–sapphire that were acquired for this projection. The HREM mode of imaging, otherwise termed 'phase contrast' or 'lattice imaging', makes use of several diffracted beams selected by a large objective aperture; the resulting interference patterns can be used to elucidate the locations of atomic columns (Fig. 17.16a). Careful simulation is required to precisely assign the atomic positions within such images, since the contrast is strongly dependent on defocus, foil thickness and sample orientation. However, the structural integrity of the interface between GaN and sapphire can be appraised, along with the presence of nanometer-scale, three-dimensional growth islands formed during the initial stages of epitaxy, prior to layer coverage. Conversely, the conventional 'many-beam, bright-field' image of heteroepitaxial GaN–sapphire (Fig. 17.16b) was created using a small objective aperture placed over the directly transmitted beam, and recorded at lower magnification. The image contrast is again complicated because there are many excited diffracted beams operating for this low-index crystal orientation and strain fields associated with the large number of threading defects ($\approx 10^{10}\,cm^{-2}$) within this sample. Despite this level of defect content, the diffraction pattern of Fig. 17.15 still reflects the high level of crystallographic perfection of the matrix. Also, rotation of this sample foil by 30° about the growth axis enabled a $\langle 1\bar{1}00\rangle_{nitride} || \langle 11\bar{2}0\rangle_{sapphire}$ projection to be established, consistent with the hexagonal symmetry of this GaN–sapphire system. Indeed, establishing different diffraction patterns for different sample projections, for known angles of sample tilt, enable the phase and structural relationship of an epilayer and substrate to be readily established.

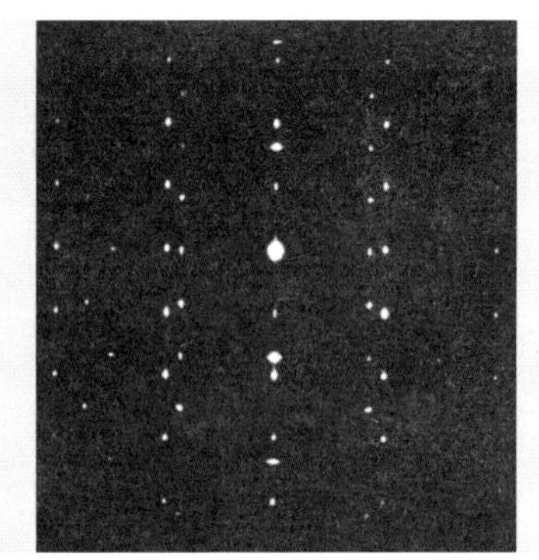

Fig. 17.15 TED pattern acquired from epitaxial GaN/sapphire{0001} viewed in cross-section, corresponding to the $\langle 11\bar{2}0\rangle_{nitride} || \langle 1\bar{1}00\rangle_{sapphire}$ axis projection

There is a need to characterize the precise nature of the fine-scale defect content within such samples. Dislocations, for example, generally act as nonradiative recombination centers, which can deleteriously affect the charge transport properties of a semiconductor [17.17]. Gaining an improved understanding of how dislocations are created and how they interact enables us to identify mechanisms that could be used to control their development, thereby improving growth of semiconductor material and the resultant properties of devices made from it.

Dislocations are one-dimensional defects that can be pure edge, pure screw or 'mixed edge-screw' in character. In certain circumstances, dislocations can dissociate

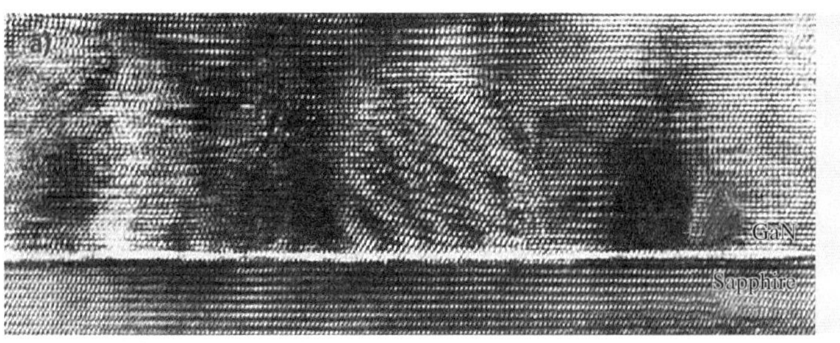

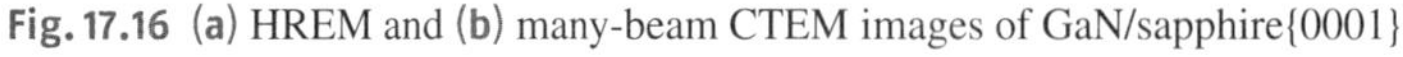

Fig. 17.16 (a) HREM and (b) many-beam CTEM images of GaN/sapphire{0001}

to form a pair of partial dislocations bounding a stacking fault ribbon, the separation of which depends on the material stacking fault energy. Dislocations can be described with reference to their line direction $\boldsymbol{u}$ and Burgers vector $\boldsymbol{b}$, representing the closure failure of a loop of equal atomic spacing around the defect core. Dislocations can move through a crystal lattice by glide or climb mechanisms; for example, when under deformation or as part of a strain relaxation mechanism during heteroepitaxial growth. Also, dislocations can interact to either self-annihilate or multiply to generate more dislocations depending on their respective type, habit plane and the slip system of the matrix.

The technique of diffraction contrast analysis allows the nature of dislocations to be ascertained. The approach used is to tilt the sample away from the highly symmetrical, low-index zone axis orientation in order to establish diffraction contrast images under selected "two-beam" conditions, with one strong diffraction spot excited ($\boldsymbol{g}$) in addition to the central transmitted beam, corresponding to one set of crystal planes at the Bragg condition. When $\boldsymbol{g}.\boldsymbol{b} = 0$, the displacement associated with a dislocation does not affect the diffracting planes used to form the image and so the defect appears invisible. In practice, two examples of this invisibility condition are generally required to determine the precise displacement associated with a given defect. The dislocation will appear visible when $\boldsymbol{g}.\boldsymbol{b} = 1$ and might show a more complex double image for the case of $\boldsymbol{g}.\boldsymbol{b} \geq 2$. Since a screw dislocation is characterized by a Burgers vector $\boldsymbol{b}$ parallel to the line direction $\boldsymbol{u}$, the defect is invisible if a diffraction vector perpendicular to the line direction is chosen. In the case of an edge or mixed dislocation where $\boldsymbol{b}$ is not parallel to $\boldsymbol{u}$, there is the stricter requirement for both $\boldsymbol{g}.\boldsymbol{b} = 0$ and $\boldsymbol{g}.(\boldsymbol{b} \times \boldsymbol{u}) = 0$ for true invisibility, otherwise residual contrast might be present that acts to confuse the image interpretation. We should also mention the deviation parameter $\boldsymbol{s}$ when establishing a diffracting condition, which represents the distance in reciprocal space from the exact Bragg condition, since this is associated with imaging artefacts such as extinction contours and thickness fringes that may act to further complicate an image.

Bright-field, dark-field or weak-beam diffraction contrast imaging techniques can be used for defect analysis, depending on the resolution required. By way of example, epitaxial GaN–$\{\overline{111}\}$B GaAs grown by molecular-beam epitaxy (MBE) at 700 °C exhibits a mosaic cell structure with subgrain boundaries delineated by predominantly mixed-type threading dislocations (typically $> 10^{11}\,\mathrm{cm}^{-2}$) [17.18]. The

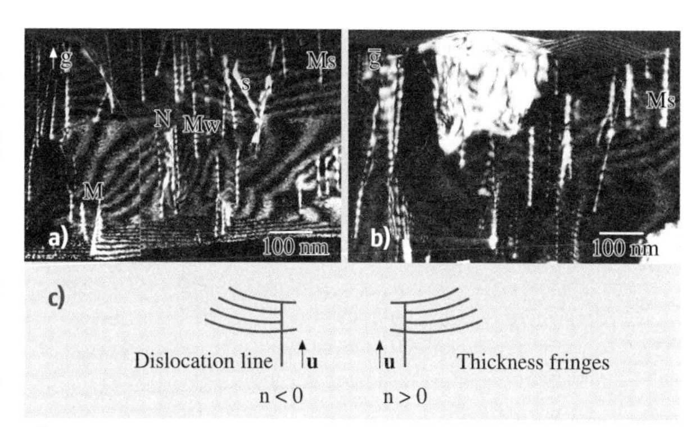

Fig. 17.17a–c Weak-beam images of the same regions of a GaN epilayer viewed in cross-section using (**a**) $\boldsymbol{g} = 000\bar{2}$ and (**b**) $\boldsymbol{g} = 1\bar{1}00$. (**c**) Schematic diagram summarizing Ishida's rule for determining $\boldsymbol{b}$ for $n = -2$ and $n = 2$

epitaxial relationship is given by $[000\bar{1}]_{\mathrm{GaN}}||[\overline{111}]_{\mathrm{GaAs}}$, $[1\bar{1}00]_{\mathrm{GaN}}||[2\bar{1}\bar{1}]_{\mathrm{GaAs}}$ and $[11\bar{2}0]_{\mathrm{GaN}}||[01\bar{1}]_{\mathrm{GaAs}}$, so the mismatch between GaN and GaAs is 38.2% between $\{1\bar{1}00\}_{\mathrm{GaN}}$ and $\{2\bar{2}0\}_{\mathrm{GaAs}}$, with the epilayer in tensile strain. Figures 17.17a,b are weak-beam images of the epilayer viewed in cross-section near the $[11\bar{2}0]$ zone axis, using $\boldsymbol{g} = 000\bar{2}$ and $\boldsymbol{g} = 1\bar{1}00$, respectively. Most of the dislocations have a line direction of $\langle 0001 \rangle$. Examples of perfect edge, $\boldsymbol{b} = 1/3\langle 11\bar{2}0 \rangle$, and screw-type dislocations, $\boldsymbol{b} = \langle 0001 \rangle$, are apparent, but mostly dislocations (typically $\approx 70\%$) can be seen in both images, hence they are mixed-type, with Burger's vector components of $\boldsymbol{a}$ and $\boldsymbol{c}$, in other words $1/3\langle 11\bar{2}3 \rangle$.

It is possible to move this analysis forwards by applying Ishida's rule, which allows the magnitude and sense of a Burgers vector to be determined if there are terminating thickness fringes at the exit of a dislocation from a wedge-shaped sample foil [17.19]. If there are n fringes terminating at one end of a dislocation, then $\boldsymbol{g}.\boldsymbol{b} = n$ and the sign of n ($n > 0$ or $n < 0$) is defined according to Fig. 17.16c. For example, the screw dislocation S (Fig. 17.17a) imaged using $\boldsymbol{g} = 000\bar{2}$ has two thickness fringes terminating on the right-hand side of the dislocation, with the dislocation line pointing towards the $[000\bar{1}]$ growth direction. Therefore $\boldsymbol{g}.\boldsymbol{b} = -2L = 2$, so $\boldsymbol{b}$ is $[000\bar{1}]$ for this screw-type dislocation. For dislocations of mixed type, Figs. 17.17a,b show that some have strong contrast in both images, while others have strong contrast when imaged using $\boldsymbol{g} = 000\bar{2}$ and weak contrast when $\boldsymbol{g} = 1\bar{1}00$. For the former, for example the dislocation pointed out as $\boldsymbol{M}_{\mathrm{s}}$ in Fig. 17.17a, thickness fringes terminate at the left-hand side of the dislocation with reference to the dislocation line pointing downwards along [0001]. Hence, $g.b = -2L = -2$ for $g = 000\bar{2}$ and $g.b = -1$ for $g = 1\bar{1}00$, so its Burgers vector is either $1/3[\bar{2}113]$ or $1/3[\bar{1}2\bar{1}3]$. For the cases of strong contrast by $g = 000\bar{2}$ and weak contrast by $g = 1\bar{1}00$, such as the dislocation labeled M_W in Fig. 17.17b, such dislocations have b of $\pm 1/3[11\bar{2}3]$ or $\pm 1/3[11\bar{2}\bar{3}]$, and the weak contrast is due to $g.(b \times u) \neq 0$. Because the thickness fringes connected with this defect type are not clear in the images, a more precise value for b cannot be determined. It is also noted that some dislocations have opposite values of b to others, such as those marked as M and N in Fig. 17.17a. These two opposite types of dislocations delineate a subgrain. The region showing bright contrast when imaged with $g = 1\bar{1}00$ in Fig. 17.17b is also a misoriented subgrain, tilted close to the Bragg condition.

17.4.2 Microdiffraction and Polarity

If the electron beam converges to form a focused probe at the sample, then there will be a range of beam directions within the incident probe and within the transmitted and diffracted beams. This leads to the formation of diffraction patterns comprising discs rather than spots. The fine detail within such microdiffraction or CBED patterns contains space and point group information. Such patterns also provide a sensitive measure of lattice parameters and strain within a sample, and can also be used to assess defect type.

Two examples showing how focused probe diffraction patterns may be used to determine the absolute polarities of sphalerite and wurtzite noncentrosymmetric crystals, in situ in the TEM, are now illustrated. It is often important to know the polar orientation of a heterostructure since this can strongly influence the mode and rate of growth, the incorporation of dopants or impurities, and the development of extended defects, and hence the functional performance of the resulting device structure. The opposite polar faces of noncentrosymmetric crystals [17.20] may be distinguished in practice using appropriate chemical etchants. However, such reagents are discriminatory in their action and need to be correlated with some other experimental technique for the purpose of absolute polarity determination. For most cases of diffraction, there is no difference between the intensities of beams scattered by the hkl and $\overline{hkl}$ reflection planes. This indeterminacy is known as Friedel's law. However, the technique of microdiffraction coupled with a breakdown in Friedel's law allows the absolute polarities of sphalerite crystals to be determined when the anion and cation sizes are very similar (as with GaAs, ZnSe and CdTe) [17.21]. The diffraction condition shown in Fig. 17.18a corresponds to a projection tilted $\approx 10°$ off a GaAs [110] zone axis, arrived at fol-

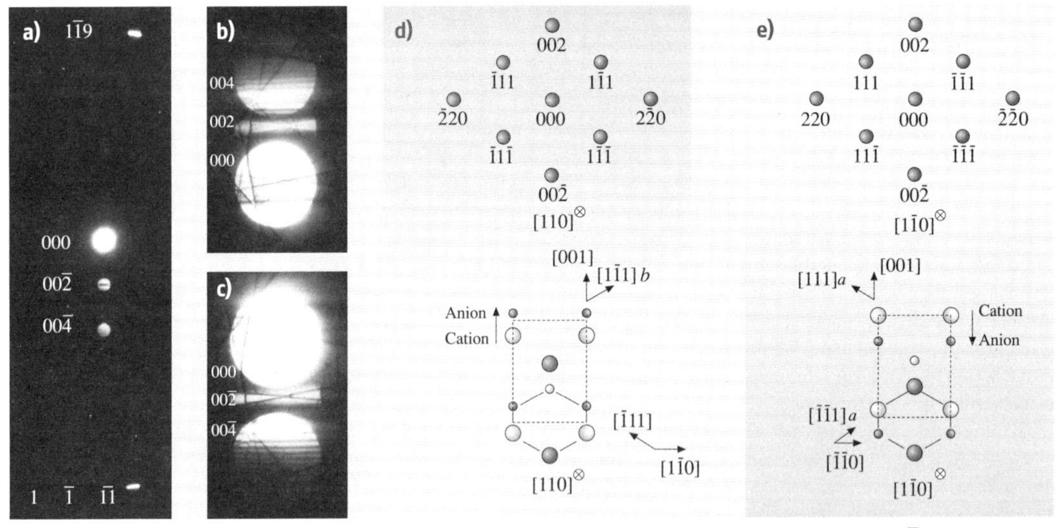

Fig. 17.18 (a)–(c) Microdiffraction patterns from [110] GaAs. (a) Destructive interference through an $00\bar{2}$ diffraction disc due to the interaction of doubly diffracted $1\bar{1}9$ and $1\bar{1}\,\bar{1}1$ beams. (b),(c) Constructive and destructive interference through 002 and $00\bar{2}$ diffraction discs. (d),(e) Sphalerite space lattice and reciprocal lattice projections for [110] and $[1\bar{1}0]$, indexed for an electron beam traveling down into the page

lowing an $\sim 00\bar{4}$ Kikuchi band. The interaction of the doubly diffracted high-order, odd-index $1\bar{1}9$ and $1\,\bar{1}\,\overline{11}$ beams with directly scattered $00\bar{2}$ reflection gives rise to destructive interference through the $00\bar{2}$ disc when all of the reflections are close to the Bragg position. Figure 17.18c, corresponding to a more highly converged incident electron beam, emphasizes the formation of the dark cross through the $00\bar{2}$ diffraction disc. Similarly, Fig. 17.18b accentuates the bright cross through the opposite 002 diffraction disc, formed by constructive interference following interaction with the corresponding $1\,\bar{1}\,11$ and $1\overline{19}$ high-order, odd-index reflections. Figures 17.18d,e, showing the [110] and $[1\bar{1}0]$ projections of the sphalerite space lattice and reciprocal lattice, respectively, are also shown here, to aid understanding of this particular sample geometry. Establishing such microdiffraction patterns requires carefully balancing crystal orientation, probe convergence and layer thickness, and the technique is found to work best with freshly plasma-cleaned sample foils to minimize the effect of extraneous hydrocarbon contamination (which diffuses out the fine-scale contrast under the imaging electron beam).

The dynamical equations of electron diffraction demonstrate that the 002 reflection, which exhibits constructive interference effects (a bright cross), always occurs in the sense of the cation to anion to bond (Fig. 17.18d), and hence is always associated with the sense of advancing $\{\overline{111}\}_b$ planes. To emphasise this, defining [001] as the growth direction of an epilayer on an (001) oriented GaAs substrate, and establishing a bright cross through an 002 diffraction disc, for example through the interaction of $1\bar{1}9$ and $1\bar{1}\,\overline{11}$ reflections, corresponds to the sense of advancing $(\bar{1}11)_b$ and $(1\bar{1}1)_b$ planes in the growth direction, which corresponds to the absolute [110] projection of the sample foil. Conversely, a dark cross through an 002 diffraction disc, such as that obtained through the interaction of $11\bar{9}$ and 1 1 11 reflections, corresponds to the sense of advancing $(111)_a$ and $(\overline{11}1)_a$ planes in the growth direction, which corresponds to the $[1\bar{1}0]$ projection of the sample foil. Since microdiffraction patterns are directly sensitive to crystallographic polarity, a qualitative interpretation of results thus allows the crystal orientation to be unambiguously determined, in situ within the TEM, without reference to any other technique.

Contrast reversals within diffraction discs of systematic row CBED patterns can similarly be used to determine absolute crystal polarity, since Friedel's law again breaks down due to dynamical scattering. In practice, this approach is found to be most effective when used with noncentrosymmetric crystals that have much larger differences in anion and cation sizes, such as wurtzite GaN [17.22]. Figure 17.19 corresponds to a thin section through a columnar defect, imaged within homoepitaxial GaN grown on an N-polar $(000\bar{1})$ GaN substrate, with two associated CBED patterns inset, recorded either side of the boundary plane corresponding to matrix and core material, respectively. Simulation of the contrast within the 0002 diffraction discs, for a known sample foil thickness, demonstrates that the central bright and dark bands correspond to N and Ga-polar growth directions, respectively. The reversal of contrast within the 0002 diffraction discs across the boundary plane therefore indicates an inversion in crystal polarity, confirming the defect to be an inversion domain in this instance.

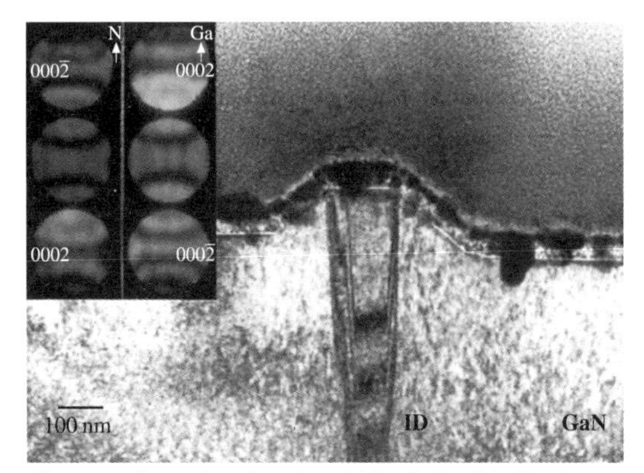

Fig. 17.19 Inversion domain within GaN with CBED patterns recorded across the boundary plane *inset*

17.4.3 Reflection High-Energy Electron Diffraction

Electron-transparent samples are required for TED investigations. In the case of semiconductor heterostructures, for example, sample preparation requires sequential mechanical polishing and ion beam thinning, which can be very time-consuming, but the advantage of this approach is that diffraction data can be directly correlated with the projected image of the internal structure of the sample. A complementary approach is to use glancing angle electron diffraction techniques to characterize the near-surface microstructure of a bulk sample or an as-grown wafer. Coupling of the electron beam with the material surface at low angle allows scattering of the electrons to occur to produce a diffraction pattern that may be viewed directly on a phosphor screen. This provides valuable information on the near-surface crystallography of the sample, which can be correlated with the growth conditions used or applied surface modifications, without the need for time-consuming sample preparation.

There are three variants of reflection electron diffraction (RED) depending on the accelerating voltage available: low ($< 1\,kV$), medium ($1-20\,kV$) or high ($20-200\,kV$) energy. Medium-energy electron diffraction (MEED) systems are commonly associated with UHV growth chambers, whilst a variant of RHEED can be performed using a conventional TEM. In general terms, coupling of an electron beam with a flat single crystal tends to be associated with the formation of diffraction streaks normal to the surface, providing information on the reconstruction of the atomic layer at the surface. More precisely, streaky RED patterns are indicative of a surface that is not quite perfectly flat, but has slight local misorientations combined with some degree of surface disorder [17.23]. When electrons are coupled with a slightly rougher surface, there is a tendency for more three-dimensional information to be obtained from the interaction with the near-surface microstructure. This leads to the production of spotty diffraction patterns from crystalline materials (half-obscured by the sample shadow edge), in an analogous fashion to TED. It is noted that UHV-MEED systems also allow the intensity fluctuations within the central beam to be monitored in real time, and this provides a way to control layer-by-layer growth.

In practice, RHEED is most effectively used for rapid comparative studies of sets of samples; for example for appraising the effect of changing the growth parameter on the structural integrity of a deposited thin film [17.24]. This approach to process mapping provides a convenient way to identify appropriate samples for more detailed TEM investigation prior to sample foil preparation. There are two variants of the RHEED technique that may be used in a TEM: one where small samples are mounted vertically in the objective lens pole piece, and the other where larger samples are mounted vertically below the projector lens; we focus on the latter variant here.

A schematic diagram of the diffraction geometry for a RHEED experiment is shown in Fig. 17.20a. The diffraction spacing R_{hkl} can be measured and the associated crystal plane spacing d_{hkl} determined using the equation, $\lambda L = R_{hkl} d_{hkl}$, where λL is known as the camera constant. Figures 17.20b,c show a RHEED stage made to interface with a TEM through an existing camera port at the base of the projector lens. The vacuum system of the TEM vents the space below the projector lens upon opening the camera chamber, and so the only waiting time is for the chamber to pump down upon changing a specimen. The RHEED stage is able to support centimeter-square sections of a crystalline specimen, whilst full tilt, twist and lateral movement enables any zone axis within the growth plane of the sample to be accessed. A shadow image of a sample, as projected onto the microscope phosphor screen, is shown in Fig. 17.20d. The area sampled by the glancing elec-

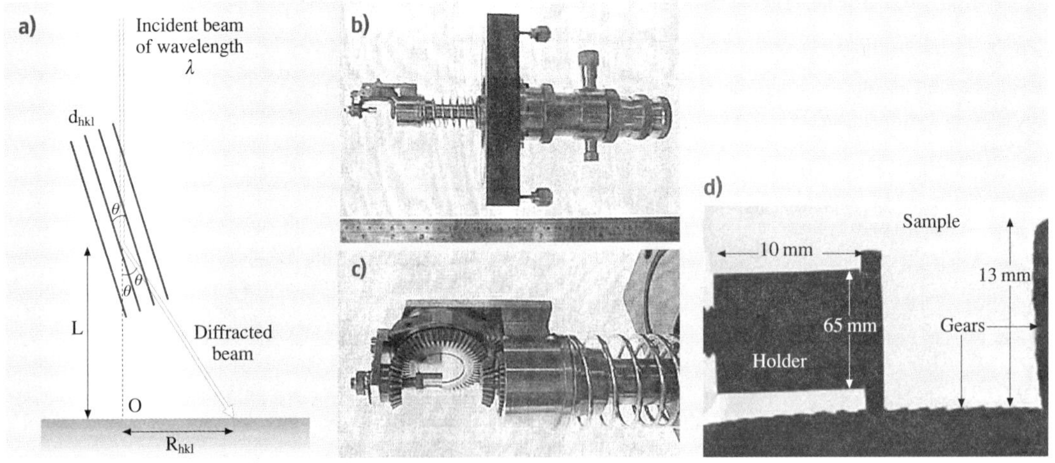

Fig. 17.20 **(a)** Schematic diagram of the RHEED diffraction geometry; **(b)** RHEED stage; **(c)** magnified view of the sample holder; and **(d)** projected shadow image of a specimen on the microscope phosphor screen

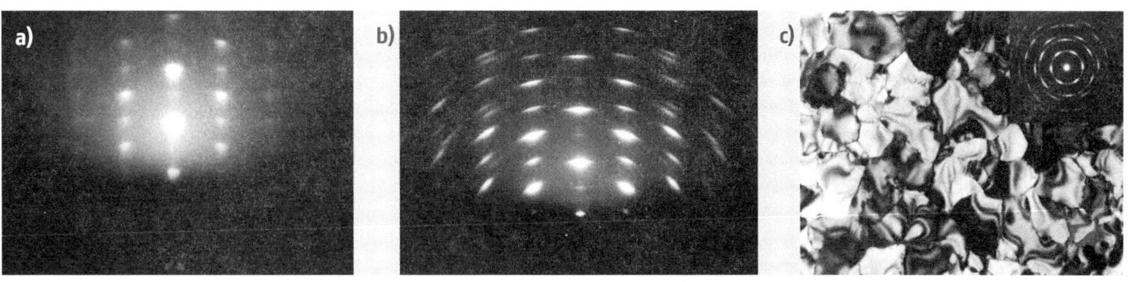

Fig. 17.21a–c RHEED patterns from Si-doped GaN/GaAs(001) (**a**) before and (**b**) after plasma cleaning. (**c**) Plan-view TEM image confirming the presence of rotated columnar grains (TED pattern *inset*)

tron beam is typically $\approx 1\,mm^2$, and so this experimental arrangement is closely associated with XRD, although the acquisition time for RHEED data is extremely short, with photographic plate exposure times of ≈ 1 s. Three or four samples can typically be examined within an hour.

The 100 kV RHEED patterns shown in Figs. 17.21a,b were acquired from a highly Si-doped GaN/GaAs(001) heterostructure, grown by MBE at 700 °C, before and after plasma cleaning. Surface hydrocarbon deposits arising from specimen handling can generate an amorphous background glow that can hinder RHEED pattern acquisition (Fig. 17.21a). The oxygen–argon plasma acts to remove such contaminants, allowing high-contrast RHEED patterns to be obtained (Fig. 17.21b). In this instance, the generated RHEED pattern remained effectively constant as the sample was rotated, indicating a random distribution of columnar grains. This predicted microstructure was subsequently confirmed by means of conventional plan-view TEM imaging (Fig. 17.21c, with TED pattern inset), which revealed a fine-scale distribution of rotated columnar grains ($\approx 20°$ of arc).

The RHEED patterns shown in Figs. 17.22a–f, acquired from a variety of III–V heterostructures grown by MBE, are presented to illustrate the variety of microstructures that can be readily distinguished. In the first instance, variations in spot, arc or ring spacings from the central beam provide evidence of the different phases present within a sample and show the general nature of the structural integrity of the near-surface layer (presumed to be representative of the bulk or thin film). Thus, polycrystalline, preferred orientation or single-crystal growth (Fig. 17.22a–c) can be rapidly distinguished, whilst embedded phases (Fig. 17.22d) and anisotropic defect distributions within zinc blende thin films (Fig. 17.22e,f) are also revealed.

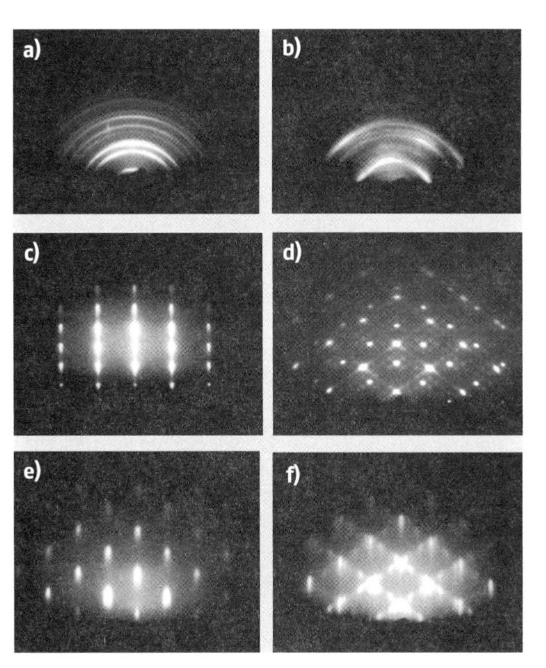

Fig. 17.22a–f RHEED patterns from (**a**) GaAs/AlN(As)/GaP(001) indicating polycrystalline growth; (**b**) GaN(As)/GaP(001) grown under low As flux at 620 °C showing disordered growth with some degree of preferred orientation; (**c**) Mg-doped GaN/GaAs$\{\overline{111}\}$B showing α-GaN single-crystal ($< 11\bar{2}0 >$ projection); (**d**) GaN(As)/GaP(001) grown under high As flux at 650 °C showing spots due to single-crystal GaAs and β-GaN; (**e**), (**f**) Be-doped GaN/GaAs(001) grown at 700 °C showing orthogonal $\langle 110 \rangle$ and $\langle 1\bar{1}0 \rangle$ projections. *Extra spots* indicate a high degree of anisotropy for the distribution of defects within this sample

17.5 Characterizing Functional Activity

There are many solid state analytical techniques that employ X-ray or electron probes, generating a variety of signals for chemical microanalysis. Techniques for performing correlated assessment of the structural and functional performance of a material are perhaps less well-covered in mainstream texts. Accordingly, we now briefly introduce the techniques of scanning transmission electron beam induced conductivity (STEBIC) and TEM-cathodoluminescence (TEM–CL), since these allow us to make correlated structure–property investigations of electrical and optical activity within a semiconductor, respectively.

As discussed earlier, when an electron beam is incident on a semiconductor specimen, electron–hole pairs are created by the excitation of crystal electrons across the band-gap. These electron–hole pairs can, for example, recombine to emit light that may be detected by a photomultiplier. A CL image can then be obtained by displaying the detected photomultiplier signal as a function of the position of the incident electron beam as it is scanned across the specimen. CL spectra can also be acquired in spot mode, which show features attributable to excitons, donor–acceptor pairs or impurities. The 'information content' of CL images and spectra therefore includes the location of recombination sites such as dislocations and precipitates, and the presence of doping-level inhomogeneities. Similarly, if the sample is configured to incorporate a collection junction, such as a Schottky-contacted semiconductor or an ohmic-contacted p–n junction, electron-hole pairs that sweep across the built-in electric field constitute current flow. This can be amplified and an image of the recombination activity displayed as the electron beam is rastered across the sample. If the dislocations within a semiconductor act as nonradiative recombination centers, then they appear as dark lines in both CL and EBIC images because of the reduced specimen luminescence or reduced current that is able to flow through the collection junction when the beam is incident at a defect.

The techniques of CL and EBIC are most commonly performed in an SEM, but this precludes the direct identification of features responsible for a given optical or electronic signature. The resolution of extended defects achieved using EBIC and CL techniques is limited by the penetration depth of the electron beam, the effect of beam spreading and the diffusion length of minority carriers. Conversely, the resolutions of the STEBIC and TEM–CL techniques, as applied to an electron-transparent sample foil, are essentially limited by specimen geometry. The constraint of minority carrier diffusion length is removed due to the close proximity of the sample foil surfaces, and resolution depends on the incident probe size, the width of the electron hole pair generation zone and the recombination velocity at the free surface. For the case of STEBIC, resolution also depends on the defect position relative to the collecting junction. The trade-off is low electrical signal and a degraded signal-to-noise ratio due to the small generation volume and surface recombination effects, in addition to the practicality of contacting and handling thin foils.

Before presenting a number of material characterization case studies based on electron beam techniques, we now discuss the preparation of electron transparent foils that are free from artefacts and suitable for TEM investigation.

17.6 Sample Preparation

We should initially consider whether destructive or nondestructive preparative techniques need to be applied. Some characterization techniques allow samples to be examined with a minimal amount of preparation, provided they are of a form and size that will fit within the apparatus. For example, the crystallography of bulk or powder samples could be directly investigated by XRD, since the penetration depth of energetic X-rays within a sample is on the scale of $\approx 100\,\mu m$. The surface morphology and near-surface bulk chemistry of a sample can be directly investigated within the SEM, noting the interaction volume of electrons [on the scale of $\approx 1\,(\mu m)^3$] associated with the EDX and WDX techniques. It might, however, be necessary to coat insulating samples with a thin layer of carbon or gold prior to SEM investigation to avoid charging effects. Similarly, minimal preparation might only be required before surface assessment using XPS or RHEED, such as cleaning using a degreasing protocol or plasma cleaning. Accordingly, the focus of this section is to introduce the techniques used to prepare samples for TEM investigation, since the requirement is for specimens that are typically submicrometer in thickness and free of preparation artefacts. For example, a complex sequence of sequential mechanical polishing, dimpling, ion beam thinning and plasma cleaning may be required to produce a pristine semiconductor heterostructure sample, with each stage of the process being designed to minimize artefacts from the previous stage of the preparation process. The idea is to minimize or eliminate artefacts from the preparation process to ensure that the sample being investigated is representative of the starting bulk material. Care is also needed to avoid artefacts that might be introduced through the interaction of the high-energy electron beam with the sample.

In this context, it is interesting to note how TEM sample preparation techniques have developed over the years. Small particles of MgO, produced by igniting the metal and drifting a specimen grid through the smoke, were typical of samples investigated in the 1940s, along with sample replicas made by a dry stripping technique using formvar film. Biological samples fashioned by enzymatic digestion, staining and microincineration were also possible by 1945. Glass and diamond knife microtomes were introduced in the 1950s and used to section soft biological materials. Advances in the controlled preparation of inorganic materials were made upon the introduction of argon ion beam thinning in the late 1960s, which allowed the cross-sectional observation of semiconductor heterostructures when combined with sequential mechanical polishing and dimpling. Significant development work in this area appears throughout the literature from the 1970s. The problem of surface amorphization, introduced by the argon sputtering process, was minimized by adopting low-voltage milling techniques to define the final electron-transparent sample foil. Low stacking fault energy semiconductors, such as II–VI compounds which are easily damaged or InP-based compounds that suffer from In droplet formation with conventional milling techniques, were also successfully prepared for TEM observation using the technique of iodine reactive ion beam etching (RIBE), otherwise known as chemically assisted ion beam etching (CAIBE), developed in the 1980s. The 1990s, however, saw the development of the most effective raft of sample preparation techniques for functional materials and complex semiconductor device structures in particular: tripod polishing, focused ion beam (FIB) milling and plasma cleaning.

The fine adjustment of micrometer supports is the key to tripod polishing that allows the direct mechanical polishing of specimens down to thickness of $< 10\,\mu m$, using diamond-impregnated polishing cloths on a stable, high-torque, low-speed polishing wheel. A brief final stage of low-voltage argon ion beam thinning (with liquid nitrogen cooling) then enables electron-transparent sample foils to be defined, free of differential mechanical polishing and shadowing artefacts, whilst minimizing remnant surface amorphization.

FIB instrumentation (now commonly integrated with an SEM) was originally developed for the semiconductor industry as a diagnostic tool for silicon chip fabrication. The application of a focused beam of gallium ions enables sample material to be sputtered away in a controlled fashion to produce an electron-transparent membrane, for example through a specific device within a complex microprocessor. Samples can be periodically observed using SEs generated by gallium ion/material interactions in order to maintain control over and the precision of the sputtering process. In particular, the ability of FIB instruments to prepare site-specific TEM membranes provides a unique opportunity to access the subsurface microstructure of complex device structures

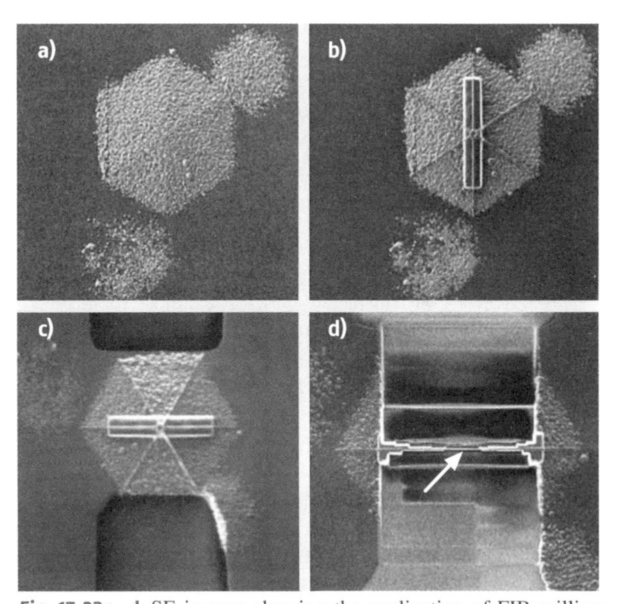

Fig. 17.23a–d SE images showing the application of FIB milling to target a source of defects buried beneath the emergent core of a growth hillock. (**a**) An etched CVD-grown GaN/sapphire hillock ($\approx 5\,\mu m$ in size). (**b**) Platimum stripes deposited along the hillock facet edges in order to retain sight of the defect core ($\approx 100\,nm$ in size). (**c**) A high Ga flux is used to create an access trench to the defect. (**d**) Decreasing the Ga flux enables an electron-transparent membrane (arrowed) to be defined at the approximate position of the defect core

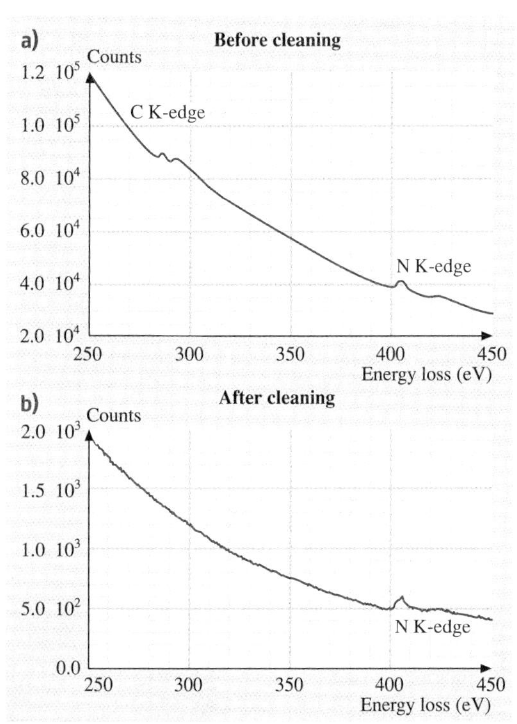

Fig. 17.24a,b EELS spectra from GaN (**a**) before and (**b**) after plasma cleaning

or specific regions within a heterostructure identified as being particularly interesting from observations of the surface. Again, by way of example, Fig. 17.23 shows a TEM membrane sectioned through the apex of a hillock identified within a sample of heteroepitaxial GaN–sapphire. Prior to sectioning, the sample was coated with a thin layer of gold and a platinum alkyl decomposed under the rastered Ga beam to define protective metallic stripes. Sequential sputtering and reduction of the incident Ga beam current allowed a supported thin membrane to be defined (Fig. 17.23d). Care is still required to minimize amorphization artefacts at the surfaces of such membranes, arising from the sputtering action of the glancing, high-energy Ga ion beam. Charging effects associated with insulating samples can act to compromise the fine-scale control of the ion milling procedure. However, this particular problem can be addressed by using low ion dose and shadowing techniques, combined with charge neutralization procedures to inhibit the deflection of the incident ion beam.

Plasma cleaning using oxygen–argon gas, commonly used for the final stage of TEM sample preparation, enables pristine electron-transparent membranes to be obtained, suitable for detailed chemical microanalysis. The disassociated oxygen component of the plasma reacts with organic surface contaminants to produce CO, CO_2 and H_2O reaction products that can be conveniently pumped away. To illustrate the effectiveness of this procedure, it was found that the quality of EELS data from GaN was significantly improved following plasma cleaning. Figure 17.24 shows the removal of the artefact carbon K-edge and enhancement of the sample nitrogen K-edge after a few minutes of exposure to the plasma. It is worth noting that in principle the argon component of the plasma permits gentle sputtering of the sample to occur if low-pressure conditions are used due to the larger mean free path and hence increased energy of the ions. This can be problematic due to the possibility of sample cross-contamination with material sputtered from the supporting sample rod. Another cautionary note on plasma cleaning relates to semi-insulating samples that can become too clean and consequently more susceptible to charging effects under the imaging electron beam.

17.7 Case Studies – Complementary Characterization of Electronic and Optoelectronic Materials

There are four general levels of interest when characterizing a given sample:

- What is it made of?
- What additional imperfections does it contain?
- How did it get to be that way?
- How does the microstructure influence the functional properties of the material/device?

For example, one might wish to identify the chemical constituents and crystal structure of a sample to start with, such as whether it is a compound or an alloy, and whether it is single-crystal, polycrystalline, exhibits a preferred orientation or is amorphous. One might then wish to characterize the additional fine-scale defect microstructure within the sample, including precipitates or extended structural defects such as dislocations, stacking faults or domain boundaries, since these could deleteriously affect the functional properties of the material. The next level of understanding focuses on how the material has formed via growth, processing or device usage, and seeks to make sense of the process of dynamic evolution, since this potentially enables us to find ways to improve the material in a controlled manner. A clear record of sample history is generally helpful in this context, particularly when trying to identify defect sources. The final level of understanding is possibly the most challenging, since it seeks to associate microstructure with the functional properties exhibited by the sample, and thereby to make sense of its structure–property–processing interrelationship at the fundamental level.

In this context, structural characterizations of semiconductor heterostructures tend to focus on the following issues:

- The integrity of the layer growth and the orientation relationship with the substrate.
- The nature of the structural defects within the epilayer, arising (for example) from the lattice mismatch or from differential thermal contraction following cool-down from the growth temperature.
- The structural integrities of the critical interfaces within the device's active region, such as multiple quantum wells, and the chemical uniformities of the associated alloy layers.
- Modification of the microstructure due to subsequent processing, such as contact formation and device usage.

XRD techniques are well-suited to assessments of the structural integrities of bulk and epitaxial thin film semiconductors. However, we should note the inherent averaging over a large number of microscopic features, such as dislocation distributions responsible for twist and tilt of mosaic grains, associated with such techniques. Conversely, TEM and related techniques are more suited to assessing the fine-scale defect microstructure of a heterostructure. For example, conventional

a)
b)
100 nm
1 µm
c)
GaN
GaN
100 nm

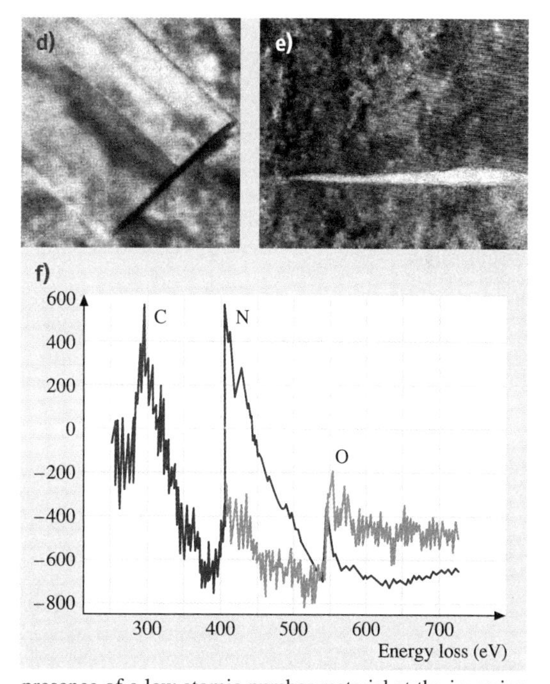

Fig. 17.25 **(a)** Optical micrograph of MOCVD-grown GaN on a KOH chemomechanically polished, N-polar, bulk GaN substrate showing a high density of hillocks. **(b)** Slightly tilted plan-view TEM image through a growth hillock revealing a central defect core (arrowed). An enlarged ⟨0001⟩ projected image of the hillock core is shown in the *inset*. **(c)** Cross-sectional, weak-beam TEM image through a hillock core. CBED analysis confirmed the feature to be an inversion domain. **(d)** HAADF image indicating the presence of a low atomic number material at the inversion domain source. **(e)** HREM image indicating the presence of a narrow band of amorphous material at the inversion domain source. **(f)** EEL second difference spectrum indicating the presence of oxygen at the nucleating event

weak-beam, HREM or CBED analysis can be used for fine-scale defect structural analysis, whilst EELS and EDX analysis can be used to profile alloy composition. However, the ability to perform atomic-level structural characterization and chemical analysis on the nanometer scale is offset by concerns about statistical significance and whether the small volume of material analyzed is truly representative of the larger object. Therefore, electron microscopy-based techniques combined with FIB procedures for site-specific sample preparation tend to be used when investigating integrated device structures.

The examples provided so far illustrate how various diffraction and imaging techniques can provide information on the structural integrity of a given sample. The following examples emphasize the need to apply complementary material characterization techniques in support of the development of semiconductor science and technology.

17.7.1 Identifying Defect Sources Within Homoepitaxial GaN

The emergence of the (In,Ga,Al)N system for short-wavelength light-emitting diodes, laser diodes and high-power field effect transistors has been *the* semiconductor success story of recent years. In parallel with the rapid commercialization of this technology, nitride-based semiconductors continue to provide fascinating problems to be solved for future technological development. In this context, a study of homoepitaxial GaN, at one time of potential interest for high-power blue–uv lasers, is presented.

The reduction in extended microstructural defects permitted by homoepitaxial growth is considered to be beneficial in the development of nitride-based technology, particularly in view of the evidence confirming that dislocations do indeed exhibit nonradiative recombinative properties. However, in the case of metalorganic chemical vapor deposition (MOCVD)-grown homoepitaxial GaN on chemomechanically polished $(000\bar{1})$, N-polar substrates, gross hexagonally shaped surface hillocks were found to develop, considered problematic for subsequent device processing [17.25]. The homoepitaxial GaN samples examined in this case study were grown at 1050 °C. The bulk GaN substrate material was grown under a high hydrostatic pressure of nitrogen (15–20 kbar) from liquid Ga at 1600 °C. Prior to growth, the $(000\bar{1})$ surfaces were mechanically polished using 0.1 μm diamond paste and then chemomechanically polished in an aqueous KOH solution. Epitaxial growth was performed using trimethylgallium and NH_3 precursors with H_2 as the carrier gas, under a total pressure of 50 mbar. Figure 17.25a shows an optical micrograph of the resultant homoepitaxial GaN/GaN$(000\bar{1})$ growth hillocks, typically 5–50 μm in size depending on the layer thickness (and therefore the time of growth).

Electron-transparent samples were prepared in plan view using conventional sequential mechanical polishing and argon ion beam thinning procedures applied from the substrate side, whilst cross-sectional samples were prepared using a Ga-source FIB workstation. As shown earlier, the selectivity of the FIB technique enables cross-sections through the emergent cores of the hillocks to be obtained, thereby allowing the nucleation events associated with these features to be isolated and characterized. When prepared in plan-view geometry for TEM observation, each hillock exhibited a small faceted core structure at the center (Fig. 17.25b), but otherwise the layers were generally found to be defect-free. Low-magnification cross-sectional TEM imaging also revealed the presence of faceted column-shaped defects beneath the apices of these growth hillocks (Fig. 17.25c). It was presumed that these features originated at the original epilayer–substrate interface since no other contrast delineating the region of this homoepitaxial interface could be discerned. A reversal of contrast within the 0002 diffraction discs from CBED patterns acquired across the boundary walls of such features (Fig. 17.19) confirmed that they were inversion domains. Thus, the defect cores were identified as having Ga-polar growth surfaces embedded within an N-polar GaN matrix. Once nucleated, the inversion domains exhibited a much higher growth rate than the surrounding matrix, being directly responsible for the development of the "circus tent" hillock structures around them. Competition between the growth and desorption rates of Ga and N-polar surfaces allowed the gross hexagonal pyramids to evolve.

This initial approach of applying electron diffraction and imaging techniques thus enabled the nature of the inversion domains to be identified and their propagation mechanism established in order to explain the development of the hillocks. However, more detailed chemical analysis was required to ascertain the nature of the source of the inversion domains and how this related to the substrate preparation and growth process. A high-angle annular dark field (HAADF) image of the inversion domain nucleation event is shown in Fig. 17.25d. HAADF is a scanned electron probe imaging technique with a resolution defined by the size of the incident probe, while the scattering (and hence contrast) is governed by the local average atomic number. In this instance, the sample, tilted slightly to minimize the effects of diffraction contrast, showed dark contrast at the position of the inversion domain source, confirming the presence of a low atomic number material associated with the nucleation event. HREM subsequently confirmed that such nucleation events were due to narrow bands of amorphous material, 2–5 nm in thickness (Fig. 17.25e), whilst EELS confirmed the presence of oxygen (Fig. 17.25f) within these narrow amorphous bands. Accordingly, these defect sources were attributed to remnant contamination from the chemomechanical polishing technique used to prepare the substrates prior to growth. The oxygen-containing residue was presumed to be gallium oxide or hydroxide – probably products of the reaction of KOH etchant with GaN. An improved surface preparation method incorporating a short, final deoxidizing polishing procedure in an aqueous solution of NaCl led to a dramatic reduction in these nucleation sources and thus allowed N-polar homoepitaxial GaN films to be grown virtually free of these gross hillock structures.

17.7.2 Cathodoluminescence/Correlated TEM Investigation of Epitaxial GaN

The CL technique is ideally suited to studies of luminescence uniformity and spectral purity. The following case study illustrates how the defect microstructure of mixed-phase epitaxial GaN/GaAs($\overline{1}\overline{1}\overline{1}$)B can be correlated with the luminescent properties of the layer [17.26].

Even though the majority of developments in GaN technology to date have come from material grown by MOCVD on sapphire and SiC substrates, there is much interest in exploring alternative growth techniques. The MBE technique offers a lower growth temperature than MOCVD and hence enables a greater range of candidate substrate materials to be investigated. One general issue for the MBE growth of heteroepitaxial GaN is the need for direct control of the process of nucleation, since this impacts on the phase and polarity of the deposit and the resultant structural integrity of the film. Epitaxial GaN preferentially adopts the wurtzite phase, with a band-gap of $3.4\,eV_{hex}$, although zinc blende inclusions, with a band-gap of $3.2\,eV_{cubic}$, are sometimes associated with MBE-grown material (for conditions of high Ga flux); see Fig. 17.26. The CL spectrum shown in Fig. 17.26a was recorded from a plan-view, electron-transparent foil of nominally single-crystal wurtzite GaN, cooled to liquid nitrogen temperatures. The peak in the CL spectrum at 386 nm (3.21 eV) was used to create the image in Fig. 17.26b indicating the distribution of cubic phase inclusions throughout the hexagonal GaN matrix. The complementary conventional TEM image of this plan-view sample foil (Fig. 17.26c) allows the nature of the fine-scale microstructure to be characterized, with diffraction patterns confirming the presence of embedded sphalerite GaN. The size and distribution of these inclusions correlated nicely with the distribution of bright spots in the CL image of 17.26b. A subsequent cross-sectional investigation of GaN/GaAs($\overline{1}\overline{1}\overline{1}$)B specimens confirmed that the cubic inclusions were nucleated at the epilayer/substrate interface (Fig. 17.26d).

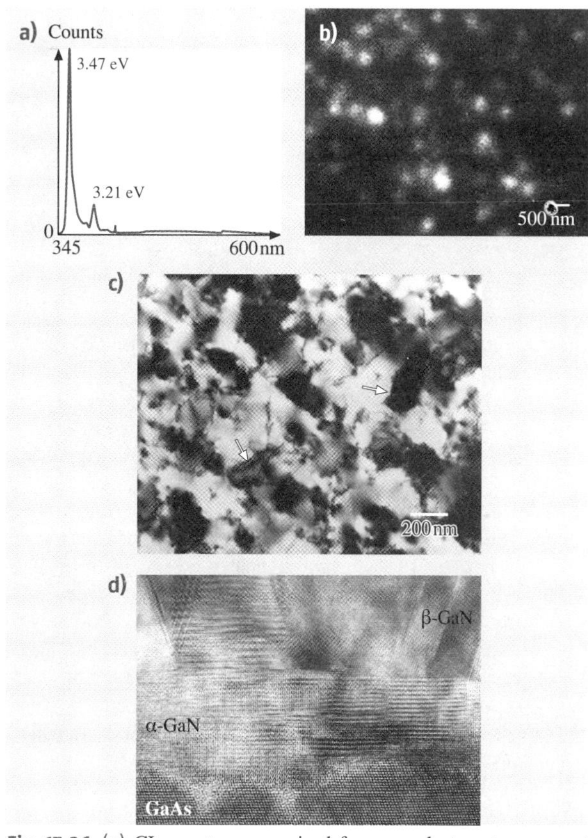

Fig. 17.26 **(a)** CL spectrum acquired from an electron-transparent foil showing peaks at 357.8 nm and 386 nm, corresponding to wurtzite and zinc blende GaN, respectively; **(b)** CL image formed at 386 nm and **(c)** complementary TEM image confirming a distribution of cubic GaN inclusions *(arrowed)* embedded within the hexagonal GaN matrix; **(d)** HREM image of epitaxial GaN/GaAs($\overline{1}\overline{1}\overline{1}$)B viewed in cross-section, indicating the nucleation of cubic phase inclusions at the epilayer–substrate interface

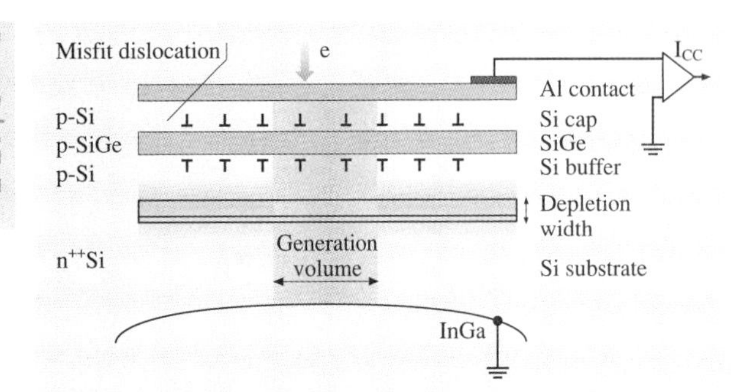

Fig. 17.27 (**a**) Schematic of the $Al/Si/Si_{1-x}Ge_x/Si(001)/InGa$ sample configuration used for STEBIC investigation.

17.7.3 Scanning Transmission Electron Beam Induced Conductivity of $Si/Si_{1-x}Ge_x/Si(001)$

The STEBIC technique was originally demonstrated in the late 1970s, using dedicated STEM instrumentation to obtain information on the electrical properties of dislocation core structures within (Ga,Al)(As,P), and thereby providing the first evidence that nonradiative recombination processes at dislocations are related to jogs and kink sites. Dissociated 60° dislocations showed the highest electrical activity, while sessile Lomer-Cottrell edge dislocations were found to be electrically neutral, indicative of reconstructed core structures. STEBIC imaging of an electron-transparent foil allows the electrical and structural properties of defects to be observed simultaneously. The availability of electron sources with high brightness in modern scanning TEM instruments compensates for the main problem of small generation volume and provides an accessible way to perform the STEBIC technique.

The next case study illustrates how the electrical nonradiative recombination properties of MBE-grown $Si/Si_{1-x}Ge_x/Si(001)$ heterostructures correlate with the distribution of interfacial misfit dislocations [17.27]. The $Si_{1-x}Ge_x$ system has potential applications in devices with high electron and hole mobilities. However, the introduction of dislocation networks, or the multiplication of existing dislocations, driven by the 4% misfit strain between Ge and Si, is generally regarded as being detrimental to device operation. With a view to gaining an improved understanding of the relationship between fine-scale structural defects and electron transport properties, structures of capped MBE-grown $Si/Si_{1-x}Ge_x/Si(001)$ were investigated using STEBIC and a range of complementary microscopies. The samples incorporated buried p–n junctions to assist with charge collection and surmount the problem of surface recombination effects.

To prepare for the STEBIC investigation of the $Si/Si_{1-x}Ge_x/Si(001)$, evaporated Al contacts were attached to the top surface prior to preparing electron-transparent foils in plan view by sequential mechanical polishing and argon ion milling of the substrate (Fig. 17.27a). Ohmic contact to the lower surface was made using an InGa eutectic. Electrical activity images were acquired using an electrical contact stage and a scanning TEM. Signal amplification was performed using an amplifier with a low noise current. By controlling the STEM scan rate, the beam could be rastered at a rate compatible with the low bandwidth constraint of the amplifier. STEBIC signals were typically $\approx 100\,pA$ for an electron-transparent $Si/Si_{1-x}Ge_x/Si$ foil imaged at a magnification of $\times1000$.

Figure 17.27a illustrates the physical parameters relevant to this STEBIC experiment. The silicon substrate was n-type $10^{18}\,cm^{-3}$, whilst the epilayer was p-type boron-doped to a level of $10^{16}\,cm^{-3}$. These

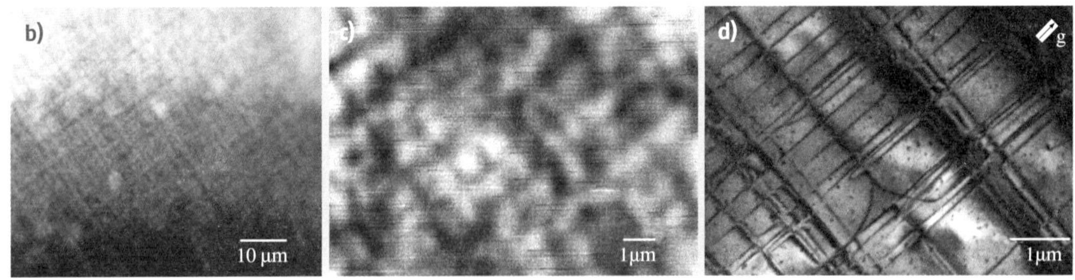

Fig. 17.27 (**b**) STEBIC image showing recombination activity due to ⟨110⟩ orthogonal arrays of misfit dislocations within relaxed $Si/Si_{0.96}Ge_{0.04}/Si(001)$. (**c**) Higher magnification STEBIC image illustrating sub-μm resolution of the electrical activity. (**d**) Bright-field TEM image demonstrating the presence of bundles of misfit dislocations ($g = 220$)

values of sample doping were chosen to create a depletion region of ≈ 30 nm width to assist with charge collection, away from the misfit dislocations delineating the $Si/Si_{1-x}Ge_x$ and $Si_{1-x}Ge_x/Si$ interfaces. The low-magnification STEBIC image shown in Fig. 17.27b illustrates the recombination activity within a relaxed $Si/Si_{0.96}Ge_{0.04}/Si(001)$ sample imaged in plan view. Submicron resolution of the recombination activity is readily achievable using this technique (Fig. 17.27c), with line scans from digitized images indicating a resolution of ≈ 0.3 μm in this case. The spacing of ≈ 1 μm striations in the STEBIC image is much greater than the spacing of individual misfit dislocations shown in the associated TEM image (Fig. 17.27d), and is more closely associated with the spacing of dislocation bundles. Thus, correlation with structural images shows that bunched arrays of orthogonal <110> misfit dislocations are primarily responsible for the enhanced recombination. For this particular sample, detailed $\boldsymbol{g}.\boldsymbol{b}$ analysis confirmed the presence of bands of predominantly 60° misfit dislocations with a few 90° segments arising from dislocation interactions.

Supporting evidence for these dislocations being dissociated and probably decorated by transition metal impurities was obtained from complementary HREM and EDX investigations of metastable and relaxed $Si/Si_{1-x}Ge_x/Si(001)$ samples from the same growth trial. As the relaxation of a low Ge content, metastable $Si/Si_{1-x}Ge_x/Si$ structure proceeds, extensive arrays of orthogonal ⟨110⟩ dislocations form and interact, with dislocations generated in the strained $Si_{1-x}Ge_x$ layer being pushed by repulsive dislocation forces into the Si substrate and cap on {111} glide planes. HREM observations of relaxed $Si/Si_{1-x}Ge_x/Si$ samples in cross-section confirmed that the misfit dislocations were dissociated, with tails associated with each of the partials, indicative of impurity decoration (Fig. 17.27e). This particular image was acquired before the development of electron beam-induced damage artefact structures within this sample foil and so is considered representative of the as-grown material. A distribution of small precipitates, of typical size 5 nm, was also identified within metastable samples prior to strain relaxation. These precipitates showed strong scattering in HREM (Fig. 17.27f) and revealed the presence of Fe when analyzed using EDX within a dedicated STEM (1 nm probe size), as shown in Fig. 17.27g. No Fe was present in spectra acquired immediately to the side of the precipitates. (The Ni signal was considered to be an artefact of EDX acquisition and attributed to fluorescence from X-rays and electrons interacting with the specimen's Ni support ring.) Hence, the suggestion is that enhanced transition metal impurity segregation at dissociated dislocations is responsible for the STEBIC contrast observed. This emphasizes the need to control dopants or impurity sources in the vicinity of the heterostructure interface during the development of functional device structures.

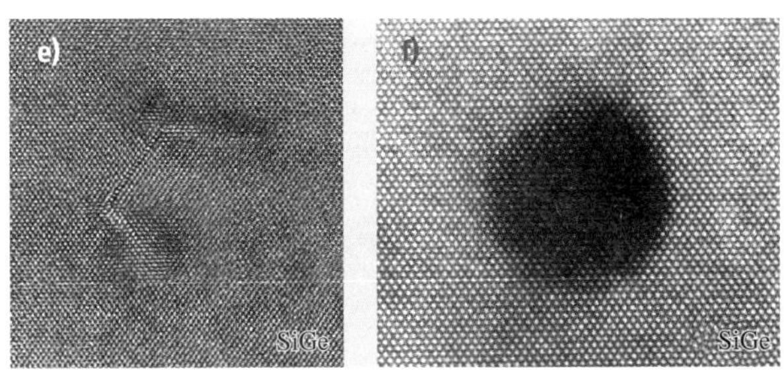

Fig. 17.27 (e) HREM image of a decorated dissociated dislocation viewed in cross-section at the $Si/Si_{1-x}Ge_x$ interface following sample annealing and relaxation. (f) HREM image of a precipitate within as-grown, metastable $Si/Si_{1-x}Ge_x/Si$.

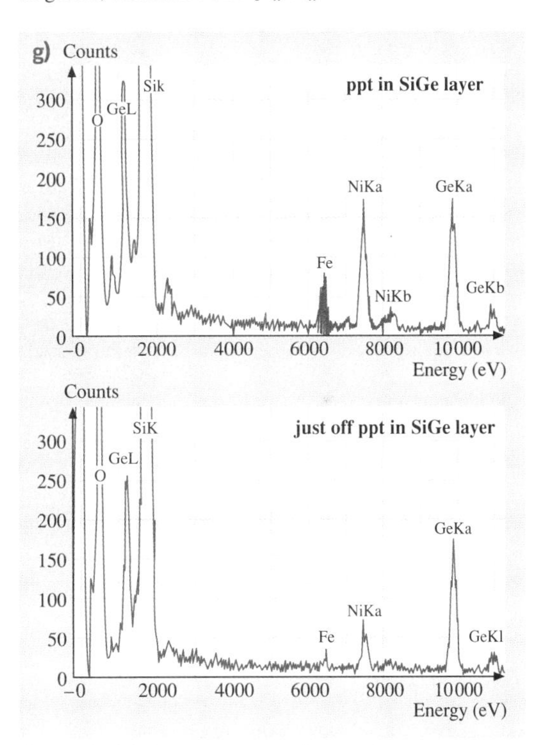

Fig. 17.27 (g) EDX spectra confirming the presence of Fe transition metal impurities

17.8 Concluding Remarks

The above commentary has attempted to convey the framework underpinning a variety of analytical techniques used to investigate the structures of semiconductors. It is emphasized that an appropriate combination of assessment techniques should generally be applied, since no single technique of assessment will provide information on the composition, morphology, microstructure and (opto)electronic properties of a given functional material or processed device structure.

This type of considered approach to materials characterization is required in order to break free of the "black-box" mentality that can develop if one is too trusting of the output generated by automated or computerized instrumentation systems. We must always bear in mind the process of signal generation that provides the information content. This in turn should help us to develop an appreciation of performance parameters such as spatial or spectral resolution, in addition to sensitivity, precision and the detection limit. We should consider technique calibration and the appropriate use of standards in order to ensure that the data acquired is appropriate (and reproducible) to the problem being addressed. Consideration should also be given to the form and structure of the data being acquired and how the data sets are analyzed. In this context, distinction should be made between the processing of analog and digital information and the consequences of data conversion. Issues regarding the interpretation (or misinterpretation) of results often stem from the handling of experimental errors. On a practical level, a rigorous experimental technique should certainly be applied to ensure that the data generated is both meaningful and representative of the sample being investigated, free from artefacts from the preparation and investigation processes. There are clearly differences between qualitative assessment and the more rigorous demands of quantitative analysis. The level of effort invested often reflects the nature of the problem that is being addressed. A comparative assessment of a number of samples may simply require a qualitative investigation (for example, in order to solve a specific materials science problem within a growth or device fabrication process). Alternatively, quantitative analysis may be required to gain a more complete understanding of the nature of a given sample, such as the precise composition. To summarize, an awareness of the methodology used in any investigation is required to establish confidence in the relevance of the results obtained. A range of complementary analysis techniques should ideally be applied to gain a more considered view of a given sample structure.

References

17.1 R. W. Cahn, E. Lifshin: *Concise Encyclopedia of Materials Characterization* (Pergamon, New York 1992)

17.2 J. M Cowley: *Electron Diffraction Techniques*, Vol. 1, 2 (Oxford Univ. Press., Oxford 1992, 1993)

17.3 B. D. Cullity, S. R. Stock: *Elements of X-Ray Diffraction*, 3rd edn. (Addison Wesley, New York 1978)

17.4 J. W. Edington: *Practical Electron Microscopy in Materials Science* (Philips Electron Optics, Eindhoven 1976)

17.5 R. F. Egerton: *Electron Energy-Loss Spectroscopy in the Electron Microscope* (Plenum, New York 1996)

17.6 P. J. Goodhew, F. J. Humphreys, R. Beanland: *Electron Microscopy and Analysis* (Taylor Francis, New York 2001)

17.7 P. J. Grundy, G. A. Jones: *Electron Microscopy in the Study of Materials* (Edward Arnold, London 1976)

17.8 P. B. Hirsch, A. Howie, R. B. Nicholson, D. W. Pashley, M. J. Whelan: *Electron Microscopy of Thin Crystals* (Butterworths, London 1965)

17.9 I. P. Jones: *Chemical Microanalysis Using Electron Beams* (Institute of Materials, London 1992)

17.10 D. C. Joy, A. D. Romig, J. I. Goldstein: *Principles of Analytical Electron Microscopy* (Plenum, New York 1986)

17.11 M. H. Loretto, R. E. Smallman: *Defect Analysis in Electron Microscopy* (Chapman Hall, London 1975)

17.12 D. Shindo, K. Hiraga: *High-Resolution Electron Microscopy for Materials Science* (Springer, Berlin, Heidelberg 1998)

17.13 J. C. H. Spence: *Experimental High-Resolution Electron Microscopy – Fundamentals and Applications* (Oxford Univ. Press, New York 1988)

17.14 G. Thomas, M. J. Goringe: *Transmission Electron Microscopy of Metals* (Wiley, New York 1979)

17.15 D. B. Williams, C. B. Carter: *Transmission Electron Microscopy: A Textbook for Materials Science* (Plenum, New York 1996)

17.16 R. Hull, J. C. Bean: Crit. Rev. Solid State **17**, 507 (1992)

17.17 T. Sugahara, H. Sato, M. Hao, Y. Naoi, S. Kurai, S. Tattori, K. Yamashita, K. Nishino, L. T. Romano, S. Sakai: Jpn. J. Appl. Phys. **37**, 398 (1997)

17.18 Y. Xin, P. D. Brown, T. S. Cheng, C. T. Foxon, C. J. Humphreys: Inst. Phys. Conf. Ser. **157**, 95 (1997)

17.19 Y. Ishida, H. Ishida, K. Kohra, H. Ichinose: Philos. Mag. A **42**, 453 (1980)

17.20 D. B. Holt: J. Mater. Sci. **23**, 1131 (1988)

17.21 K. Ishizuka, J. Taftø: Acta Cryst. B **40**, 332 (1984)

17.22 D. Cherns, W. T. Young, M. Saunders, J. W. Steeds, F. A. Ponce, S. Nakamura: Philos. Mag. **A77**, 273 (1998)

17.23 J. M. Cowley: *Electron Diffraction: An Introduction*, Vol. 1 (Oxford Univ. Press, Oxford 1992)

17.24 G. J. Russell: Prog. Cryst. Growth Ch. **5**, 291 (1982)

17.25 J. L. Weyher, P. D. Brown, A. R. A. Zauner, S. Muller, C. B. Boothroyd, D. T. Foord, P. R. Hageman, C. J. Humphreys, P. K. Larsen, I. Grzegory, S. Porowski: J. Cryst. Growth **204**, 419 (1999)

17.26 P. D. Brown, D. M. Tricker, C. J. Humphreys, T. S. Cheng, C. T. Foxon, D. Evans, S. Galloway, J. Brock: Mater. Res. Soc. Symp. Proc **482**, 399 (1998)

17.27 P. D. Brown, C. J. Humphreys: J. Appl. Phys. **80**, 2527 (1996)

18. Surface Chemical Analysis

The physical bases of surface chemical analysis techniques are described in the context of semiconductor analysis. Particular emphasis is placed on the SIMS (secondary ion mass spectrometry) technique, as this is one of the more useful tools for routine semiconductor characterization. The practical application of these methods is addressed in preference to describing the frontiers of current research.

Surface chemical analysis is a term that is applied to a range of analytical techniques that are used to determine the elements and molecules present in the outer layers of solid samples. In most cases, these techniques can also be used to probe the depth distributions of species below the outermost surface. In 1992 the International Standards Organisation (ISO) established a technical committee on surface chemical analysis (ISO TC 201) to harmonize methods and procedures in surface chemical analysis. ISO TC 201 has a number of subcommittees that deal with different surface chemical analytical techniques and this chapter will discuss the applications of these different methods, defined by ISO TC 201, in the context of semiconductor analyses. In particular, this discussion is intended to deal with practical issues concerning the application of surface chemical analysis to routine measurement rather than to the frontiers of current research. Standards relating to surface chemical analysis developed by the ISO TC201 committee can be found on the ISO TC201 web site www.iso.org (under "standards development").

Traditional surface chemical analysis techniques include the electron spectroscopy-based methods Auger electron spectroscopy (AES or simply Auger) and X-ray photoelectron spectroscopy (XPS, once also known as ESCA – electron spectroscopy for chemical analysis), and the mass spectrometry method SIMS (secondary ion mass spectrometry). The ISO TC 201 committee also has a subcommittee that deals with glow discharge spectroscopies. Whilst these latter methods have been used more for bulk analysis than surface analysis, the information they produce comes from the surface of the sample as that surface moves into the sample, and so they have been finding applications in depth profiling studies.

One thing that is common to all of these surface chemical analysis techniques is that they are vacuum-based methods. In other words, the sample has to be loaded into a high or ultrahigh vacuum system for the analysis to be carried out. With the one exception of glow discharge optical emission spectroscopy (GDOES), where the analysis relies upon the detection of photons, all of the techniques also depend upon the detection of charged particles. This requirement for vacuum operation necessarily imposes limits on the types and sizes of samples that can be analyzed, although of course instruments capable of handling semiconductor wafers do exist. The quality of the vacuum environment around the sample can also affect the quality of the analysis, especially with regard to the detection of elements that exist in the atmosphere around us. The size and complexity of surface chemical analysis equipment has arguably tended to limit the wider use of these powerful methods.

18.1 Electron Spectroscopy

In the electron spectroscopies, Auger and XPS, the surface of the sample is probed by an exciting beam which causes electrons to be ejected from the atoms in the sample. These electrons are collected and their energies analyzed. The two techniques are similar but subtly different.

18.1.1 Auger Electron Spectroscopy

In Auger, a beam of electrons is used to excite the sample. During the interaction of the primary electron beam with the sample atoms, core electrons are knocked out, creating vacancies in the inner electron shells. Electrons from outer shells can fall into the vacancy, thus leaving the atom in an unstable state and, in order to return to equilibrium, the excess energy the atom possesses can be dissipated in one of two ways: either by the emission of an X-ray photon with a characteristic energy (the basis of energy- or wavelength-dispersive X-ray analysis), or by the emission of a third electron (the Auger electron), with an energy determined by the difference in energy between the original core state and those of the two other levels involved. Clearly these energies are uniquely determined by the energy levels in the atom and thus provide a route for analysis. As three electrons are necessary for the Auger process to occur, AES is incapable of detecting hydrogen or helium, but all other elements produce characteristic Auger electrons. The energies of the Auger electrons can range from a few tens of electron volts to a few thousand electron volts. In this energy regime, electrons can only travel of the order of monolayers through a solid before an interaction occurs which causes a loss of energy, thus destroying the analytical information the electron possessed. It is this property which gives the Auger technique its surface sensitivity – although Auger electrons will be produced as far into the material as the primary beam can penetrate, only those produced in the top few monolayers can escape with their characteristic energy intact. The Auger electrons excited deeper into the sample lose energy before escaping and contribute to a background signal, as do the scattered primary electrons and the initial core electrons ejected at the start of the process. These scattered electrons produce a background signal against which the Auger electrons must be detected; this limits the analytical sensitivity that can be achieved by Auger. This is also why Auger spectra are sometimes displayed as the differential of the number of electrons against energy, because the small Auger peaks are more rapidly varying functions than the larger, slowly changing background signal and hence are enhanced by the differentiation process. Figure 18.1 shows an Auger spectrum from GaInAsP, where the spectrum plotted as the number of electrons as a function of energy is compared to the differential of the number of electrons as a function of energy.

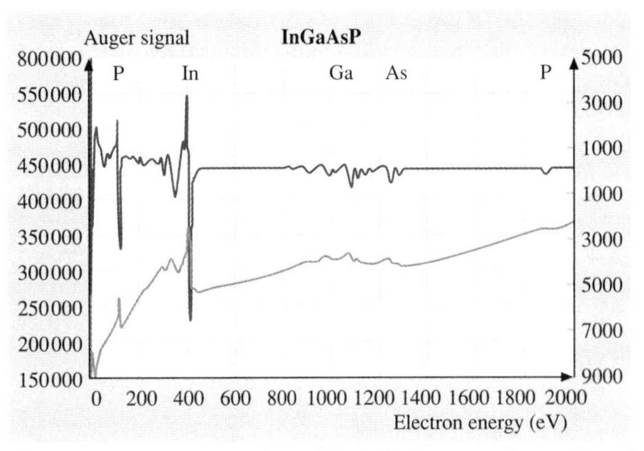

Fig. 18.1 Auger spectrum from GaInAsP, shown as the number of electrons and the differential of the number of electrons as a function of energy

As the number of Auger electrons produced is proportional to the number of atoms excited, Auger offers the ability to perform quantitative as well as qualitative analyses, although some form of calibration is required, either through the use of local reference materials or instrument calibration and standard databases. So Auger offers quantitative analysis for all elements from lithium to uranium from layers only a few atoms in thickness. What gives Auger an extra dimension is the ability to profile into the sample by removing the outermost surface layers with an inert gas ion beam (usually argon) in order to expose the layers below. This sputter depth profiling is the inverse of the sputter deposition widely used to deposit thin layers of material. Argon is the most widely used ion beam in Auger depth profiling, as its effects are physical rather than chemical, although it should be noted that some chemicals can be modified by the sputtering process. For example, some metal oxides can be reduced by sputtering while others are not, and so any chemical state data inferred from atomic compositions in sputter depth profiles should be treated with caution. Physical effects can also occur, for example atomic mixing and the development of surface topography, which can distort the shape of buried features.

One of the great advantages of Auger is that, as the excitation is provided by electrons, the primary beam can be easily focused and scanned over the surface of the sample. By detecting the scattered electrons, a physical image of the sample is produced as in the scanning electron microscope, and maps of elemental distributions can be obtained by detecting the Auger signal as

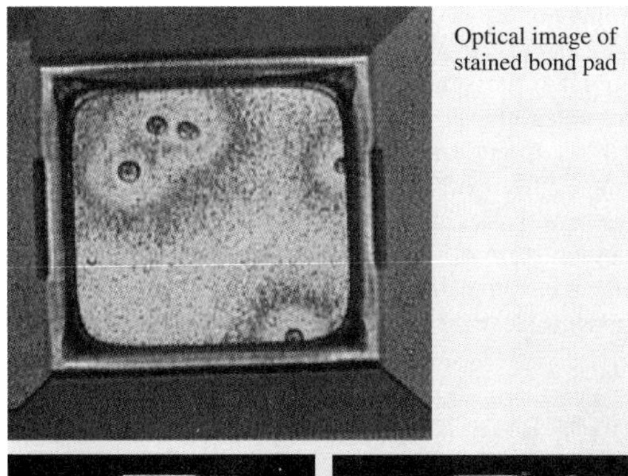

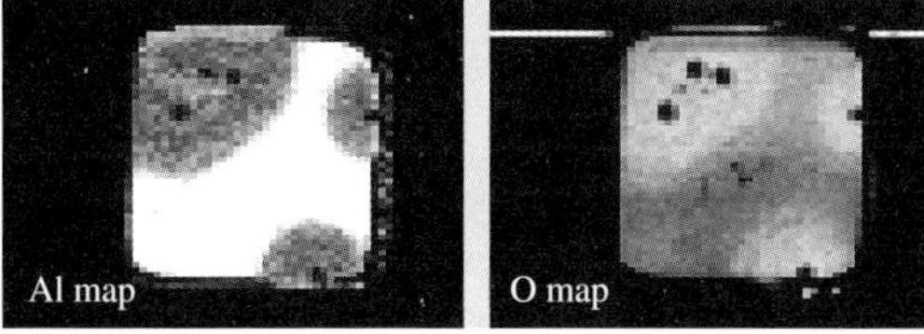

Fig. 18.2 Optical image and Auger maps of a stained aluminium bond pad

a function of beam position. These elemental maps can be time-consuming to acquire but can be useful when making a point or illustrating a book chapter; however, for practical analyses it is often adequate to identify the feature of interest from the physical image. Figure 18.2 shows aluminium and oxygen Auger maps from a contaminated bond pad. In the region of the contamination, the aluminium signal is reduced and the oxygen signal is higher compared with the uncontaminated regions.

In terms of hardware, Auger systems can be standalone systems comprising an electron beam column, an electron energy analyzer, an inert gas ion gun and sample handling stage and an associated vacuum chamber (or chambers), or form a part of a multitechnique system with X-ray sources for XPS analysis or a mass spectrometer for basic SIMS studies.

18.1.2 X-Ray Photoelectron Spectroscopy (XPS)

XPS is very similar to Auger in terms of the instrumentation and physics involved. The primary excitation, as the name implies, is, in this case, a beam of X-rays. The X-rays, often magnesium or aluminium Kα, eject core electrons from the surface atoms by the photoelectron effect. The kinetic energy of the emitted photoelectrons will be equal to the difference between the X-ray photon energy and the core level binding energy, and thus will be less than ≈ 1100 eV or ≈ 1400 eV for magnesium and aluminium Kα respectively. So, as with Auger, the mean free paths of the photoelectrons are of the order of monolayers in solid materials. As a core level vacancy is produced by the excitation, Auger electrons will also be present in the XPS spectra but, by convention, Auger spectroscopy refers to the electron-excited situation. Whereas in Auger electron spectroscopy the electron energy is usually referred to in terms of the electron's kinetic energy, in XPS the electron binding energy is usually plotted as the ordinate in the spectra. This means that, whatever X-ray excitation is used, be it magnesium, aluminium or a more exotic material, the photoelectrons will appear at the same binding energy in the spectra but the apparent positions of the Auger peaks will change (on the binding energy scale) as their kinetic energy is independent of the excitation source. There are no scattered primary electrons (which are always present in Auger spectra) in the XPS spectra, so these spectra have better signal-to-background, and the photoelectron peaks are easily distinguished against the background arising from scattered photoelectrons produced deeper into the sample and other secondary processes.

As the problem of focusing electron beams is much simpler than focusing X-ray beams, XPS is perceived as a large-area technique, whereas Auger is the technique of choice for small area analysis. However, the relentless advances made in the performance and design of instruments means that XPS instruments can achieve spatial resolutions of the order of 1 to 10 μm. In XPS no charge is brought to the sample by the primary excitation, and so insulating samples are easier to analyze with XPS than with Auger; also, the photoelectron peaks show small energy differences in the peaks positions depending upon the local chemical environment of the atom from which they originated, the so-called chemical shift. While chemical shifts are present in some Auger peaks, this is the exception rather than the rule, and XPS is the technique of choice where information on the local chemical state of the surface is required. The physics of the XPS process is probably even better understood than the Auger process, and quantification of the spectra is relatively routine.

As with Auger, composition depth profiles can be produced by sputtering the surface of the sample with an inert gas ion beam. Again caution is advised when interpreting chemical state information from a surface that has been subject to ion bombardment. With both of the electron spectroscopies it is the surface of the sample that remains after sputtering that is analyzed in a depth profile, and there are two factors to be aware of (if not more). Once the passivating surface layer has been sputtered away, the surface of the sample may become chemically active and getter residual gas from the vacuum system. If the sample is a multicomponent material, one component may have a higher sputtering rate than the other, so as the sputtering process proceeds the surface will become depleted in the higher sputtering rate material. This process will continue until an equilibrium state is reached where the material leaving the surface is in the same ratio as the bulk composition; the corollary of this is that the surface will be enriched in the lower sputter rate material and so the composition of the material as measured by either XPS or Auger will be in error unless this effect is understood and accounted for.

Both Auger and XPS are capable of detecting all elements from lithium to uranium (and beyond), and have sensitivities in the parts per hundred to parts per thousand regime. The responses vary from element to element but typically sensitivities remain within an order of magnitude or so between elements.

18.2 Glow-Discharge Spectroscopies (GDOES and GDMS)

These are two apparently similar but quite unrelated techniques that rely upon glow discharges as the excitation source. In glow discharge optical emission spectroscopy (GDOES) a high-pressure glow discharge is used to sputter material from the surface of a sample, and this sputtered material is detected by the optical emissions it produces in the glow discharge. In glow discharge mass spectrometry (GDMS), a low pressure dc glow discharge is used to sputter material from the sample surface and ionized material from the discharge is extracted into a mass spectrometer for analysis.

GDOES is probably the simplest of all the surface chemical analysis techniques, at least as far as the vacuum requirements are concerned. There is no complex vacuum system, as needed for all of the other methods, and the sample itself sits with atmospheric pressure on one side of it while the opposite face acts as one electrode of a glow discharge cell that is pumped by a simple vacuum pump. A flow of high-purity argon gas flows through the cell, providing the sputtering and discharge gas and purging the cell of impurities and material removed from the sample. A window at the other end of the discharge cell transmits light from the discharge into an optical spectrometer. By using a spectrometer with a number of photomultiplier detectors, prepositioned at the known wavelengths of the expected element emission lines, data from a large number of elemental channels can be collected in parallel, making GDOES an extremely efficient analytical system. A scanning spectrometer can also be included in the instrument to provide a continuous spectral scan to detect emission lines from elements other than those built into the instrument, but this is, of course, a serial detection device and the advantages of parallel acquisition are lost.

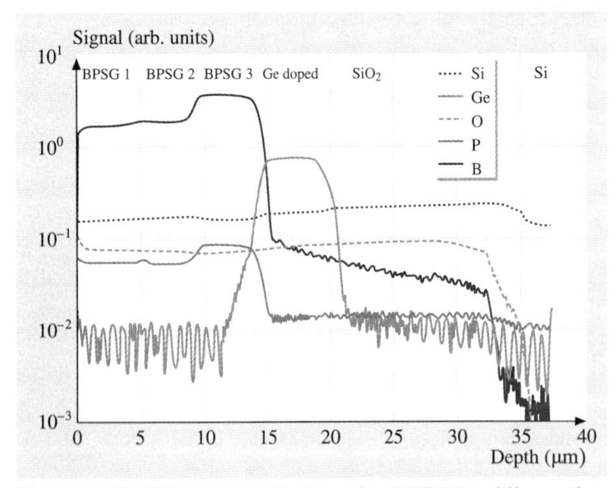

Fig. 18.3 GDOES depth profile through a DWDM multilayer glass structure

Traditionally GDOES has been widely used for the analysis of metals and coatings on metals, but it is currently also finding application in the area of semiconductor materials. With the development of radio frequency glow discharge sources, the technique is capable of analyzing insulators, and with the high sputtering rates available, it can depth profile tens of μm into dielectric layers. With fast electronics and parallel detection, thin oxide layers can also be profiled. The sensitivity of the GDOES technique lies between that of Auger or XPS and that of SIMS. One potential weakness of the method is that the technique has no spatial resolution and the analysis area is millimeters in diameter. It is thus useful for large-area plain samples, but cannot be used with patterned material or to probe small features. The technique is useful for bulk analysis, but it is the ability to depth-profile into material, providing an insight into layer structures, that is its greatest appeal. However, because the technique has no ability to discriminate where the analytical signal is coming from, the quality of the depth profiles produced will be compromised by crater edge effects. In other words, while most of the analytical signal will originate from the bottom of the sputtered crater, there will always be some information that comes from the crater side wall. The consequence of this is that, with layered structures, layers closer to the surface will appear to tail into layers beneath them, even though the interface between the layers is abrupt. This effect can be seen in the depth profile shown in Fig. 18.3, which shows a GDOES profile into a DWDM structure.

Glow discharge mass spectrometry (GDMS) is a considerably more complex technique, at least from an instrumental point of view. Originally developed as a method of bulk analysis, GDMS is probably the most sensitive, in terms of the detection limit achievable, of all of the techniques being considered here. As with GDOES, in GDMS the sample forms one electrode in a simple glow discharge cell. However, in the case of GDMS, the discharge cell is mounted within a high-vacuum system. In its original form, the sample (typically be 1 mm^2 by about 15 mm long) is placed in the center of a cylindrical cell into which argon is leaked at low pressure. By applying a dc voltage between the sample and the cell, an argon plasma is created which sputters the outside of the sample, removing material. This material, some of which is ionized but the majority of which is neutral as it leaves the surface, is ionized by a variety of processes as it passes through the glow discharge plasma. These ions are then accelerated into a high-resolution magnetic sector mass spectrometer where they are mass-analyzed and counted. Instruments can also be based on quadrupole mass spectrometers, but it is the magnetic sector instruments which offer the greater sensitivity. By sweeping the mass spectrometer through a range of masses, which can cover the entire periodic table, the major, minor and trace elements present in a sample can be determined. GDMS is a particularly powerful method of detecting the trace elements present in bulk semiconductor materials at levels down to parts per billion.

It is also possible to analyze flat, rather than matchstick-shaped, samples in GDMS. Just as in GDOES, the flat sample is positioned at the end of the discharge cell, and a cylindrical crater is etched into the sample surface. As with GDOES, with GDMS there is no spatial resolution, and the depth information from layered structures will be distorted by crater edge effects and loss of crater base flatness as it is not possible to discriminate between ions produced from the base of the crater and those produced from the sidewalls.

18.3 Secondary Ion Mass Spectrometry (SIMS)

SIMS is probably the most powerful and versatile of all of the surface analysis techniques and comes in the widest variety of instrumentations, from big, stand-alone instruments to bench-top instruments and add-ons to electron spectrometers. SIMS can offer chemical identification of submonolayer organic contamination, measurement of dopant concentrations, and can produce maps and depth profile distributions from nanometers to tens of μm in depth. However, no one instrument is going to be capable of all of these tasks, and even if it could it would not be able to achieve all of them at the same time.

SIMS, in its simplest form, requires an ion gun and a mass spectrometer. The sample is placed in a vacuum chamber and ions from the ion gun sputter the sample surface. Material is sputtered from the sample surface and some of this will be ionized, although in most cases the major part of the sputtered material will be in the form of a neutral species. The ionized component of the sputtered material is mass-analyzed with the mass spectrometer.

The technique has evolved in various directions from this common origin to produce a variety of subtly different variants of the SIMS technique, including dynamic SIMS (DSIMS), static SIMS (SSIMS) and time of flight SIMS (ToFSIMS), each of which has its own distinct attributes. There are three main types of mass spectrometer used for SIMS analysis: the magnetic sector, the quadrupole and the time of flight, ToF. Dedicated depth-profiling SIMS machines, dynamic SIMS instruments, tend to employ either magnetic sector or quadrupole mass spectrometers. Magnetic sector instruments offer high transmission and high mass resolution capabilities, useful for separating adjacent mass peaks with a very small mass difference, for example ^{31}P from ^{30}SiH. Quadrupole mass spectrometers offer ultrahigh vacuum compatibility and, as well as being used in DSIMS instruments, smaller versions are also found as add-ons to Auger/XPS instruments and bench-top instruments. Time of flight instruments are remarkably efficient in their use of material in that the entire mass spectrum is sampled in parallel, whereas in the

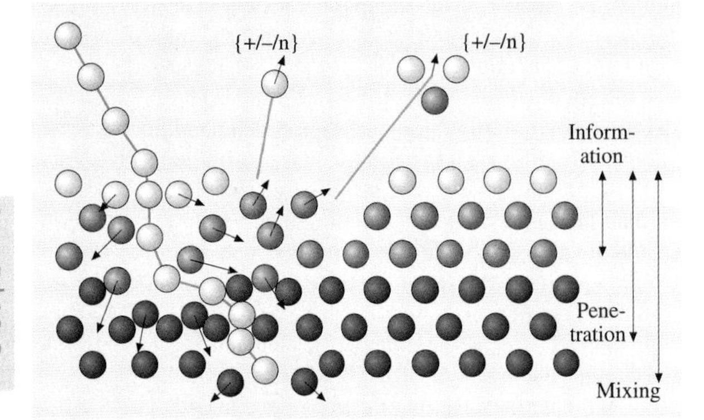

Fig. 18.4 Schematic diagram of the SIMS process

magnetic sector and quadrupole instruments the spectrum is produced by sequentially scanning through the mass range of interest. ToFs can also offer high mass resolution but profile relatively slowly because it is necessary to use a pulsed ion beam with a low duty cycle. Whilst static SIMS can be carried out on either a quadrupole or magnetic sector instrument, the ToF-based instruments are more suited to the task as less material is used in the course of the analysis. However, when dynamic SIMS, where only a few elements need to be monitored, is the goal, the magnetic sector or quadrupole instrument is the appropriate choice, although ToF machines can be used for shallow profiling.

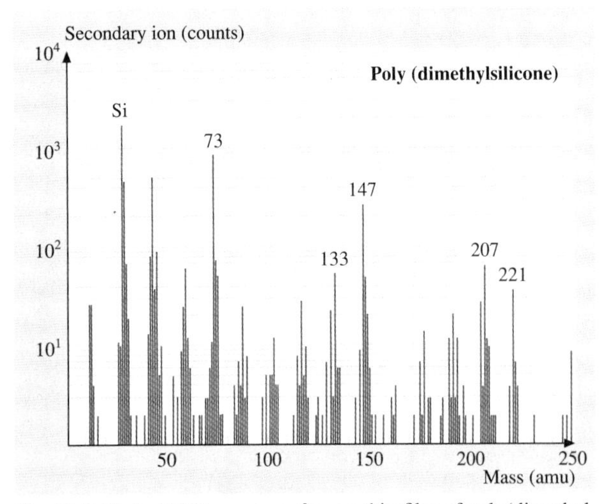

Fig. 18.5 Static SIMS spectrum from a thin film of poly(dimethylsilicone)

Figure 18.4 shows a schematic diagram of the basis of these SIMS techniques. Ions from the ion gun bombard the sample surface and transfer momentum to the atoms in the sample creating a collision cascade. This cascade distributes the energy of the incoming ion amongst the atoms in the sample, causing them to be displaced from their original sites, possibly breaking some bonds between atoms whilst creating others. The bombarding ion also can become implanted into the target material, modifying the chemistry of the material in the process. None of this is peculiar to SIMS – it happens in any sputtering process, be it in Auger depth-profiling, GDMS analysis or sputter deposition systems. It is probable that some of the energy deposited in the sample will cause atoms or molecules to be ejected from the surface of the sample, and those atoms or molecules that leave the surface as ions can be collected and detected by the mass spectrometer.

In the early stages of sputtering, the atoms and molecules sputtered from the surface will originate from areas of virgin surface – in other words from sites that have not yet been damaged by the primary ion – and thus carry with them information about the chemistry of the outer molecular layers of the sample. As the sputtering process proceeds, the probability of the sputtered particles being emitted from an area that has been modified by earlier ion impacts increases. Thus, at the start of the process the sputtered particles are characteristic of the virgin surface, but they will eventually become characteristic of the ion beam-modified surface.

Static SIMS is concerned with the measurement of the sputtered molecules produced at the start of the process, where information about the surface chemistry of the sample can be obtained. In SSIMS a very low dose of primary ions is used, typically less than 10^{13} primary ions per square centimeter, so that there is a very low probability that the ions that are detected come from damaged material. Used in this way, SIMS can be used to identify organic contamination on surfaces and obtain information about the molecular structure of the sample. Figure 18.5 shows a typical SSIMS mass spectrum from poly(dimethyl silicone), a common surface contaminant, which can be recognized by prominent peaks in the mass spectrum at masses 73, 133, 147, 207 and 221, which originate from fragmentation of the parent molecule $(CH_3-Si(CH_3)_2-O-Si(CH_3)_2-O-\ldots -Si(CH_3)_2-CH_3)$.

Whilst the abundance of molecular fragments that can be produced by the sputtering process is of great value for revealing chemical information about the sample surface in the SSIMS context, it can also be something of a problem when carrying out elemental analysis of trace impurities in simple matrices. For example, the spectrum shown in Fig. 18.6a, from an unknown Ga-based material, illustrates the number and complexity of molecular species produced by the SIMS process, which can make interpretation difficult. Fortunately, the distribution of the number of secondary ions with energy is different between atomic ions and molecular ions. The atomic ions, in general, have a broader energy distribution than the molecular ions, and its energy distribution tends to become sharper as the complexity of the molecular ion increases. For example, Fig. 18.7 shows the ion energy distributions of the Si^+, Si_2^+ and Si_3^+ ions, illustrating the narrower energy distributions of the molecular ions. Thus, by selecting the ion energy range from which the spectrum is recorded, the molecular information can be suppressed, allowing the elements present in the material to be identified, see Fig. 18.6b.

The intensities of the peaks in the SIMS mass spectra reveal more about the relative ionization probability and instrument transmission function than the number of species of that mass on the sample surface. The signals that are measured in SIMS are proportional to the numbers of ions produced, and this is not simply related to the amount of material present. The degree of ionization can vary enormously from element to element and matrix to matrix. For example, the numbers of positive ions formed by the inert gas bombardment of clean metal samples can be several orders of magnitude lower than the numbers produced from oxidized surfaces of the same metals under identical bombardment conditions. Clearly it is not the concentration of the metal atoms on the surface that is important, as there will be fewer metal atoms in the oxide layer than in the pure metal. It is the presence of oxygen that increases the ionization probability for material leaving the surface as positive ions.

In SSIMS, the chemical nature of the primary ion species is of relatively little consequence in terms of its effect on the number of ions produced by the surface, but in DSIMS, where it is the deliberately modified surface that is of interest, oxygen ion beams are widely used for analyses of electropositive species. Figure 18.8 shows how the silicon matrix signal from a silicon wafer with a native oxide layer behaves as the sputtering process proceeds. Initially there is a strong signal from silicon as the surface oxide layer is still present on the sample surface. The signal then falls as the oxide layer is sputtered away, and then slowly recovers in intensity as oxygen from the primary ion beam is implanted into the silicon surface. Eventually a steady state is reached where the material that is being sputtered is constant-composition oxygen-implanted silicon. The thickness of this transient region will depend upon the energy and angle of incidence of the primary oxygen ion beam. The higher the ion beam energy, the thicker the transient region. In the transient region the sputtering rate of the material and the ionization probability may well be changing, and precise quantification in this part of the

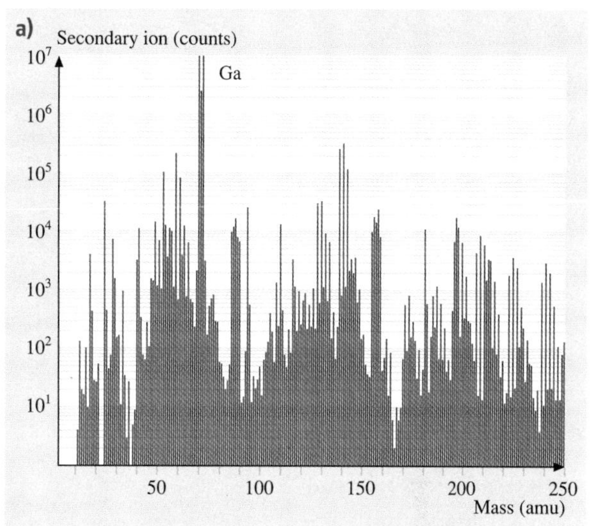

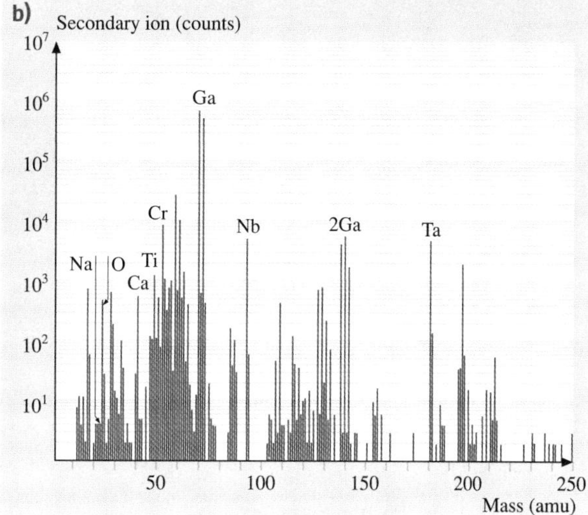

Fig. 18.6 SIMS mass spectrum from an unknown sample *top panel* without energy filtering, *bottom panel* with energy filtering

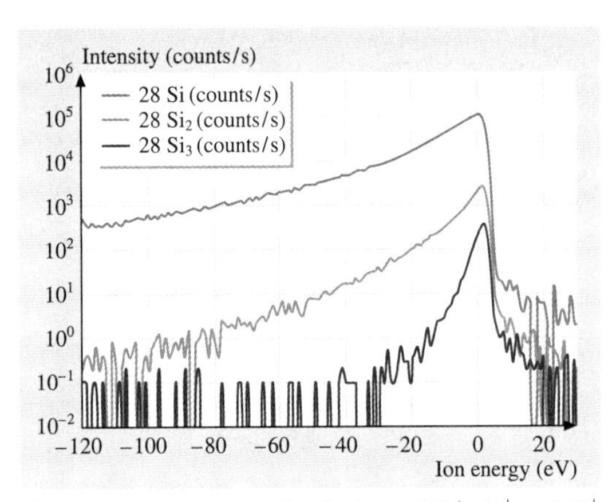

Fig. 18.7 The ion energy distributions of Si^+, Si_2^+ and Si_3^+ ions

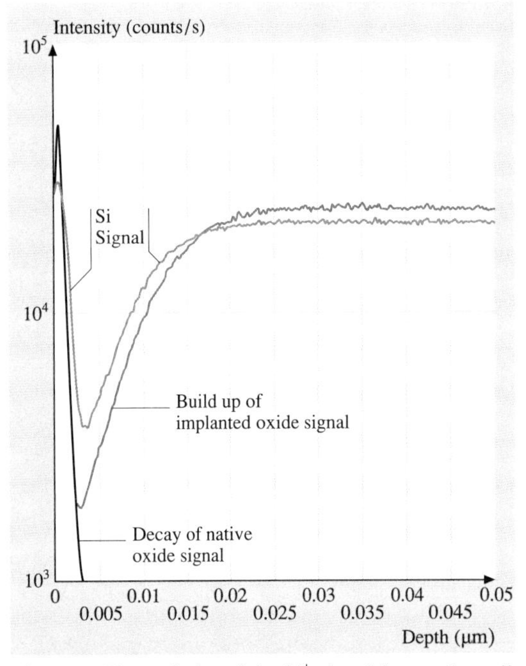

Fig. 18.8 The evolution of the Si^+ signal from a clean silicon wafer as a function of sputtering time with an oxygen primary ion beam, showing the decay of the surface oxide signal and the build-up of the implanted oxide layer

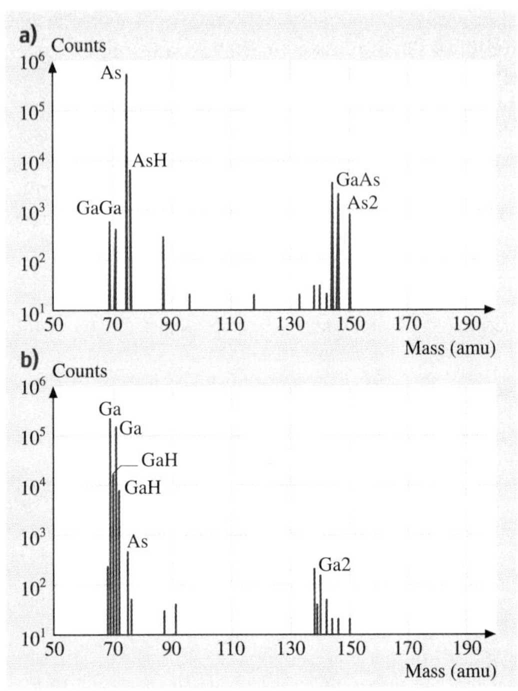

Fig. 18.9a,b Mass spectra from a GaAs wafer recorded (a) using Cs^+ primary ion bombardment and negative secondary ion detection, and (b) using O_2^+ primary ion bombardment and positive secondary ion detection

profile may be difficult. This is of particular importance when the features of interest are very close to the surface and within the transient region. The technological importance of shallow structures has driven the development of very low energy primary ion beam columns to enable the characterization of shallow implants.

Clearly, as is illustrated in Fig. 18.8, the number of positive ions produced in the sputtering process depends upon the concentration of oxygen in the sample surface. In order to maximize the positive secondary ion yield, the sample surface needs to be saturated with oxygen, and this can be achieved with normal-incidence primary ion beams or by deliberately allowing oxygen to flood the sample surface as well as using an oxygen primary ion beam. In the example shown in Fig. 18.8, the primary ion beam was incident at approximately 45° and so complete oxidation of the silicon surface was not achieved, meaning that the silicon and oxygen signals in the equilibrium region are not as high as from the fully oxidized native oxide layer.

Not all species prefer to form positive ions in the sputtering process; some prefer to produce negative ions. Oxygen is a case in point, and indeed it would be difficult to conduct an analysis for oxygen using an oxygen beam. For electronegative species, it turns out that high

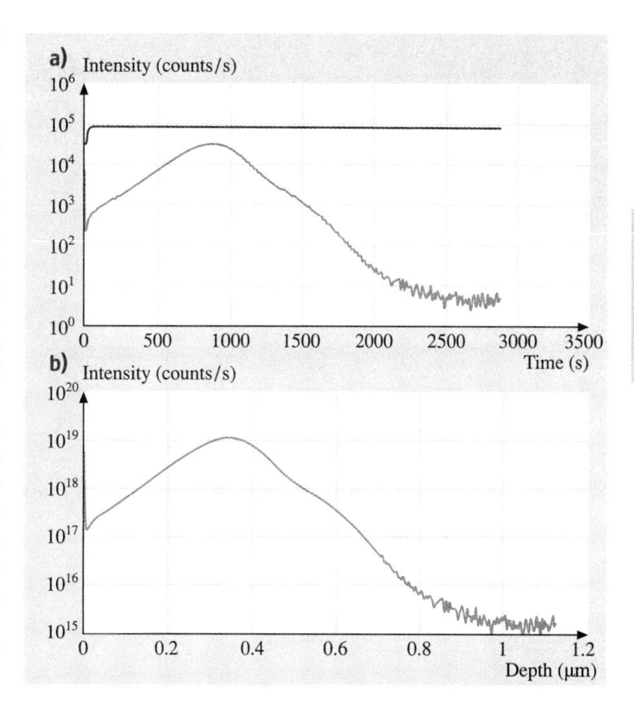

Fig. 18.10a,b Reference sample of boron ion-implanted silicon, (**a**) raw data profile and (**b**) quantified depth profile

negative secondary ion yields can be achieved if the surface is sputtered using cesium ions. An illustration of the relative ion yields of gallium and arsenic as positive and negative ions from a gallium arsenide surface with oxygen and cesium ion bombardment are shown in Fig. 18.9. Some species can be reluctant to produce either positive or negative secondary ions, but in some cases (such as zinc), good sensitivity can be achieved by using cesium ion bombardment and monitoring the cesium–element molecular ion, $CsZn^+$ in the case of zinc.

Despite the wide variations in secondary ion yields from element to element and matrix to matrix, the quantification of impurities in semiconductors can be relatively straightforward when reference materials are used. For example, Fig. 18.10a shows a raw data depth profile of ^{11}B ions implanted in silicon; knowing the areal dose of ions implanted into the sample and the depth of the crater sputtered into the sample, it is a relatively simple task (usually buried in the instrument software) to convert boron counts into concentrations

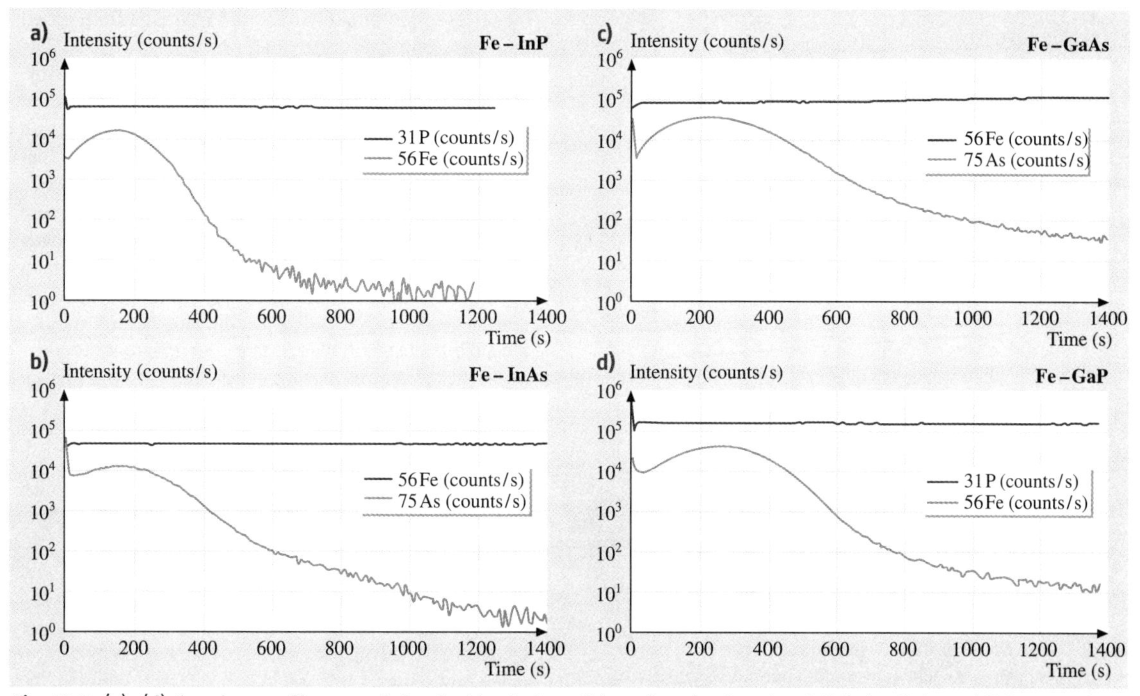

Fig. 18.11 (**a**)–(**d**) show iron profiles, recorded under identical conditions, from implants into InP, InAs, GaAs and GaP, respectively

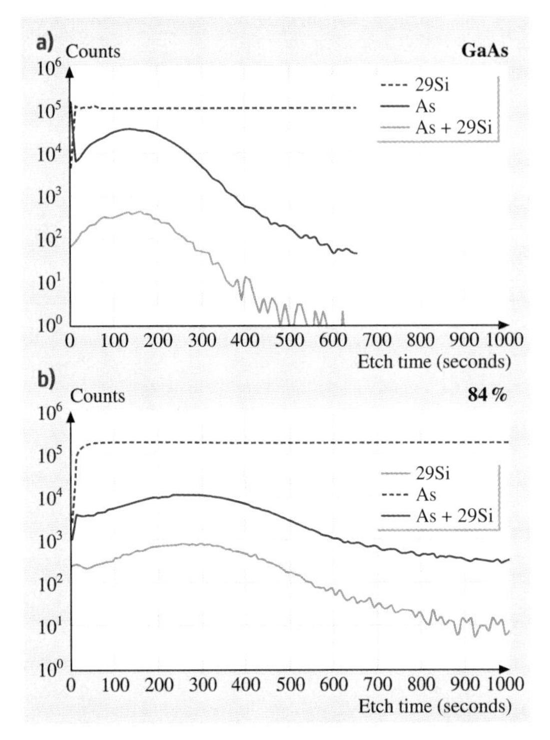

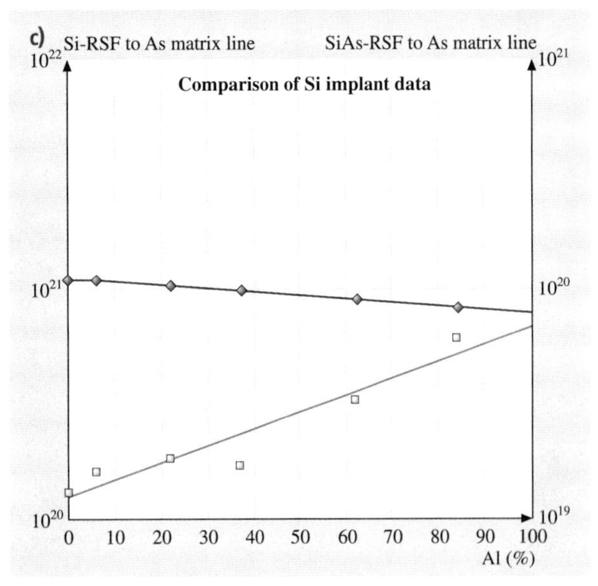

Fig. 18.12a–c Depth profiles of silicon as Si^- and $AsSi^-$ in (**a**) GaAs and (**b**) $Ga_{0.16}Al_{0.84}As$, and (**c**) variation in relative sensitivity factor as a function of aluminium content

(Fig. 18.10b). By recording a signal from the matrix element, either during the profile as in Fig. 18.10a, or in the crater after the profile, a relative sensitivity factor for the analyte (boron) in the matrix of interest (silicon) can be obtained which can then be used to determine the concentration of that analyte in another sample of the same matrix. Errors or uncertainties can arise from both the crater depth measurement and the implanted dose measurement. At the time of writing, there are only three metrologically traceable reference samples for SIMS (boron, arsenic and phosphorus in silicon) produced by NIST in the USA. For other species and other matrices, the analyst must rely upon locally produced materials with no traceability.

Care must be exercised when the analyte of interest is present in different matrices. For example, Figs. 18.11a–c and d show iron profiles, recorded under identical conditions (positive ion detection with O_2^+ primary ion bombardment), from implants into InP, InAs, GaAs and GaP produced in the same implant run. While the relative sensitivity of iron to the phosphorus matrix signal is the same for InP and GaP matrices, there is a variation of the order of 20% in the arsenide matrices. The useful ion yield of iron (the total number of ions produced as a fraction of the number of atoms present) increases approximately four-fold in the gallium-based matrices compared to the indium-based matrices. The profile in the GaAs sample, Fig. 18.11c, also shows a further complication: the arsenic matrix signal increases at about 1000 s into the profile. This could be interpreted as an upwards drift in the primary ion beam current, but it is actually caused by roughening of the GaAs surface as a result of the sputtering process. This effect is more common with oxygen primary ion bombardment than with cesium bombardment, and it has been shown that it is possible to reduce such effects by rotating the sample during analysis. Not only does the roughening modify the analytical signal, it also causes an apparent broadening of sharp features present in the sample as the surface roughness is convoluted with the feature width. The problems associated with sputter-induced roughness can become more severe when dealing with metal films but, once again, can be overcome by sample rotation.

Another example of how relative sensitivity factors vary with the matrix is shown in Fig. 18.12. Here a comparison is shown between the behavior of the Si^- and $AsSi^-$ signals as the matrix is changed from GaAs to $Ga_{0.16}Al_{0.84}As$. Molecular signals such as $AsSi^-$ are sometimes used to give increased signal compared to the atomic signal but, as in the example shown, the backgrounds are often similarly increased. What is inter-

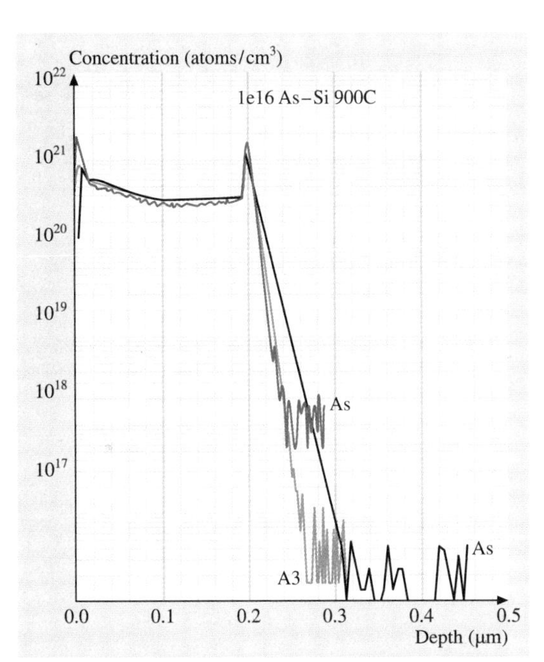

Fig. 18.13 Depth profiles of As in Si for different analytical conditions

esting here is that whilst the ion yields from the Si^- and As^- signals increase, but at different rates with increasing Al content, the $AsSi^-$ signals decrease. Ideally the analyst needs to be aware of, and able to take account of, such variations in response. Simply using the relative sensitivity factor derived for a GaAs matrix and applying this to high Al content material using a point-by-point normalization to the As^- signal can lead to a 30% overestimate of the silicon content. However, monitoring the $AsSi^-$ signal instead and using the same normalization procedure can underestimate the silicon content by over 350%.

With an ideal analytical technique the analytical signal will vary linearly with concentration over all concentrations. With SIMS, for uniform matrices, the measured signal is indeed linear with concentration for dilute systems. However, once the analyte concentration begins to exceed a few percent of the material, the signal may begin to vary in a nonlinear way with concentration. Thus, whilst SIMS is the ideal technique for quantifying minor and trace contaminants, it is less than ideal for determining matrix element concentrations. The most successful method of measuring elemental concentrations at high levels was achieved by monitoring the cesium molecular species ($CsAl^+$ and $CsGa^+$ and $CsAs^+$) in GaAlAs. In GaAlAs the ($CsAl^+/CsAs^+$) and ($CsGa^+/CsAs^+$) ratios do vary linearly with the Al content.

Returning to practical analysis, another factor that needs to be considered in relation to DSIMS, and one that is related to the primary bombarding particles, is atomic mixing. The primary ions used to bombard the surface transfer momentum to the atoms in the sample surface. While some of the atoms will be sputtered from the surface, others will be driven deeper into the sample, thus broadening any concentration distributions that were present within the original sample. The higher the primary ion beam energy, the more severe the atomic mixing. Consequently, as the need to analyze shallower and shallower structures has increased, there has been a move to develop instruments that use lower bombardment energies and sources producing polyatomic ions. A simple example of how the analytical conditions can cause profile distortion is shown in Fig. 18.13. These profiles were recorded from an arsenic-implanted and annealed polysilicon layer on a thin oxide on silicon. Using an oxygen primary ion beam striking the surface at oblique incidence with an impact energy of 5.5 keV (a) produces a profile that shows a spike followed by a rapid fall at the polysilicon/oxide interface. Using a cesium primary ion beam striking the surface at close to normal incidence with an impact energy of 14.5 keV (b) gives a much better detection limit with smaller interface spikes but with a much less abrupt fall in the arsenic signal. Profile (c), which uses the same bombardment conditions as profile (a) but combined with oxygen flooding of the sample surface (and consequently took longer to acquire), shows good agreement with profile (b) in terms of the size of the interface spike – which is exaggerated in profile (a) because of the increased ion yield from the oxide layer compared to the polysilicon – and good agreement with profile (a) in terms of the decay of the arsenic signal in the interface region, and has better sensitivity than profile (a) but less than profile (b). These profiles demonstrate the trade-off that the analyst must make when choosing analysis conditions.

18.4 Conclusion

The various surface chemical analysis techniques have their own strengths and weaknesses. No one method is suitable for all of the tasks the analyst faces; sometime one technique is sufficient to address the problem at hand, sometimes a combination of them is required. However, the approach should be successful if the technique(s) is (are) fit for the purpose of the task.

19. Thermal Properties and Thermal Analysis: Fundamentals, Experimental Techniques and Applications

The chapter provides a summary of the fundamental concepts that are needed to understand the heat capacity C_P, thermal conductivity κ, and thermal expansion coefficient α_L of materials. The C_P, κ, and α of various classes of materials, namely, semiconductors, polymers, and glasses, are reviewed, and various typical characteristics are summarized. A key concept in crystalline solids is the Debye theory of the heat capacity, which has been widely used for many decades for calculating the C_P of crystals. The thermal properties are interrelated through Grüneisen's theorem. Various useful empirical rules for calculating C_P and κ have been used, some of which are summarized. Conventional differential scanning calorimetry (DSC) is a powerful and convenient thermal analysis technique that allows various important physical and chemical transformations, such as the glass transition, crystallization, oxidation, melting etc. to be studied. DSC can also be used to obtain information on the kinetics of the transformations, and some of these thermal analysis techniques are summarized. Temperature-modulated DSC, TMDSC, is a relatively recent innovation in which the sample temperature is ramped slowly and, at the same time, sinusoidally modulated. TMDSC has a number of distinct advantages compared with the conventional DSC since it measures the complex heat capacity. For example, the glass-transition temperature T_g measured by TMDSC has almost no dependence on the thermal history, and corresponds to an almost step life change in C_P.

The new Tzero DSC has an additional thermocouple to calibrate better for thermal lags inherent in the DSC measurement, and allows more accurate thermal analysis.

The selection and use of electronic materials, one way or another, invariably involves considering such thermal properties as the specific heat capacity (c_s), thermal conductivity (κ), and various thermodynamic and structural transition temperatures, for example, the melting or fusion temperature (T_m) of a crystal, glass transformation (T_g) and crystallization temperature (T_c) for glasses and amorphous polymers. The thermal expansion coefficient (α) is yet another important material property that comes into full play in applications of electronic materials inasmuch as the thermal expansion mismatch is one of the main causes of electronic device failure. One of the most important thermal characterization tools is the differential scanning calorimeter (DSC), which enables the heat capacity, and various structural transition temperatures to be determined. Modulated-temperature DSC in which the sample temperature is modulated sinusoidally while being slowly ramped is a recent powerful thermal analysis technique that allows better thermal characterization and heat-capacity measurement. In addition, it can be used to measure the thermal conductivity. The present review is a selected overview of thermal properties and the DSC technique, in particular MTDSC. The overview is written from a materials science perspective with emphasis on phenomenology rather than fundamental physics.

The thermal properties of a large selection of materials can be found in various handbooks [19.1, 2]. In the case of semiconductors, *Adachi*'s book is highly recommended [19.3] since it provides useful relationships between the thermal properties for various group IV, II–V and II–VI semiconductors.

19.1 Heat Capacity

19.1.1 Fundamental Debye Heat Capacity of Crystals

The heat capacity of a solid represents the increase in the enthalpy of the crystal per unit increase in the temperature. The heat capacity is usually defined either at constant volume or at constant pressure, C_V and C_P, respectively. C_V represents the increase in the internal energy of the crystal when the temperature is raised because the heat added to the system increases the internal energy U without doing mechanical work by changing the volume. On the other hand C_P represents the increase in the enthalpy H of the system per unit increase in the temperature. Thus,

$$C_V = \left(\frac{\partial H}{\partial T}\right)_V = \left(\frac{\partial U}{\partial T}\right)_V \quad \text{and} \quad C_P = \left(\frac{\partial H}{\partial T}\right)_P . \tag{19.1}$$

The exact relationship between C_V and C_P is

$$C_V = C_P - \frac{T\alpha^2}{\rho K} , \tag{19.2}$$

where T is the temperature, ρ is the density, α is the linear expansion coefficient and K is the compressibility. For solids, C_V and C_P are approximately the same. The increase in the internal energy U is due to an increase in the energy of lattice vibrations. This is generally true for all solids except metals at very low temperatures where the heat capacity is due to the conduction electrons near the Fermi level becoming excited to higher energies. For most practical temperature ranges of interest, the heat capacity of most solids is determined by the excitation of lattice vibrations. The molar heat capacity C_m is the increase in the internal energy U_m of a crystal of Avogadro's number N_A atoms per unit increase in the temperature at constant volume, that is, $C_m = (dU_m/dT)_V$. The Debye heat capacity is still the most successful model for understanding the heat capacity of crystals, and is based on the thermal excitation of lattice vibrations, that is phonons, in the crystal [19.4]; it is widely described as a conventional heat capacity model in many textbooks [19.5,6]. The vibrational mean energy at a frequency ω is given by

$$\bar{E}(\omega) = \frac{\hbar\omega}{\exp(\frac{\hbar\omega}{k_B T}) - 1} , \tag{19.3}$$

where k_B is the Boltzmann constant. The energy $\bar{E}(\omega)$ increases with temperature. Each phonon has an energy of $\hbar\omega$ so that the phonon concentration in the crystal increases with temperature. To find the internal energy due to all the lattice vibrations we must also consider how many vibrational modes there are at various frequencies. That is, the distribution of the modes over the possible frequencies: the spectrum of the vibrations. Suppose that $g(\omega)$ is the number of modes per unit frequency, that is, $g(\omega)$ is the vibrational density of states or modes. Then $g(\omega)d\omega$ is the number of vibrational states in the range $d\omega$. The internal energy U_m of all lattice vibrations for 1 mole of solid is,

$$U_m = \int_0^{\omega_{max}} \bar{E}(\omega) g(\omega) d\omega . \tag{19.4}$$

The integration is up to certain allowed maximum frequency ω_{max}. The density of states $g(\omega)$ for the lattice vibrations in a periodic three-dimensional lattice, in a highly simplified form, is given by

$$g(\omega) \approx \frac{3}{2\pi^2} \frac{\omega^2}{v^3} , \tag{19.5}$$

where v is the mean velocity of longitudinal and transverse waves in the solid. The maximum frequency is ω_{max} and is determined by the fact that the total number of modes up to ω_{max} must be $3N_A$. It is called the *Debye frequency*. Thus, integrating $g(\omega)$ up to ω_{max} we find,

$$\omega_{max} \approx v(6\pi^2 N_A)^{1/3} . \tag{19.6}$$

This maximum frequency ω_{max} corresponds to an energy $\hbar\omega_{max}$ and to a temperature T_D defined by,

$$T_D = \frac{\hbar\omega_{max}}{k} , \tag{19.7}$$

and is called the *Debye temperature*. Qualitatively, it represents the temperature above which all vibrational frequencies are executed by the lattice waves.

Thus, by using (19.3) and (19.5) in (19.4) we can evaluate U_m and hence differentiate U_m with respect to temperature to obtain the molar heat capacity at constant volume,

$$C_m = 9R\left(\frac{T}{T_D}\right)^3 \int_0^{T_D/T} \frac{x^4 e^x \, dx}{(e^x - 1)} , \tag{19.8}$$

which is the well-known Debye heat capacity expression.

Figure 19.1 represents the constant-volume Debye molar heat capacity C_m for a perfect crystal (19.8) as a function of temperature, normalized with respect to the Debye temperature. The well-known classical *Dulong–Petit rule* ($C_m = 3R$) is only obeyed when $T > T_D$. The Dulong–Petit (DP) rule is a direct consequence of the applications of Maxwell's theorem of equipartition of energy and the classical kinetic molecular theory to vibrations of atoms in a crystal. Notice that C_m at $T = 0.5T_D$ is $0.825(3R)$, whereas at $T = T_D$ it is $0.952(3R)$. For most practical purposes, C_m is within 6% of $3R$ when the temperature is $0.9T_D$. For example, for copper $T_D = 315$ K and above about $0.9T_D$, that is above 283 K (or 10 °C), $C_m \approx 3R$, as borne out by experiments. Table 19.1 provides typical values for T_D, and heat capacities for a few selected elements. At the lowest temperatures, when $T \ll T_D$, (19.8) predicts that $C_m \propto T^3$ and this is indeed observed in low-temperature heat capacity experiments on a variety of crystals. On the other hand, well-known exceptions are glasses, noncrystalline solids, whose heat capacity is proportional to $a_1 T + a_2 T^3$, where a_1 and a_2 are constants (e.g. [19.6, 7]).

It is useful to provide a physical picture of the Debye model inherent in (19.8). As the temperature increases from near zero, the increase in the crystal's vibrational energy is due to more phonons being created and higher frequencies being excited. The phonon concentration increases as T^3 and the mean phonon energy increases as T. Thus, the internal energy increases as T^4. At temperatures above T_D, increasing the temperature creates more phonons but does not increase the mean phonon energy and does not excite higher frequencies. All frequencies up to ω_{max} have been now excited. The internal

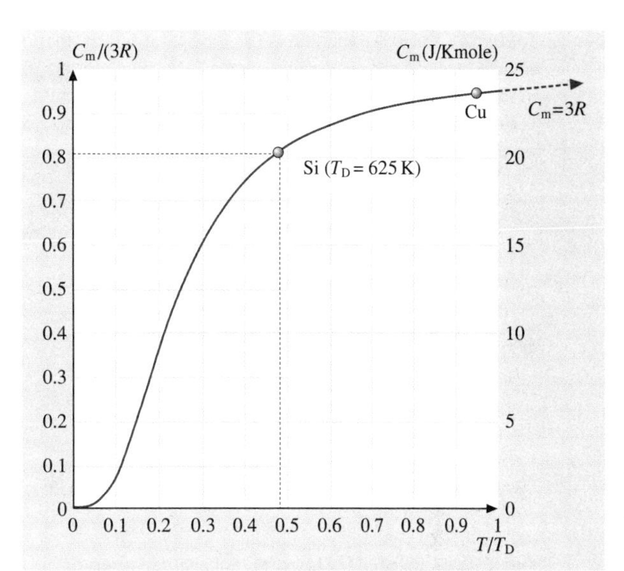

Fig. 19.1 Debye constant-volume molar heat capacity curve. The dependence of the molar heat capacity C_m on temperature with respect to the Debye temperature: C_m versus T/T_D. For Si, T_D = 625 K, so at room temperature (300 K), $T/T_D = 0.48$ and C_m is only 0.81 ($3R$)

energy increases only due to more phonons being created. The phonon concentration and hence the internal energy increases as T; the heat capacity is constant, as expected from (19.8).

In general terms, elements with higher atomic mass tend to have lower T_D, and their molar heat capacities are closer to the DP limit. Usually T_D is taken to depend on the temperature to accurately model the experimental C_P–T behavior; the dependence of T_D on T is only important at low temperatures, and can be neglected in most practical applications in engineering.

There are two important assumptions in the Debye theory. First is that all thermal excitations are lattice vibrations (i.e., involving phonons), which is usually the case in solids. Secondly, the interatomic potential-energy (PE) curve as a function of interatomic displacement x from equilibrium, is assumed to be parabolic, that is *harmonic* or symmetrical: $PE = PE_{min} + (1/2)\beta x^2$, where β is a constant (the spring constant in the equation of net force versus displacement, $F = -\beta x$). Within the harmonic approximation, the lattice vibrations are *plane waves* that are independent of each other; they are vibrational normal modes of the lattice. Further, a symmetrical PE curve does not allow thermal expansion, which means that C_V

Table 19.1 Debye temperatures (T_D), heat capacities, thermal conductivities and linear expansion coefficients of various selected metals and semiconductors. C_m, c_s, κ, and α are at 25 °C. For metals, T_D is obtained by fitting the Debye curve to the experimental molar heat capacity data at the point $C_m = \frac{1}{2}(3R)$. T_D data for metals from [19.8]. Other data from various references, including [19.2] and the Goodfellow metals website

Metals	Ag	Al	Au	Bi	Cu	Ga	Hg	In	Pd	W	Zn
T_D (K)	215	394	170	120	315	240	100	129	275	310	234
C_m (J/K mol)	25.6	24.36	25.41	25.5	24.5	25.8	27.68	26.8	25.97	24.45	25.44
c_s (J/K g)	0.237	0.903	0.129	0.122	0.385	0.370	0.138	0.233	0.244	0.133	0.389
κ (W/m K)	420	237	317	7.9	400	40.6	8.65	81.6	71.8	173	116
α $(K^{-1}) \times 10^{-6}$	19.1	23.5	14.1	13.4	17	18.3	61	24.8	11	4.5	31
Semiconductors	**Diamond**	**Si**	**Ge**	**AlAs**	**CdSe**	**GaAs**	**GaP**	**InAs**	**InP**	**ZnSe**	**ZnTe**
T_D (K)	1860	643	360	450	135	370	560	280	425	340	260
C_m (J/K mol)	6.20	20.03	23.38	43.21	53.77	47.3	31.52	66.79	46.95	51.97	49.79
c_s (J/K g)	0.540	0.713	0.322	0.424	0.281	0.327	0.313	0.352	0.322	0.360	0.258
κ (W/m K)	1000	156	60	91	4	45	77	30	68	19	18
α $(K^{-1}) \times 10^{-6}$	1.05	2.62	5.75	4.28	7.43	6.03	4.89	5	4.56	7.8	8.33

and C_P are identical. However, the actual interatomic PE is anharmonic, that is, it has an additional x^3 term. It is not difficult to show that in this case the vibrations or phonons interact. For example, two phonons can *mix* to generate a third phonon of higher frequency or a phonon can decay into two phonons of lower frequency etc. Further, the anharmonicity also leads to thermal expansion, so that C_V and C_P are not identical as is the case in the Debye model. As a result of the anharmonic effects, C_P continues to increase with temperature beyond the $3R$ Dulong–Petit rule, though the increase with temperature is usually small.

19.1.2 Specific Heat Capacity of Selected Groups of Materials

Many researchers prefer to quote the heat capacity for one mole of the substance, that is quote C_m, and sometime express C_m in terms of R. The limit $C_m = 3R$ is the DP rule. It is not unusual to find materials for which C_m can exceed the $3R$ limit at sufficiently high temperatures for a number of reasons, as discussed, for example, by *Elliott* [19.6]. Most applications of electronic materials require a knowledge of the specific heat capacity c_s, the heat capacity per unit mass. The heat capacity per unit volume is simply c_s/ρ, where ρ is the density. For a crystal that has only one type of atom with an atomic mass M_{at} (g/mol) in its unit cell (e.g. Si), c_s is C_m/M_{at} expressed in J/K g.

While the Debye heat capacity is useful in predicting the molar heat capacity of a crystal at any temperature, there are many substances, such as metals, both pure metals and alloys, and various semiconductors (e.g. Ge, CdSe, ZnSe etc.) and ionic crystals (e.g. CsI), whose room-temperature heat capacities approximately follow the simple DP rule of $C_m = 3R$, the limiting value in Fig. 19.1. For a metal alloy, or a compound such as $A_xB_yC_z$, that is made up of three components A, B and C with molar fractions x, y and z, where $x+y+z=1$, the overall molar heat capacity can be found by adding individual molar heat capacities weighted by the molar fraction of the component,

$$C_m = xC_{mA} + yC_{mB} + xC_{mA} , \tag{19.9}$$

where C_{mA}, C_{mB} and C_{mA} are the individual molar heat capacities. Equation (19.9) is *the additive rule of molar heat capacities*. The corresponding specific heat capacity is

$$c_s = 3R/\bar{M}_{at} , \tag{19.10}$$

where $\bar{M}_{at} = xM_A + yM_B + zM_C$ is the mean atomic mass of the compound, and M_A, M_B and M_C are the atomic masses of A, B and C. For example, for ZnSe, the average mass $\bar{M}_{at} = (1/2)(78.96 + 65.41) = 72.19$ g/mol, and the DP rule predicts $c_s = 3R/\bar{M}_{at} = 0.346$ J/Kg, which is almost identical to the experimental value at 300 K.

The modern Debye theory and the classical DP rule that $C_m = 3R$ are both based on the addition of heat increasing the vibrational energy of the atoms or molecules in the solid. If the molecules are able to rotate, as in certain polymers and liquids, then the molar heat capacity will be more than $3R$. For example, the heat capacity of molten Bi–BiI_2 mixture is more than the DP rule, and the difference is due to the rotational energies associated with various molecular units in the molten state [19.9].

The heat capacity of polymers are usually quite different than the expected simple Debye heat capacity behavior. Most polymers are not fully crystalline but either amorphous or semicrystalline, a mixture of amorphous and crystalline regions. Secondly, and most significantly, the polymer structures usually have a main chain (a backbone) and various side groups attached to this backbone. Stretching and possible bending vibrations and possible partial rotations of these side groups provide additional contributions to the overall heat capacity. The extent of freedom the main chain and side groups have in executing stretching and bending vibrations and rotations depends on whether the structure is above or below the glass-transition temperature T_g. (The glass-transition temperature is defined and discussed later.) Above T_g, the polymer structure is floppy and has sufficient free volume for various molecular motions to be able to absorb the added heat. The heat capacity is therefore larger above T_g than that below T_g, where the structure is frozen.

The heat capacity of most polymers increases with temperature. For most solid polymers (below the glass-transition temperature), the fractional rate of increase in the heat capacity with temperature follows

$$(1/C_P)(dC_P/dT) \approx 3 \times 10^{-3}\,K^{-1} \quad (T < T_g) \tag{19.11}$$

to within about 5%. This empirical rule is particularly useful for estimating C_P at a desired temperature from a given value at another temperature e.g. at room temperature. For liquid polymers (above T_g) on the other hand, $(1/C_P)(dC_P/dT)$ is less ($\approx 1 \times 10^{-3}\,K^{-1}$) and has a large variability (more than 30%) from polymer to polymer. There are various empirical rules for predicting the heat capacity of a polymer based on the backbone chain structure and types of side groups, as discussed by *van Krelen* and *Hoftyzer* [19.10]. The most popular is to simply sum heat capacity contributions per mole from each distinguishable molecular unit in the polymer structure's repeat unit. The heat capacity contribution from each type of identifiable molecular unit (e.g. $-CH_2-$ in the backbone or $-CH_3$ side group) in polymer structures are listed in various tables. For example, the molar heat capacity of polypropylene can be calculated approximately by adding contributions from three distinguishable units, CH_2, CH and CH_3 in the repeating unit that makes up the polypropylene structure, as illustrated schematically in Fig. 19.2. Once the contributions are added, the sum is 71.9 J/K mol. The molecular mass of the repeat unit corresponds to three C atoms and six H atoms, or to 42 g/mol so that the specific heat is 71.9 J/K mol/42 g/mol or 1.71 J/Kg, which is close to the experimental value. (Polypropylene is normally semicrystalline and its specific heat also depends on the crystalline-to-amorphous phase ratio.) The term *heat capacity per mole* in polymers normally refers to a mole of repeat units or segments in the polymer.

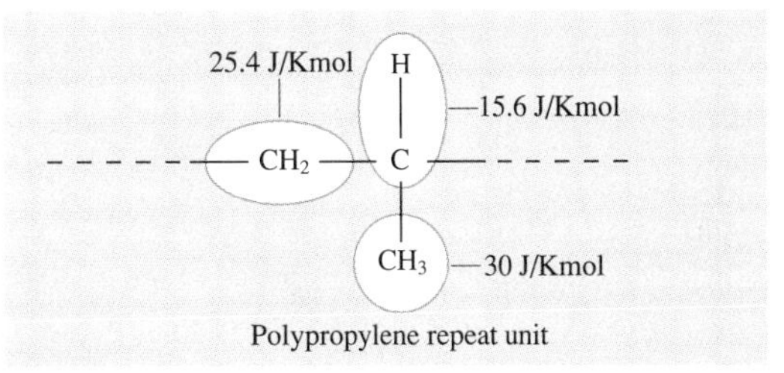

Fig. 19.2 Molar heat capacity of solid polypropylene calculated from various distinguishable units in the repeat unit

More rigorous characterization of the heat capacity of polymers tends to be complicated by the fact that most polymers are a mixture of the crystalline and amorphous phases, which have different C_P versus T characteristics, and the exact volume fractions of the phases may not be readily available. There have been a number of examples recently where the heat capacity characteristics of semicrystalline polymers have been critically examined [19.11]. A comprehensive overview of the heat capacity of polymers has been given by *Wunderlich* [19.12], which is highly recommended.

The overall or effective heat capacity of polymer blends can be usually calculated from the individual capacities and the volume fractions of the constituents. The simplest is a linear mixture rule in which individual capacities are weighted by the volume fraction and then summed.

The heat capacity of glasses depends on the composition and increases with temperature. There is a sharp change in the dependence of C_P on T through the glass-transition temperature T_g. For example, for zirconium, barium, lanthanum, aluminum, sodium (ZBLAN) glasses, C_m is 24 J/K mol below T_g and slightly less than $3R$. On the other hand, C_m is 38.8 J/K mol above T_g, and greater than $3R$. (The C_P change at T_g is discussed below.) The dependence of C_P on T below T_g has been modeled by a number of expressions dating back to the 1950s. For example, *Sharp* and *Ginther* [19.13] in 1951 proposed the following empirical expressions

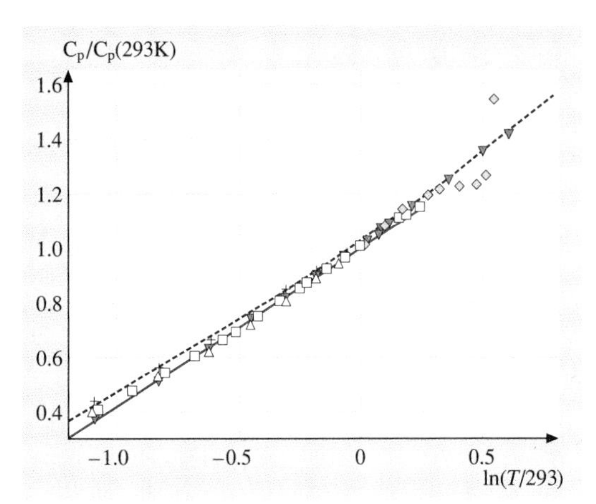

Fig. 19.3 Normalized heat capacity $C_P(T)/C_P(298)$ versus $\ln(T/298)$ for various glasses. *Open square:* $12.5Na_2O$, $12.5Al_2O_3$, $75SiO_2$; *solid circle:* 25MgO, 25CaO, $50SiO_2$; *open triangle:* $33Na_2O$, $67B_2O_3$; *solid triangle:* 24BaO, $1Fe_2O_3$, $75B_2O_3$; *open diamond:* B_2O_3 (includes T_g); $+30K_2O$, $70GeO_2$. Adapted from *Khalimovskaya-Churkina* and *Priven* [19.14]. The *solid straight line* is (19.14), whereas the *dashed line* is the experimental trend line

for the mean and true specific heat capacity c_m and c respectively, of oxide glasses in the range 0–1300 °C,

$$c_m = \frac{H_2 - H_1}{T_2 - T_1} = \frac{aT + c_o}{0.001\,46T + 1} \quad \text{and} \tag{19.12a}$$

$$c = \frac{dH}{dT} = \frac{aT + c_m}{0.001\,46T + 1} \tag{19.12b}$$

in which T is in °C, and a and c_o are calculated from the glass composition as described in the original paper [19.13]. The mean heat capacity in (19.12)a then predicts the heat capacity to within about 1%. More recently, for various oxide glasses from 300 K to T_g, *Inaba* et al. [19.16] have proposed an exponential empirical expression of the form,

$$C_P = 3R\left[1 - \exp\left(-1.5\frac{T}{T_D}\right)\right], \tag{19.13}$$

where T_D is the Debye temperature. Equation (19.13) predicts a lower C_P than the Debye equation itself at the same temperature for the same T_D. There are also other empirical expressions for C_P (for example, a three-term polynomial in T) that typically involve at least three fitting parameters. *Khalimovskaya-Churkina* and

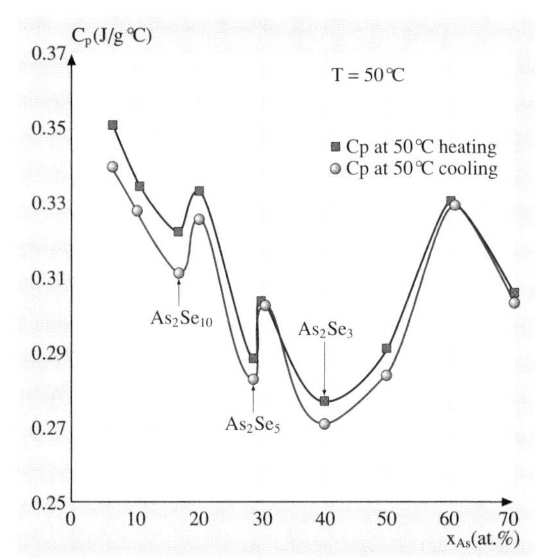

Fig. 19.4 The specific heat capacity versus As content in As_xSe_{1-x}. (After *Wagner* and *Kasap* [19.15])

Priven [19.14] have shown that a simple two-parameter empirical expression of the form

$$C_P = a + b\ln T\,, \tag{19.14}$$

where a and b are composition-dependent parameters, can be fruitfully used to represent the dependence of C_P on T from 100 K to the lower boundary of T_g for various oxide glasses to within about 2.5% error as shown in Fig. 19.3. The authors further show that, to within about 4%,

$$C_P = C_P(293\,\text{K})\,[1 + 0.52\ln(T/293)]\,. \tag{19.15}$$

What is significant in this equation is that the composition dependence of C_P has been incorporated into $C_P(293\,\text{K})$, and once the latter is calculated, the temperature dependence of C_P is then given by (19.15).

Various empirical rules have been proposed for approximately calculating the heat capacity of oxide glasses from its components. The simplest is the additive rule stated in (19.9), in which the overall heat capacity per mole is found from adding individual molar heat capacities of the constituents, so-called partial molar heat capacities, weighted by their molar fractions. However, for a heterogeneous mixture, that is for a solid in which there are two or more distinct phases present, the specific heat capacity may involve more complicated calculations.

Figure 19.4 shows the heat capacity of an As_xSe_{1-x} glass as the composition is varied. The specific heat capacity changes with composition and shows special features at certain critical compositions that correspond to the appearance of various characteristic molecular units in the glass, or the structure becoming optimally connected (when the mean coordination number $\langle r \rangle = 2.4$). With low concentrations of As, the structure is Se-rich and has a floppy structure, and the heat capacity per mole is about $3R$. As the As concentration is increased, the structure becomes more rigid, and the heat capacity decreases, and eventually at 40 at % As, corresponding to As_2Se_3, the structure has an optimum connectivity ($\langle r \rangle = 2.4$) and the heat capacity is minimum and, in this case, close to $2.51R$. There appears to be two minima, which probably correspond to As_2Se_{10} and As_2Se_5, that is to $AsSe_5$ and $AsSe_{5/2}$ units within the structure.

19.2 Thermal Conductivity

19.2.1 Definition and Typical Values

Heat conduction in materials is generally described by *Fourier's heat conduction law*. Suppose that J_x is the heat flux in the x-direction, defined as the quantity of heat flowing in the x-direction per unit area per unit second: the thermal energy flux. Fourier's law states that the heat flux at a point in a solid is proportional to the temperature gradient at that point and the proportionality constant depends on the material,

$$J_x = -\kappa \frac{dT}{dx} , \tag{19.16}$$

where κ is a constant that depends on the material, called the thermal conductivity (W/m K or W/m °C), and dT/dx is the temperature gradient. Equation (19.16) is called Fourier's law, and effectively defines the thermal conductivity of a medium. Table 19.2 provides an overview of typical values for the thermal conductivity of various classes of materials.

The thermal conductivity depends on how the atoms in the solid transfer the energy from the hot region to the cold region. In metals, the energy transfer involves the conduction electrons. In nonmetals, the energy transfer involves lattice vibrations, that is atomic vibrations of the crystal, which are described in terms of phonons.

The thermal conductivity, in general, depends on the temperature. Different classes of materials exhibit different κ values and dependence of κ on T, but we can generalize very roughly as follows:

Most pure metals: $\kappa \approx 50$–400 W/m K. At sufficiently high T, e.g. above ≈ 100 K for copper, $\kappa \approx$ constant. In magnetic materials such as iron and nickel, κ decreases with T.

Most metal alloys: κ lower than for pure metals; $\kappa \approx 10$–100 W/mK. κ increases with increasing T.

Most ceramics: Large range of κ, typically 10–200 W/m K with diamond and beryllia being exceptions with high κ. At high T, typically above ≈ 100 K, κ decreases with increasing T.

Most glasses: Small κ, typically less than ≈ 5 W/mK and increases with increasing T. Typical examples are borosilicate glasses, window glass, soda-lime glasses, fused silica etc. Fused silica is noncrystalline SiO_2 with $\kappa \approx 2$ W/mK.

Most polymers: κ is very small and typically less than 2 W/mK and increases with increasing T. Good thermal insulators.

19.2.2 Thermal Conductivity of Crystalline Insulators

In nonmetals heat transfer involves lattice vibrations, that is phonons. The heat absorbed in the hot region increases the amplitudes of the lattice vibrations which is the same as generating more phonons. These new phonons travel towards the cold regions and thereby transport the lattice energy from the the hot to cold region. The thermal conductivity κ measures the rate at which heat can be transported through a medium per unit area per unit temperature gradient. It is proportional to the rate at which a medium can absorb energy, that is, κ is proportional to the heat capacity. κ is also proportional to the rate at which phonons are transported which is determined by their mean velocity v_{ph}. In addition, of

Table 19.2 Typical thermal conductivities of various classes of materials at 25 °C

Pure metals	Nb	Sn	Fe	Zn	W	Al	Cu	Ag
κ (W/m K)	53	64	80	116	173	237	400	420
Metal alloys	Stainless steel	55Cu−45Ni	Manganin (86Cu−12Mn −2Ni)	70Ni−30Cu	1080 Steel	Bronze (95Cu−5Sn)	Brass (63Cu−37Zn)	Dural (95Al−4Cu −1Mg)
κ (W/m K)	12−16	19.5	22	25	50	80	125	147
Ceramics and glasses	Glass-borosilicate	Silica-fused (SiO_2)	S_3N_4	Alumina (Al_2O_3)	Magnesia (MgO)	Saphire (Al_2O_3)	Beryllia (BeO)	Diamond
κ (W/m K)	0.75	1.5	20	30	37	37	260	1000
Polymers	Poly-propylene	Polystyrene	PVC	Poly-carbonate	Nylon 6,6	Teflon	Polyethylene low density	Polyethylene high density
κ (W/m K)	0.12	0.13	0.17	0.22	0.24	0.25	0.3	0.5

course, κ is proportional to the mean free path Λ_{ph} that a phonon has to travel before losing its momentum, just as the electrical conductivity is proportional to the electron's mean free path. A rigorous classical treatment gives κ as,

$$\kappa = \tfrac{1}{3} C_V v_{ph} \Lambda_{ph} , \tag{19.17}$$

where C_V is the heat capacity per unit volume. The mean phonon velocity v_{ph} is constant and approximately independent of temperature so that κ is governed primarily by C_V and Λ_{ph}. The mean free path Λ_{ph} depends on various processes that can scatter the phonons and hinder their propagation along the direction of heat flow. Phonons can collide with other phonons, crystal defects, impurities and crystal surfaces.

At high temperatures, κ is controlled by phonon–phonon scattering. The phonon–phonon collisions that are normally responsible for limiting the thermal conductivity, that is scattering the phonon momentum in the opposite direction to the heat flow, are due to the anharmonicity (asymmetry) of the interatomic potential energy curve. Stated differently, the net force F acting on an atom is not simply βx (where β is a spring constant) but also has an x^2 term, i.e., it is nonlinear. The greater the asymmetry or nonlinearity, the larger is the effect of such momentum-flipping collisions. The same asymmetry that is responsible for thermal expansion of solids is also responsible for determining the thermal conductivity. When two phonons 1 and 2 interact in a crystal region as in Fig. 19.5, the nonlinear behavior and the periodicity of the lattice, cause a new phonon 3 to be generated. This new phonon 3 has the same energy as the sum of 1 and 2 but it is traveling in the wrong direction. (The frequency of 3 is the sum of frequencies of 1 and 2.) The flipping of the phonon momentum is called an *Umklapp process* and is a result of the anharmonicity in the interatomic bond. Figure 19.6 shows the dependence of κ on T for three semiconductors. We can identify three types of behavior.

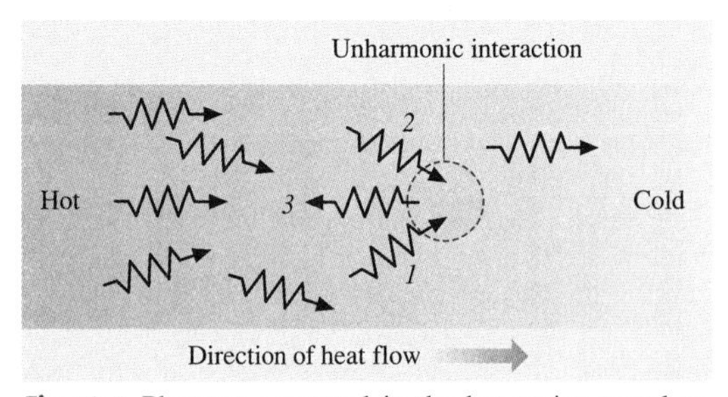

Fig. 19.5 Phonons generated in the hot region travel towards the cold region and thereby transport heat energy. Phonon–phonon anharmonic interaction generates a new phonon whose momentum is towards the hot region. (After [19.5])

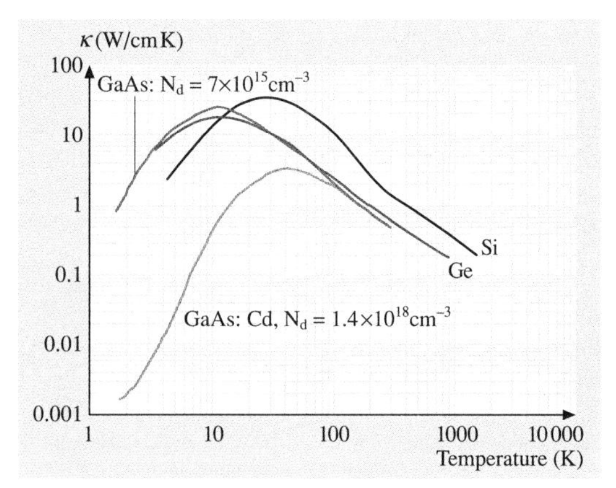

Fig. 19.6 Thermal conductivity of Si, Ge and GaAs. GaAs for two different levels of doping. Data extracted and replotted: Si [19.17]; Ge [19.18]; GaAs [19.19]

At low temperatures there are two factors that control κ. The phonon concentration is too low for phonon–phonon collisions to be significant. Instead, the mean free path Λ_{ph} is determined by phonon collisions with crystal imperfections, most significantly, crystal surfaces and grain boundaries. Thus, Λ_{ph} depends on the sample geometry and crystallinity. Further, as we expect from the Debye model, C_v depends on T^3 so that κ has the same temperature dependence as C_v, that is $\kappa \propto T^3$. Figure 19.6 shows the dependence of κ on T for Ge, Si and GaAs. Notice that κ increases with increasing T at low temperatures, and that κ also depends on the concentration of dopants (GaAs) or impurities in the crystal, which can scatter phonons. An extensive treatise on the thermal conductivity of semiconductors is given by *Bhandari* and *Rowe* [19.20]

As the temperature increases, the temperature dependence of κ becomes controlled by Λ_{ph} rather than C_v, which changes only slowly with T. The mean free path becomes limited by phonon–phonon collisions that obey the Umklapp process. These phonons have energies $\approx \frac{1}{2}k_B T_D$. The concentration of such phonons is therefore proportional to $\exp(T_D/2T)$. Since Λ_{ph} is inversely proportional to the Umklapp-obeying phonon concentration, κ decreases with increasing temperature, following a $\kappa \propto \exp(T_D/2T)$ type of behavior. At temperatures above the Debye temperature, C_v is, of course, constant, and the phonon concentration n_{ph} increases with temperature, $n_{ph} \propto T$. Thus, the mean free path decreases as $\Lambda_{ph} \propto 1/T$, which means that $\kappa \propto T^{-1}$, as observed for most crystals at sufficiently high temperatures. Figure 19.7 illustrates the expected dependence of κ on T for nonmetals, with three temperature regimes:

$$\kappa \propto \begin{cases} T^3 ; & \text{Low } T \\ \exp(-T_D/2T) ; & \text{Intermediate } T \\ T^{-1} ; & \text{High } T . \end{cases} \qquad (19.18)$$

Most crystalline semiconductors tend to follow this type of behavior. The thermal conductivity of important semiconductors has been discussed by *Adachi* [19.3]. In the temperature range above the peak of κ with T, Adachi used a $\kappa = AT^n$ type of power law to model the κ–T data, and has provided an extensive table for A and n for various group IV, III–V and II–V semiconductors [19.3]. For example, for GaAs, over $150 \le T \le 1500$, $A = 750$ and $n = -1.28$ so that $\kappa\,(\text{W/cm K}) = (750)T^{-1.28}$, where T is in Kelvin. Further, Adachi has been able to correlate κ with the lattice parameter a, mean atomic mass $\bar{M}$ in the unit cell and the Debye temperature T_D.

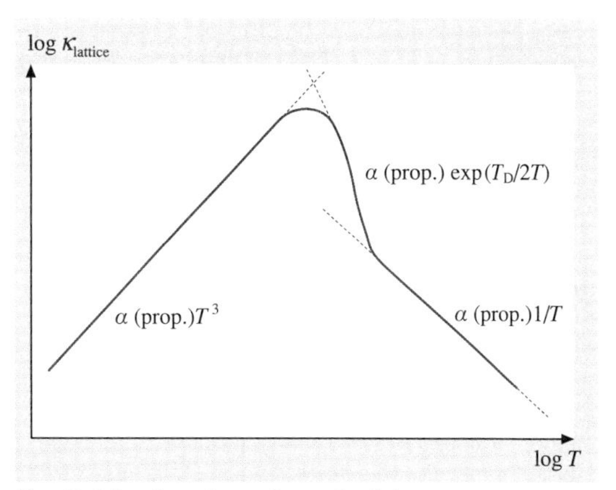

Fig. 19.7 Lattice thermal conductivity versus temperature (K) on a log–log plot. κ is due to phonons

19.2.3 Thermal Conductivity of Noncrystalline Insulators

Glasses

The thermal conductivity of glasses tends to be substantially smaller than their crystalline counterparts, and tends to increase with the temperature almost monotonically. The most striking difference at room temperature is that, as the temperature is decreased, κ for a glass decreases whereas for a crystal it increases. This dif-

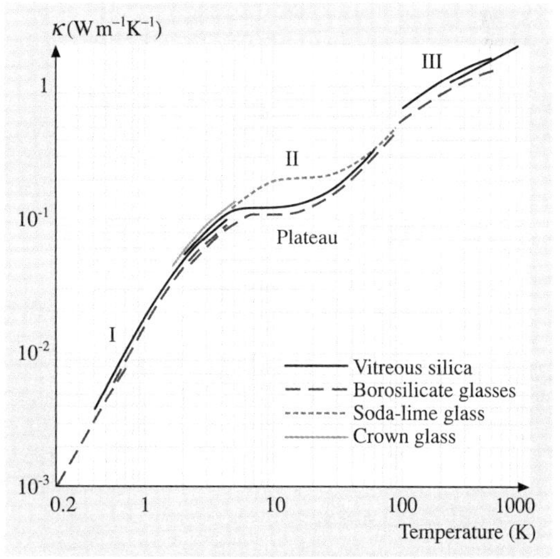

Fig. 19.8 The temperature dependence of κ for four types of glasses. Data in Figure 2 of *Zeller* and *Pohl* [19.21] replotted

ference has been qualitatively attributed to the phonon mean free path Λ_{ph} in a glass remaining roughly constant as the temperature is lowered and being limited by the range of disorder in the structure, which is a few atomic spacings [19.22, 23]. While intuitively this explanation seems reasonable, a more rigorous derivation excludes (19.17) since concepts like phonon velocity and wavelength tend to become meaningless over very short distances [19.24]. Most of the theories that have been proposed for describing κ more rigorously in the noncrystalline state are quite mathematical, and beyond the scope of this review based on materials science. (These models include the Kubo transport equation [19.25], hopping of localized vibrations [19.26], simulations based on molecular dynamics [19.27] etc.) Figure 19.8 shows κ versus T for a few selected glasses [19.21, 28]. Other glasses tend to follow this type of behavior in which there are three regions of distinct behavior. At very low temperatures (I in Fig. 19.8, below 1–2 K, $\kappa \approx T^n$ where $n = 1.8$–2.0. In the intermediate temperature regime, from about 2 K to about 20–30 K, κ exhibits a plateau. At high temperatures, above 20–30 K, beyond the plateau, κ increases with T, with a decreasing slope.

Polymers

Crystalline polymers tend to follow the general predictions in Fig. 19.7 that at high temperatures κ decreases with the temperature. As the degree of crystallinity decreases, κ also decreases, and eventually for amorphous polymers κ is substantially smaller than the crystalline phase, and κ increases with increasing temperature. If ρ_c and ρ_a are the densities of a given polymer in the crystalline (c) and amorphous (a) phase, then various empirical rules have been proposed to relate the conductivities of c- and a-phases through the ratio ρ_c/ρ_a of their densities. For example, in one empirical rule [19.29] $\kappa_c/\kappa_a = 1 + 1.58[(\rho_c/\rho_a) - 1]$. This means that, if we know κ_c, and ρ_c/ρ_a, we can calculate κ_a. If we know the density ρ of the partially crystalline polymer or the volume fraction v_a of the amorphous phase, then we can use an appropriate mixture rule to find the overall κ. (A solved problem is given as an example in chapter 17 of [19.10].) Figure 19.9 shows the dependence of κ on T for three different degrees of crystallinity in the structure; clearly κ versus T is strongly influenced by the crystallinity of the structure.

Figure 19.10 shows the temperature dependence of κ when it has been normalized to its value $\kappa(T_g)$ at the glass-transition temperature T_g against the normalized temperature, T/T_g for a number of amorphous polymers. While there are marked differences between various polymers, there is nonetheless an overall trend in which κ increases up to T_g and thereafter decreases with T. The solid curve in Fig. 19.10 indicates an overall trend. Notice how similar the κ values are.

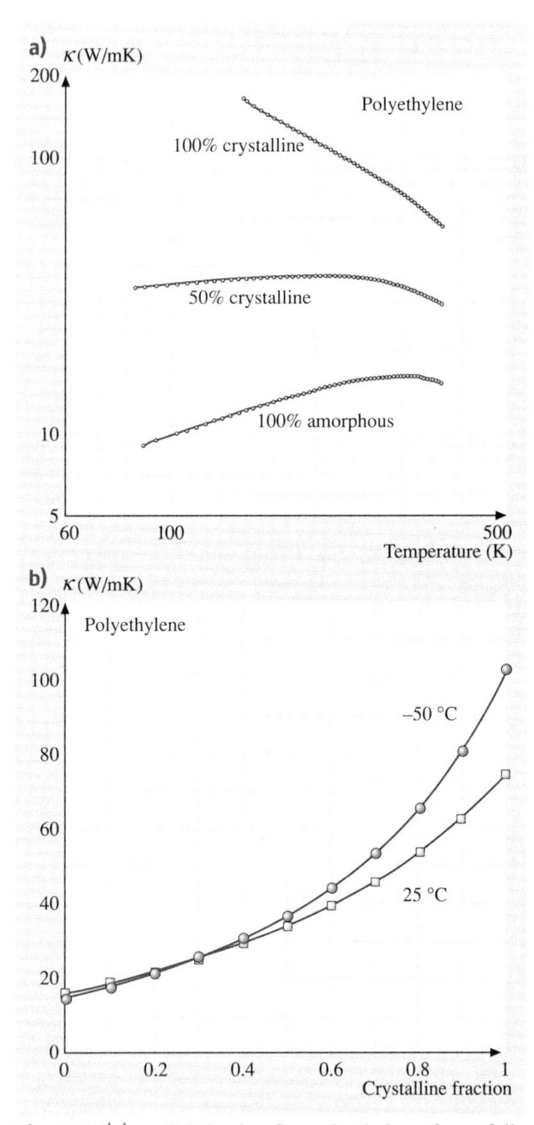

Fig. 19.9 (**a**) κ–T behavior for polyethylene for a fully crystalline, fully amorphous and 50% crystalline structures. (**b**) Dependence of κ on the crystallinity of polyethylene. Data extracted and replotted from [19.29]

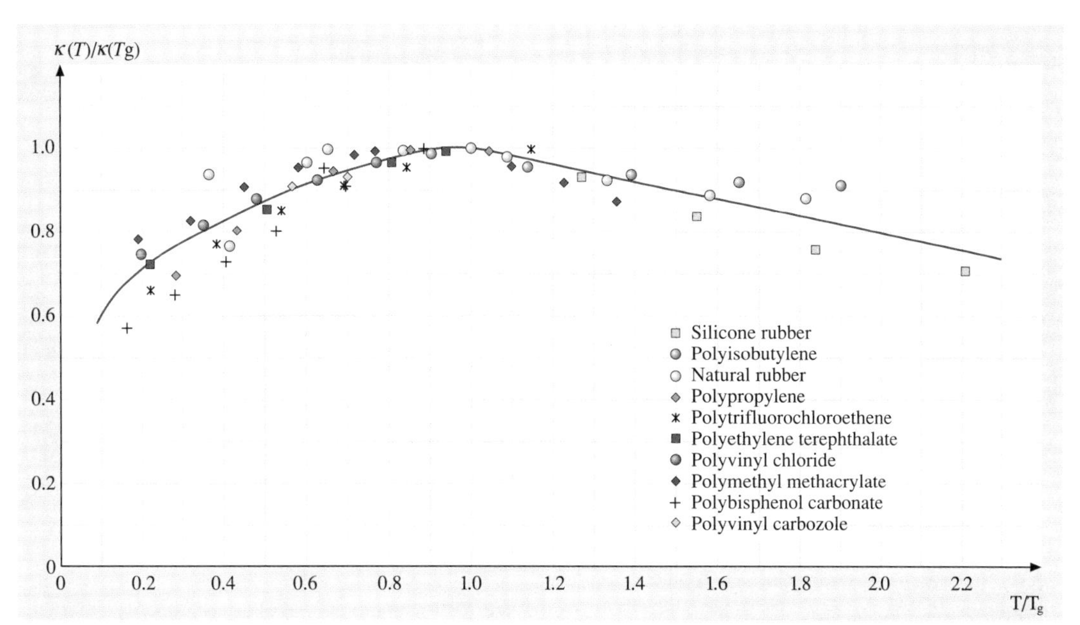

Fig. 19.10 Normalized thermal conductivity, $\kappa(T)/\kappa(T_g)$ versus normalized temperature, T/T_g, for various polymers. (Figure 17.2 in [19.10])

The overall or effective thermal conductivity of polymer blends can be usually estimated from the individual conductivities, the volume or weight fractions of the constituents, and the crystallinity of the structure; the problem can be quite complicated. A good example in which κ has been examined for a polymer blend as a function of blend composition has been given by *Agari* et al. [19.30], who also test various mixture rules.

The thermal conductivity of polymeric composite materials (as opposed to blends) can be estimated from various mixture rules [19.31, 32]. For example, if a matrix with κ_1 contains spherical particles with κ_2, then the composite has a thermal conductivity,

$$\kappa = \kappa_1 \frac{2\kappa_1 + \kappa_2 - 2v_2(\kappa_1 - \kappa_2)}{2\kappa_1 + \kappa_2 + v_2(\kappa_1 - \kappa_2)} , \quad (19.19)$$

where v_2 is the volume fraction of the particles (with κ_2).

19.2.4 Thermal Conductivity of Metals

Although the heat capacity of metals is due the lattice vibrations, the thermal conductivity is due to the free conduction electrons. Both charge transport and heat transport are carried by the conduction electrons, and hence the electrical and thermal conductivity, σ and κ respectively, are related. Normally only electrons around the Fermi level contribute to conduction and energy transport. In general κ and σ can be written as

$$\kappa = \frac{1}{3} C_e v_F \ell_e \text{ and } \sigma = \frac{1}{3} e^2 v_F \ell_e g(E_F) , \quad (19.20)$$

where C_e is the electronic heat capacity per unit volume, v_F is the Fermi velocity, $g(E_F)$ is the density of states at E_F (Fermi level), and ℓ_e is the electron mean free path. The scattering of electrons limits ℓ_e, which then limits κ and σ. If we then apply the free-electron theory to substitute for C_e and $g(E_F)$ then the ratio of κ to σT turns out to be a constant, which depends only on the constants k and e; ℓ_e cancels out in this ratio. A proper derivation involves solving the Boltzmann transport equation in the presence of a temperature gradient [19.33], and leads to the well-known *Wiedemann–Franz–Lorenz* (WFL) law

$$\frac{\kappa}{\sigma T} = C_{WFL} = \frac{\pi^2 k^2}{3e^2} = 2.44 \times 10^{-8}\,\mathrm{W\,\Omega/K^2} , \quad (19.21)$$

where C_{WFL} is called the *Lorentz number*.

As shown in Fig. 19.11, the WFL law is well obeyed at room temperature for most pure metals and their alloys. The WFL law is normally well obeyed over a wide

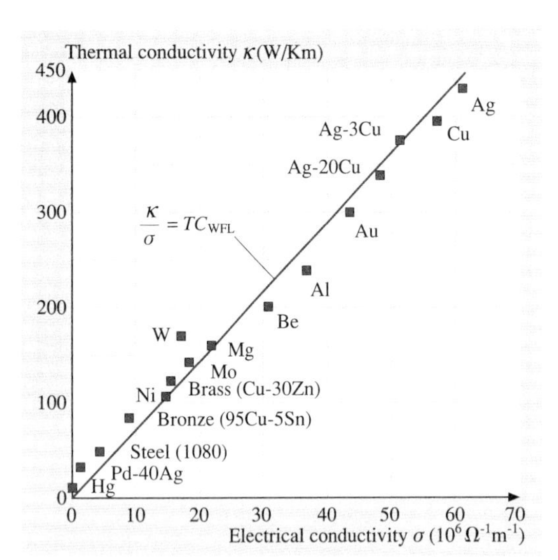

Fig. 19.11 Thermal conductivity versus electrical conductivity for various metals (both pure metals and alloys) at 20 °C. (From [19.5])

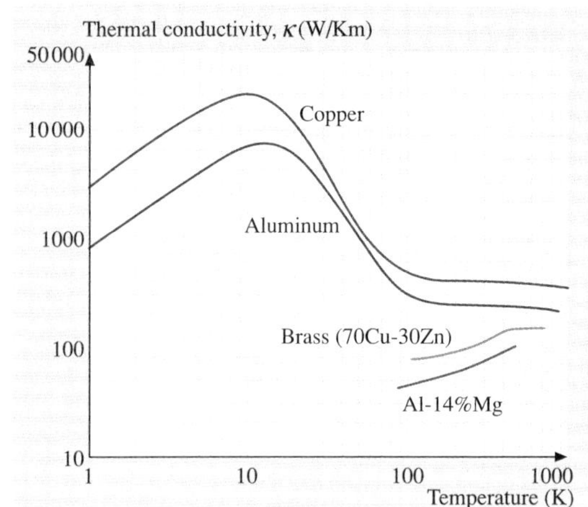

Fig. 19.12 Thermal conductivity versus temperature for two pure metals (Cu and Al) and two alloys (brass and Al–14% Mg). (From [19.5])

temperature range, and also by many liquid pure metals. Consequently the thermal conductivity κ of a metal or an alloy can be readily calculated given the conductivity or resistivity ρ. For example, we can easily calculate the resistivity of a solid solution alloy by using Matthiessen's and Nordheim's rules (as explained in this handbook by *Koughia* et al. in Chapt. 2) and then simply calculate κ from (19.21). Figure 19.12 shows the typical dependence of κ on T for two pure metals (Cu and Al) and two alloys (brass and Al–14%Mg). Above the Debye temperature, for pure metals, $\sigma \propto 1/T$, so that the thermal conductivity remains almost constant, as in Fig. 19.12. For alloys, σ values are smaller than the corresponding pure metals and σ exhibits a weaker temperature dependence. Consequently κ is smaller and increases with the temperature, as shown in Fig. 19.12.

While for pure metals and many alloys κ is due to the conduction electrons, the phonon contribution can nonetheless become comparable to the electron contribution for those alloys that have substantial resistivity or at low temperatures. Glassy metals (or metallic glasses) usually have two or more elements alloyed; one of the elements is usually a nonmetal or a metalloid. Their resistivities tend to be higher than usual crystalline or polycrystalline alloys. Both conduction electrons and also phonons contribute to the thermal conductivity of these glassy metals. For example, for $Fe_{80}B_{20}$, $\rho = 1220\,\mathrm{n\Omega\,m}$ (compare with $17\,\mathrm{n\Omega\,m}$ for Cu), and the phonon contribution is about 15% at 273 K and 61% at 100 K [19.34].

19.3 Thermal Expansion

19.3.1 Grüneisen's Law and Anharmonicity

Nearly all materials expand as the temperature increases. The thermal coefficient of linear expansion, or simply, the thermal expansion coefficient α is defined as the fractional change in length per unit temperature increase, or

$$\alpha = \frac{1}{L_0} \cdot \left(\frac{\mathrm{d}L}{\mathrm{d}T}\right)_{T_0} , \tag{19.22}$$

where L_0 is the original length of the substance at temperature T_0, $\mathrm{d}L/\mathrm{d}T$ is the rate of change of length with temperature at the reference temperature T_0.

The principle of thermal expansion is illustrated in Fig. 19.13, which is the potential-energy curve $U(r)$ for two atoms separated by a distance r in a crystal. At temperature T_1, the atoms would be vibrating about their equilibrium positions between B and C; compressing (B) and stretching (C) the bond between them. The line BC corresponds to the total energy E of the pair

Table 19.3 The Grüneisen parameter for some selected materials with different types of interatomic bonding (FCC = face-centered cubic; BCC = body-centered cubic)

Material	ρ (g/cm^3)	α ($\times 10^{-6}$ K^{-1})	K (GPa)	c_s (J/kg K)	γ
Iron (metallic, BCC)	7.9	12.1	170	444	0.20
Copper (metallic, FCC)	8.96	17	140	380	0.23
Germanium (covalent)	5.32	6	77	322	0.09
Glass (covalent–ionic)	2.45	8	70	800	0.10
NaCl (ionic)	2.16	39.5	28	880	0.19
Tellurium (mixed)	6.24	18.2	40	202	0.19
Polystyrene (van der Waals)	1.05	100	3	1200	0.08

of atoms. The average separation at T_1 is at A, halfway between B and C. The PE curve $U(r)$ is asymmetric, and it is this asymmetry that leads to the phenomenon of thermal expansion. When the temperature increases from T_1 to T_2 the atoms vibrate between B' and C' and the average separation between the atoms also increases, from A to A', which is identified as thermal expansion. If the PE curve were symmetric, then there would be no thermal expansion. The extent of expansion (A to A') depends on the amount of increase from BC to $B'C'$ per degree of increase in the temperature. α must therefore also depend on the heat capacity. When the temperature increases by a small amount δT, the energy per atom increases by $(C_V \delta T)/N$, where C_V is the heat capacity per unit volume and N is the number of atoms per unit volume. If $C_V \delta T$ is large then the line $B'C'$ in Fig. 19.13 will be higher up on the energy curve and the average separation A' will therefore be larger. Thus, $\alpha \propto C_V$. Further, the average separation, point A, depends on how much the bonds are stretched and compressed. For large amounts of displacement from equilibrium, the average A will be greater as more asymmetry of the PE curve is used. Thus, smaller is the elastic modulus K, the greater is α; clearly $\alpha \propto C_V/K$.

If we were to expand $U(r)$ about its minimum value $U_{\min}$ at $r = r_0$, we would obtain the Taylor expansion to the cubic term as,

$$U(r)=U_{\min}+(1/2)\beta(r-r_0)^2-(1/3)g(r-r_0)^3\ , \tag{19.23}$$

where β and g are coefficients related to second and third derivatives of U at r_0. The term $(r-r_0)$ is missing because we are expanding a series about $U_{\min}$ where $\mathrm{d}U/\mathrm{d}r = 0$. The $U_{\min}$ and the $\beta_2(r-r_0)^2$ term give a parabola about $U_{\min}$ which is a symmetric curve around r_0 and therefore does not lead to thermal expansion. It is the cubic term that gives the expansion because it leads to asymmetry. Thus, the amount of expansion α

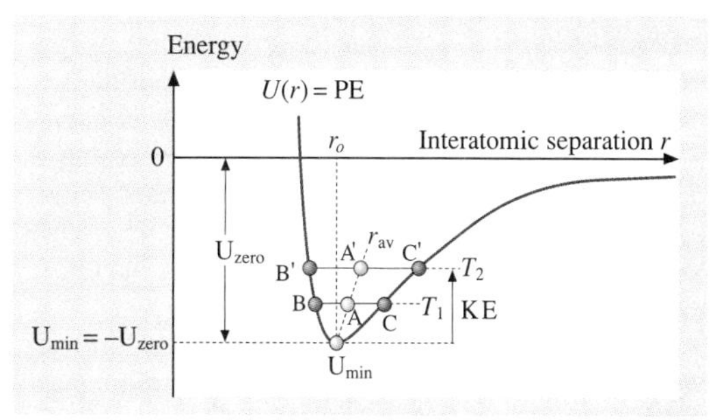

Fig. 19.13 Interatomic potential energy (PE) U curve, which has a minimum when the atoms in the solid attain the interatomic separation at $r = r_0$

also depends on the amount of asymmetry with respect to symmetry that is β/g. Thus, α is proportional to both C_V/K and g/β. The third term in (19.23) is called *anharmonicity* in the interatomic PE curve. The ratio of g to β depends on the nature of the bond. A simplified analytical treatment [19.35] gives α as,

$$\alpha \approx 3\gamma\frac{C_V}{K}\ ;\quad \gamma=\frac{r_0 g}{3\beta}\ , \tag{19.24}$$

where γ is a constant called the *Grüneisen parameter*, which represents the relative asymmetry of the energy curve, g/β. γ is of the order of unity for many materials; experimentally, $\gamma = 0.1-1$. We can also write the Grüneisen law in terms of the molar heat capacity C_m [heat capacity per mole or the specific heat capacity c_s (heat capacity per unit mass)]. If ρ is the density, and M_{at} is the atomic mass of the constituent atoms of the crystal, then

$$\lambda=3\gamma\frac{\rho C_m}{M_{at}K}=3\gamma\frac{\rho c_s}{K}\ . \tag{19.25}$$

Given the experimental values for α, K, ρ and c_s, the Grüneisen parameters have been calculated from (19.25) and are listed in Table 19.3. An interesting feature is that the experimental γ values, within a factor of 2–3, are about the same, at least to an order of magnitude. Equation (19.24) also indicates that the dependence of α on T should resemble the dependence of C_v on T, which is borne out by experiments.

19.3.2 Thermal Expansion Coefficient α

The thermal expansion coefficient normally depends on the temperature, $\alpha = \alpha(T)$, and typically increases with increasing temperature, except at the lowest temperatures. Figure 19.14 shows α versus T for selected materials and also compares α for a wide range of materials at 300 K. In very general terms, except at very low (typically below 100 K) and very high temperatures (near the melting temperature), for most metals α does not depend strongly on the temperature; many engineers take α for a metal to be approximately temperature independent. Further, higher-melting-point temperature metals tend to have lower α.

To calculate the final length L from the original length L_0 that corresponds to the temperature increasing from T_0 to T, we have to integrate $\alpha(T)$ with respect to temperature from T_0 to T. If $\bar{\alpha}$ is the mean value for the expansion coefficient from T_0 to T, then

$$L = L_0[1 + \bar{\alpha}(T - T_0)] \text{ and} \tag{19.26}$$

$$\bar{\alpha} = \frac{1}{(T - T_0)} \int_{T_0}^{T} \alpha(T) \mathrm{d}T \tag{19.27}$$

For example, the expansion coefficient of silicon over the temperature range 120–1500 K is given by [19.36]

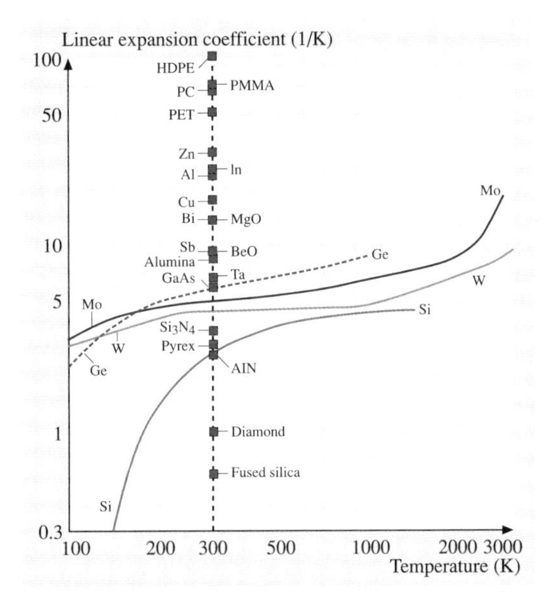

Fig. 19.14 α versus T for selected materials and comparison of α for a range of materials at 300 K. (After [19.5])

as

$$\alpha = 3.725 \times 10^{-6} \left(1 - e^{-5.88\times10^{-3}(T-124)}\right) + 5.548 \times 10^{-10} T \,, \tag{19.28}$$

where T is in Kelvin. At 20 °C, $\alpha = 2.51 \times 10^{-6}\,\mathrm{K}^{-1}$, whereas from 20 °C to 320 °C, $\bar{\alpha} = 3.35 \times 10^{-6}\,\mathrm{K}^{-1}$. Most $\alpha(T)$ expressions have exponential-type expressions as in (19.28) but they can always be expanded in terms of T or $(T - T_0)$ up to the third term, beyond which errors become insignificant.

19.4 Enthalpic Thermal Properties

19.4.1 Enthalpy, Heat Capacity and Physical Transformations

Consider the behavior of the enthalpy $H(T)$ versus temperature T of a typical material as it is cooled from a liquid state starting at O as shown in Fig. 19.15 (the sketch is idealized to highlight the fundamentals). For crystalline solids, at the melting or fusion temperature T_m, the liquid L solidifies to form a crystalline solid C. Some degree of undercooling is necessary to nucleate the crystals but we have ignored this fact in Fig. 19.15 so that T_m is the thermodynamic melting temperature where the Gibbs free energies of the liquid (L) and the crystal (C) are the same. The heat capacity $C_{\mathrm{liquid}} = \mathrm{d}H_{\mathrm{liquid}}/\mathrm{d}T$ for the liquid is greater than that for the crystal, $C_{\mathrm{crystal}} = \mathrm{d}H_{\mathrm{crystal}}/\mathrm{d}T$, as apparent from the H-T slopes in Fig. 19.15. The fusion enthalpy ΔH_f corresponds to the enthalpy change between the liquid and crystalline phases, L and C, in Fig. 19.15.

When a glass-forming liquid is cooled, it does not crystallize at T_m; its crystallization kinetics are too slow or negligible. Its enthalpy H continues to decrease along

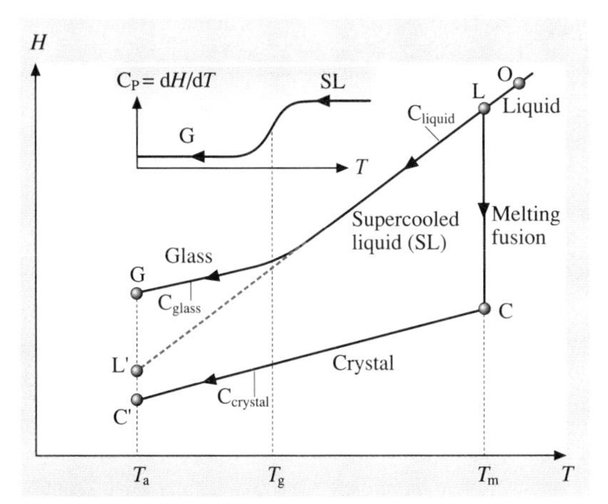

Fig. 19.15 Enthalpy H versus temperature T

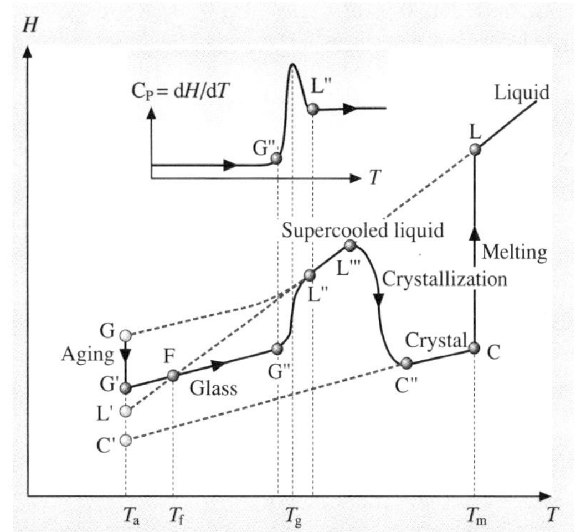

Fig. 19.16 Isothermal aging at T_a changes the glass enthalpy. When the glass is heated at a constant rate r, it undergoes glass transition at around T_g, and then ideally crystallization and then melting

the liquid H versus T line with a slope C_{liquid} and it becomes a supercooled liquid. However, as the cooling proceeds, eventually, over a certain small temperature range around a temperature marked as T_g in Fig. 19.15, the atomic motions become too sluggish, and the viscosity becomes too high to follow the liquid-like behavior. The material cools along a solid-like H–T curve that is almost parallel to the crystalline or solid H–T line. This region where the slope or the heat capacity is similar to that of the solid or the crystal ($C_{crystal}$) is called the glass region, and the material is said to be in the *glassy state*. The intersection of the liquid-like (supercooled) and solid-like (glass) H–T lines defines a *glass-transition temperature* T_g as observed under cooling at a particular cooling rate q. T_g depends on the cooling rate, and obviously on the structure. C_P changes in a step-like fashion over a small interval around T_g during cooling as shown in Fig. 19.15. The shift in T_g with q can provide valuable information on the kinetics of the glass-transformation process as discussed below. The glass enthalpy is greater than the crystalline enthalpy due to the disorder in the glass structure.

Suppose we stop the cooling at an ambient temperature T_a, corresponding to the glass state G. The glass enthalpy $H_{glass}(T_a)$ will relax via structural relaxations toward the enthalpy of the metastable liquid-like equilibrium state $H_{liquid}(T_a)$ at L′. This process is called aging or annealing. When we start heating a glass, its initial enthalpy can be anywhere between G and L′, depending on the thermal history (including aging) and the cooling rate.

When the sample at $H(T_a)$ at G′ is heated at a constant rate r, as shown in Fig. 19.16, the enthalpy of the system, as a result of long structural relaxation times at these low temperatures, follows the glass H–T curve until the structural relaxation rate is sufficiently rapid to allow the system to recover toward equilibrium $H_{liquid}(T)$; the transition is from G″ to L″. The sigmoidal-like change in $H(T)$ leads to the glass-transformation endotherm observed for many glasses when they are heated through T_g. This endothermic heat, called the enthalpy of relaxation, depends on the duration of aging, and the structural relaxation processes, which in turn depend on the structure (e.g. [19.37]). The observed glass-transition temperature T_g during heating depends not only on the heating rate r but also on the initial state $H(T_a)$ or G′. On the other hand, during cooling from the melt down to low temperatures, one observes only a change in the heat capacity. As we further heat the sample, the supercooled liquid state can crystallize, from L‴ to C″ and form a true solid, usually polycrystalline. This transformation from L‴ to C″ releases heat; it is exothermic. At T_m, the solid melts to form the liquid, C changes to L, which means that heat is absorbed.

It is apparent from the simple H–T behavior in Fig. 19.16 that there are at least several key parameters of interest in characterizing the thermal properties of materials: the heat capacity of the liquid, crystal and glass phases, C_{liquid}, $C_{crystal}$, C_{glass}, their temperature dependences, melting or fusion temperature T_m, fusion enthalpy ΔH_f, crystallization onset temperature (L‴), exothermic enthalpy of crystallization ($L''' - C'''$), T_g (which depends on various factors), and the endothermic glass-transformation enthalpy ($L'' - G''$). These properties are most conveniently examined using a *differential scanning calorimeter* (DSC).

19.4.2 Conventional Differential Scanning Calorimetry (DSC)

Principles

Differential scanning calorimetry (DSC) is a method which is extensively used to measure heats and temperatures of various transitions and has been recognized as a very useful tool for the interpretation of thermal events as discussed and reviewed by a number of authors (see for example [19.38–44]). In essence, the DSC measures the net rate of heat flow $Q' = dQ/dT$, into a sample with respect to heat flow into a reference inert sample, as the sample temperature is ramped. Figure 19.17 shows a simplified schematic diagram of the heat-flux DSC cell. The constantan disk allows heat transfer to and from the reference and sample pans by thermal conduction. Due to symmetry, the thermal resistance of the heat paths from the heater to the reference and sample pans is approximately the same. The constantan disk has two slightly raised platforms on which the reference and sample pans are placed. Chromel disks fixed under these platforms form area thermocouple junctions with the constantan platforms. The constantan-chromel thermocouples are in series and read the temperature difference ΔT between the reference and sample pans. There are also alumel–chromel thermocouples attached to the chromel disks to measure the individual reference and sample temperatures.

The measured temperature difference ΔT corresponds to a differential heat flow $Q' = dQ/dt$ between the sample and the reference pans by virtue of Fourier's law of heat conduction,

$$Q' = \Delta T/R\,, \tag{19.29}$$

where R is the effective thermal resistance of the heat flow path to the sample (or reference) through the constantan disk. In conventional thermal analysis, the sample temperature is either ramped linearly at a constant heating or cooling rate or kept constant (as in isothermal experiments).

A typical example of a conventional DSC is shown in Fig. 19.18 where the thermogram evinces three distinct regions that correspond to the glass transition (T_g), crystallization and melting. T_g in DSC analysis is normally defined in terms of an onset temperature as in Fig. 19.18. The crystallization exotherm is normally used to obtain a crystallization onset temperature T_c and a peak crystallization rate temperature T_p that corresponds to the location of the peak crystallization. ΔH_c and fusion ΔH_f enthalpies are obtained by integrating the corresponding peaks as shown in Fig. 19.18 with an appropriate baseline. The base line represents the overall heat capacity of the sample, which changes during the transformation since the sample is gradually transformed into the new phase. Usually a simple sigmoidal, or even a straight line between the onset and the end points, is sufficient for an accurate determination of the enthalpy.

Conventional DSC has been used to study and characterize a wide range of physical and chemical transformations. When the DSC is used in a heating or cooling scan, the temperature of the sample is ramped along

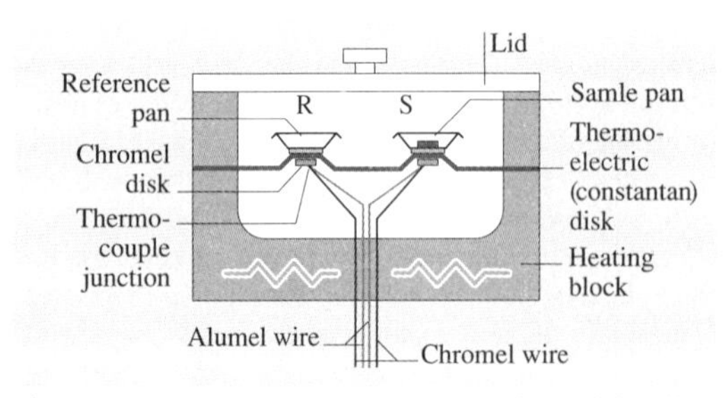

Fig. 19.17 Schematic illustration of heat-flux DSC cell. R and S refer to reference and sample pans

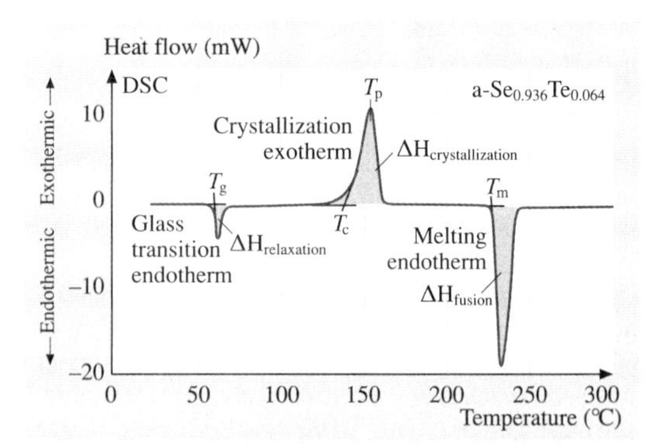

Fig. 19.18 Typical DSC signal for a chalcogenide $Se_{0.936}Te_{0.064}$ glass sample (well relaxed or aged). From left to right: the glass transition, crystallization and the melting temperatures and corresponding enthalpies of transitions for this material. $\Delta H_c = \Delta H_{crystallization}$; $\Delta H_f = \Delta H_{fusion}$

$T = T_0 + rt$, where T_0 is the initial temperature. Consequently, the transformations in DSC are carried out under non-isothermal conditions, and well-known isothermal rate equations cannot be directly applied without some modification.

Glass Transformation

There are extensive discussions in the literature on the meaning of the glass-transition region and the corresponding T_g (e.g. [19.45–50]). The most popular interpretation of T_g is based on the fact that this transformation is a kinetic phenomenon. The glass-transformation kinetics have been most widely studied by examining the shift in T_g with the heating or cooling rate in a so-called T_g-shift technique. The relaxation process can be modeled by assuming that the glass structure has a characteristic structural relaxation time that controls the rate at which the enthalpy can change. It is well recognized that the glass-transformation kinetics of glasses are nonlinear. In the simplest description, the relaxation can be conveniently described by using a single phenomenological relaxation time τ (called the Narayanaswamy or *Tool–Narayanaswamy–Moynihan relaxation time*) that depends not only on the temperature but also on the glass structure through the fictive temperature T_f as

$$\tau(T, T_f) = \tau_0 \exp[x\Delta h^*/RT + (1-x)\Delta h^*/RT_f] , \quad (19.30)$$

where Δh^* is the activation enthalpy, T_f is the fictive temperature and x is the partition parameter which determines the relative contributions of temperature and structure to the relaxation process. T_f is defined in Fig. 19.16 as the intersection of the glass line passing through the starting enthalpic state G′ and the extended liquid H–T lines. It depends on the starting enthalpy G′ so that T_f is used as a convenient temperature parameter to identify the initial state at G′. Due to the presence of the structural parameter x, the activation energies obtained by examining the heating and cooling rate dependences of the glass-transition temperature are not the same. If the shift in T_g is examined as a function of the cooling rate q starting from a liquid-like state (above T_g) then a plot of $\ln q$ versus $1/T_g$ (called a *Ritland plot* [19.51]) should yield the activation enthalpy Δh^* in (19.30) [19.52–54]. In many material systems, the relaxation time τ in (19.30) is proportional to the viscosity, $\tau \propto \eta$ [19.55–57] so that Δh^* from cooling scans agrees with the activation energy for the viscosity [19.58–62] over the same temperature range. The viscosity η usually follows either an Arrhenius temperature dependence, as in oxide glasses, or a Vogel–Tammann–Fulcher behavior, $\eta \propto \exp[A/(T - T_0)]$, where A and T_0 are constants, as in many polymers and some glasses, e.g. chalcogenides.

The relaxation kinetics of various structural properties such as the enthalpy, specific volume, elastic modulus, dielectric constant etc. have been extensively studied near and around T_g, and there are various reviews on the topic (e.g. [19.63]). One particular relaxation kinetics that has found widespread use is the stretched exponential in which the rate of relaxation of the measured property is given by

$$\text{Rate of relaxation} \propto \exp\left[-\left(\frac{t}{\tau}\right)^{\beta}\right] , \quad (19.31)$$

where β (< 1) is a constant that characterizes the departure from the pure exponential relaxation rate. Equation (19.31) is often referred to as the *Kohlrausch–Williams–Watts* (KWW) [19.64] stretch exponential relaxation function. β depends not only on the material but also on the property that is being studied. In some relaxation processes, the whole relaxation process over a very long time is sometimes described by two stretched exponentials to handle the different fast and slow kinetic processes that take place in the structure [19.65].

The kinetic interpretation of T_g implies that, as the cooling rate is slowed, the transition at T_g from the supercooled liquid to the glass state is observed at lower temperatures. There is however a theoretical thermodynamic boundary to the lowest value of T_g. As the supercooled liquid is cooled, its entropy decreases faster than that of the corresponding crystal because $C_{liquid} > C_{crystal}$. Eventually at a certain temperature T_0, the relative entropy lost $\Delta S_{liquid\text{-}crystal}$ by the supercooled liquid with respect to the crystal will be the same as the entropy decrease (latent entropy of fusion) $\Delta S_f = \Delta H_f/T_m$ during fusion. This is called *Kauzmann's paradox* [19.66], and the temperature at which $\Delta S_{liquid\text{–}crystal} = \Delta S_f$ is the lowest theoretical boundary for the glass transformation; $T_g > T_0$.

The changes in T_g with practically usable heating or cooling rates are usually of the order of 10 °C or so. There have been various empirical rules that relate T_g to the melting temperature T_m and the glass structure and composition. Since T_g depends on the heating or cooling rate, such rules should be used as an approximation; nonetheless, they are extremely useful in engineering as a guide to the selection and use of materials.

Table 19.4 Some selected examples of T_g dependences on various factors

Rule	Notation	Comment
$T_g \approx (2/3)T_m$	T_g and T_m in K. T_m = melting temperature of corresponding crystalline phase.	Kauzmman's empirical rule [19.66]. Most glass structures including many amorphous polymers [19.67]. Some highly symmetrical polymers with short repeat units follow $T_g \approx (1/2)T_m$ [19.10]
$\ln(q) \approx -\Delta h^*/RT_g + C$	q = cooling rate; Δh^* = activation energy in (19.30); C = constant.	Dependence of T_g on the cooling rate. Δh^* may depend on the range of temperature accessed. Bartenev–Lukianov equation [19.51, 68]
$T_g \approx T_g(\infty) + C/M_n$	M_n = average molecular weight of polymer; C = constant; $T_g(\infty)$ is T_g for very large M_n	Dependence of T_g on the average molecular weight of a polymer [19.69, 70]. Tanaka's rule
$\ln(T_g) \approx 1.6Z + C$	Z = mean coordination number, C = constant (≈ 2.3)	Network glasses. Dependence of T_g on the mean coordination number. Neglects the heating rate dependence. [19.71]
$T_g(x) = T_g(0) - 626x$	T_g in K; x is atomic fraction in $a: (Na_2O + MgO)_x(Al_2O_3 + SiO_2)_{1-x}$ $b: (PbO)_x(SiO_2)_{1-x}$ $c: (Na_2O)_x(SiO_2)_{1-x}$ $T_g(0) = 1080$ K for a; 967 K for b; 895 K for c.	±5%. Silicate glasses [19.72]; x is network modifier. $0.01 < x < 0.6$

Non-Isothermal Phase Transformations

The crystallization process observed during a DSC heating scan is a non-isothermal transformation in which nucleation and growth occur either at the same time as in homogenous nucleation or nucleation occurs before growth as in heterogenous nucleation. In the case of isothermal transformations by nucleation and growth, the key equation is the so-called *Johnson–Mehl–Avrami equation*,

$$x = 1 - \exp(-Kt^n) , \qquad (19.32)$$

where $K \propto \exp(-E_A/k_B T)$ is the thermally activated rate constant, and n is a constant called the *Avrami index* whose value depends on whether the nucleation is heterogeneous or homogenous, and the dimensionality m of growth ($m = 1, 2$ or 3 for one-, two- or three-dimensional growth). For example, for growth from preexisting nuclei (heterogeneous nucleation) $n = m$, and for continuing nucleation during growth, $n = m + 1$. A detailed summary of possible n and m values has been given by *Donald* [19.73]

DSC studies however are conventionally non-isothermal. There have been numerous papers and discussions on how to extract the kinetic parameters of the transformation from a DSC non-isothermal experiment [19.74–78]. It is possible to carry out a reasonable examination of the crystallization kinetics by combining a single scan experiment with a set of multiple scans; there are many examples in the literature (e.g. [19.78, 79]). Suppose that we take a single DSC scan, as in Fig. 19.18, and calculate the fraction of crystallized material x at a temperature T. The plot of $\ln[-\ln(1-x)]$ versus $1/T$ (*Coats–Redfern–Sestak plot* [19.80, 81]) then provides an activation energy E'_A from a single scan. For heterogenous nucleation, E'_A is mE_G, whereas for homogenous nucleation it is $E_N + mE_G$, where E_G is the activation energy for growth and E_N is the activation energy for nucleation. Clearly, we need to know something about the nucleation process and dimensionality of growth to make a sensible use of E'_A. Thus,

$$\ln[-\ln(1-x)] = -\frac{E'_A}{RT} + C' , \qquad (19.33)$$

where C' is a constant.

Suppose we then examine how the peak rate temperature T_p shifts with the heating rate r. Then a plot of $\ln(r/T_p^2)$ versus $1/T_p$ is called a *Kissinger plot* [19.82,83], and gives an activation energy E''_A. Thus,

$$\ln\left(\frac{r}{T_p^2}\right) = -\frac{E''_A}{RT_p} + C'' , \qquad (19.34)$$

where C'' is a constant. In heterogenous nucleation E''_A simply represents the activation energy of growth E_G, whereas if the nucleation continues during growth, $E''_A = (E_N + mE_G)/(m+1)$. The ratio E'_A/E''_A represents the non-isothermal Avrami index n. Table 19.5 provides an overview of various thermal analysis techniques that have been used for characterizing non-isothermal phase transformations.

Table 19.5 Typical examples of studies of transformation kinetics. Usual interpretation of E'_A and E''_A are $E'_A = mE_G$ or $(E_N + mE_G)$ and $E''_A = E_G$ or $(E_N + mE_G)/(m+1)$; $\dot{x}$ is the rate of crystallization

Study	Method and plot	Slope provides	Method
Single scan	$\ln[-\ln(1-x)]$ versus $1/T$	E'_A	*Coats-Redfern-Sestak* [19.80,81]
Single scan	$\frac{\dot{x}}{(1-x)[-\ln(1-x)]^{(n-1)/n}}$	E''_A	If n chosen correctly, this agrees with the Kissinger method [19.77,79]. Independent of the initial temperature
Multiple scan	$\ln(r/T_p^2)$ versus $1/T_p$	E''_A	*Kissinger* [19.82,83]. Initial temperature effect in [19.77]
Multiple scan	$\ln(r^n/T_p^2)$ versus $1/T_p$	mE''_A	Modified Kissinger; *Matusita, Sakka* [19.89,90]
Multiple scan	$\ln r$ versus $1/T_c$	E''_A	*Ozawa method* [19.74–76]
Multiple scan	$\ln[r/(T_p - T_c)]$ versus $1/T_p$; T_c = initial temperature	E''_A	*Augis, Bennett* [19.91]
Multiple scan	$[d\Delta H/dt]_{max}$ versus $1/T_p$; $\dot{x} = [d\Delta H/dt]_{max}$	E''_A	*Borchardt–Pilonyan* [19.92,93]
Multiple scan	$\ln(r/T_1^2)$ versus $1/T_1$; when $x = x_1$, $T = T_1$; T_1 depends on r	E''_A	*Ozawa–Chen* [19.94,95]
Multiple scan	$\ln[-\ln(1-x)]$ versus $\ln r$; x is the crystallized amount at $T = T_1$; T_1 is constant	$-n$	*Ozawa* method [19.90,94]

DSC has been widely used to study the kinetics of crystallization and various phase transformations occurring in a wide range of material systems; there are numerous recent examples in the literature [19.84–88]. Equations (19.33) and (19.34) represent a simplified analysis. As emphasized recently [19.73], a modified Kissinger analysis [19.89, 90] involves plotting $\ln(r^n/T_p^2)$ versus $1/T_p$, the slope of which represents an activation energy, mE''_A; however, the latter requires some knowledge of n or m to render the analysis useful. n can be obtained examining the dependence of $\ln[-\ln(1-x)]$ on $\ln r$ at one particular temperature, which is called the Ozawa method as listed in Table 19.5 [19.90]. It is possible to combine the modified Kissinger analysis with an isothermal study of crystallization kinetics to infer n and m given the type of nucleation process (heterogeneous or homogenous) that takes place.

19.5 Temperature-Modulated DSC (TMDSC)

19.5.1 TMDSC Principles

In the early 1990s, a greatly enhanced version of the DSC method called temperature-modulated differential scanning calorimetry (MDSC^{TM}) was introduced by the efforts of *Reading* and coworkers [19.96–98]. The MDSC method incorporates not only the ability of conventional DSC but it also provides significant and distinct advantages over traditional DSC. The benefits of the MDSC technique have been documented in several recent papers, and include the following: separation of complex transitions, e.g. glass transition, into easily interpreted components; measurement of heat flow and heat capacity in a single experiment; ability to determine more accurately the initial crystallinity of the studied material; increased sensitivity for the detection of weak transitions; increased resolution without the loss of sensitivity; measurement of thermal conductivity [19.99, 100]. One of the most important benefits is the separation of complex transitions such as the glass transition into more easily interpreted components. Recent applications of MDSC to glasses has shown that it can be very useful for the interpretation of thermal properties, such as the heat capacity, in relation to the structure as, for example, in the case of chalcogenide glasses (e.g. [19.15, 101, 102]).

The MDSC^{TM} that is currently commercialized by TA Instruments uses a conventional heat-flux DSC cell whose heating block temperature is sinusoidally modulated. In MDSC, the sample temperature is modulated sinusoidally about a constant ramp so that the temperature T at time t is,

$$T = T_0 + rt + A\sin(\omega t)\,, \tag{19.35}$$

where T_0 is the initial (or starting) temperature, r is the heating rate (which may also be a cooling ramp q), A is the amplitude of the temperature modulation, $\omega = 2\pi/P$ is the angular frequency of modulation and P is the modulation period. It should be emphasized that (19.35) is a simplified statement of the fact that the cell has reached a steady-state operation and that the initial temperature transients have died out. The resulting instantaneous heating rate, $\mathrm{d}T/\mathrm{d}t$, therefore varies sinusoidally about the average heating rate r, and is given by

$$\mathrm{d}T/\mathrm{d}t = r + A\omega\sin(\omega t)\,. \tag{19.36}$$

At any time, the apparatus measures the sample temperature and the amplitude of the instantaneous heat flow (by measuring ΔT) and then, by carrying out a suitable Fourier deconvolution of the measured quantities, it determines two quantities (which have been termed by TA Instruments):

1. Reversing heat flow (RHF),
2. Nonreversing heat flow (NHF).

Fourier transforms are made on one full cycle of temperature variation, which means that the average quantities refer to moving averages. The average heat flow, which corresponds to the average heating rate (r), is called the total heat flow (HF). Total heat flow is the only quantity that is available and hence it is the only quantity that is always measured in conventional DSC experiments. MDSC determines the heat capacity using the magnitudes of heat flow and heating rate obtained by averaging over one full temperature cycle. If triangular brackets are used for averages over one period P, then $\langle Q'\rangle$ is the average heat flow per temperature cycle. The heat capacity per cycle is then calculated from

$$mC_\mathrm{P} = \langle\text{Heat flow}\rangle/\langle\text{Heating rate}\rangle\,, \tag{19.37}$$

where m is the mass of the sample. This C_P has been called the *reversing heat capacity*, though Schawe has defined as the complex heat capacity [19.12]. The reversing in this context refers to a heat flow that is reversing over the time scale of the modulation period. Furthermore, it is assumed that C_P is constant, that is, it does not change with time or temperature over the modulation period. The reversing heat flow is then obtained by

$$\mathrm{RHF} = C_\mathrm{P}\langle \mathrm{d}T/\mathrm{d}t\rangle\,. \tag{19.38}$$

The nonreversing heat flow (NHF) is the difference between the total heat flow and the reversing heat flow and represents heat flow due to a kinetically hindered process such as crystallization. There are a few subtle issues in the above condensed qualitative explanation. First is that the method requires several temperature cycles during a phase transition to obtain RHF and NHF components, which sets certain requirements on r, A and ω. The second is that the phase difference between the heat flow and the heating rate oscillations is assumed to be small as it would be the case through a glass-transition region or crystallization; but not through a melting process. There are a number of useful discussions and reviews on the MDSC technique and its applications in the literature [19.103–107].

19.5.2 TMDSC Applications

At present there is considerable scientific interest in applying TMDSC measurements to the study of glass-transformation kinetics in glasses and polymers (e.g. [19.15, 101, 108–112]). The interpretation of TMDSC measurements in the glass-transition region has been recently discussed and reviewed by *Hutchinson* and *Montserrat* [19.113–116]. The reversing heat flow (RHF) through the T_g region exhibits a step-like change and represents the change in the heat capacity. The hysteresis effects associated with thermal history seem to be less important in the RHF but present in the NHF. The measurement of T_g from the RHF in TMDSC experiments shows only a weak dependence on glass aging and thermal history [19.117–119], which is a distinct advantage of this technique. The interpretation of the NHF has been more difficult but it is believed that it provides a qualitative indication of the enthalpy loss during the annealing period below T_g [19.103] though more research is needed to clarify its interpretation. Figure 19.19 shows a typical MDSC result through the glass-transition region of $\mathrm{Se_{99.5}As_{0.5}}$ glass.

It is important to emphasize that ideally the underlying heating rate in TMDSC experiments should be as small as possible. In this way we can separate the conventional DSC experiment, which also takes place during TMDSC measurements, from the dynamic, frequency-controlled TMDSC experiment. The oscillation amplitude A in the TMDSC must be properly chosen so that the C_P measurements do not depend on A; typically $A = \pm 1.0\,°\mathrm{C}$ [19.61]. The oscillation period P should be chosen to ensure that there are at least four full modulations within the half width of the temperature transition, that is, a minimum of eight oscillations over

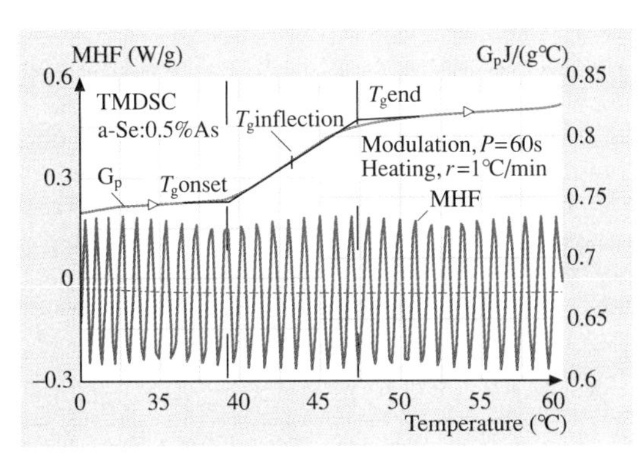

Fig. 19.19 A typical TMDSC scan

the T_g range, as can be seen in the example in Fig. 19.19 which has a particularly narrow glass-transition range and required a $P = 60$ s. The difference between the average specific capacity, as determined from the conventional DSC heat flow, and the complex heat capacity, as determined from TMDSC RHF, in the T_g region is substantial [19.115]. The influence of experimental conditions on TMDSC measurements have been well discussed in the literature [19.120].

The glass transition T_g as observed in DSC scans is well known to depend on thermal history. The classical well-known example is the fact that the T_g value determined by DSC under a heating scan at a given rate r is different from that determined under a cooling scan at a rate $q = r$. Further, T_g determined in a DSC heating scan also depends on aging. The case for As_2Se_3 is well documented [19.60]. Figure 19.20 shows an example in which the T_g values for vitreous As_2Se_3 determined in TMDSC heating and cooling scans are approximately the same for a given modulation

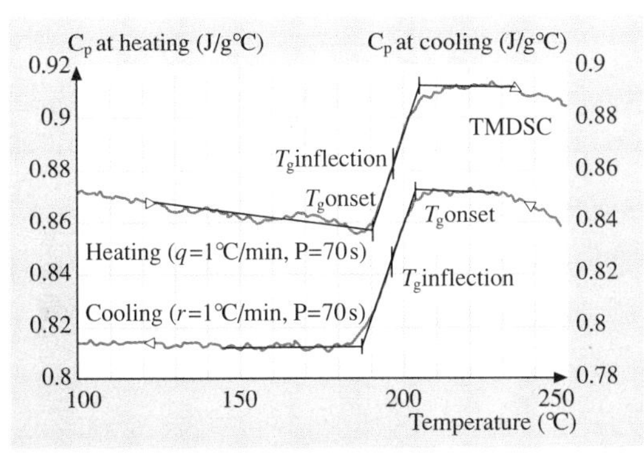

Fig. 19.20 Typical TMDSC heating and cooling scans on As_2Se_3 glass

period P, within experimental errors. T_g depends only on the modulation frequency $\omega(= 2\pi/P)$ provided that the underlying heating or cooling scan is sufficiently slow, usually less than 1 °C/min. Other examples are also available in the literature for both glasses and polymers [19.121, 122]. The T_g measurement in the TMDSC experiments represents the relaxation processes associated with the temperature modulation, and not the underlying heating rate as in the conventional DSC. For example, T_g in TMDSC represents a temperature where the relaxation time τ is comparable with the modulation period P. Above T_g, $\tau \ll P$ and below T_g, $\tau \gg P$, and in the T_g region $\tau \approx P$. As P (or ω) is varied, from experiment to experiment, different values of τ satisfy $\tau \approx P$ and hence T_g depends on ω. It is possible to examine the shift in T_g with ω, and then plot $\ln \omega$ versus $1/T_g$ to extract the activation energy of the relaxation time τ, as shown for As_2Se_3 [19.61]. T_g that is observed in conventional DSC experiments is associated with the devitrification (on heating) and vitrification (on cooling) of the structure, whereas T_g in TMDSC experiments represents a dynamic glass-transition temperature; it is important to arrange the experimental conditions so that the two processes are separated in the measurement [19.123].

While a single phenomenological relaxation time provides a convenient description of the overall glass-transformation behavior, a more complete description necessarily involves incorporating a distribution of relaxation times, as discussed in [19.116].

19.5.3 Tzero Technology

It is well known that there is a thermal lag between the sample and temperature-control sensors in the DSC and most DSCs require calibration at the heating rate to be used. Alternatively, calibration at multiple rates is required to correct for the heating rate dependence of the thermal lag. Without this calibration, there is considerable error in the temperature data in changing from one heating rate to another. This problem is especially acute in cooling experiments due to the thermal lag being in the opposite direction to that of heating, thereby doubling the lag effect. The lag problem has been solved in a new DSC called the *Tzero technology*, as developed by TA Instruments [19.124, 125]. With the Q-series DSC system, an additional thermocouple on the DSC sensor allows a complete calibration of the DSC cell and compensation for all thermal lags, including those caused by the encapsulated sample. This means that, for all practical purposes, the DSC can be calibrated at any reasonable heating rate, and the DSC temperature data will be correct within a few tenths of a Celsius degree for data taken at other heating or cooling rates. Inasmuch as advanced Tzero compensates for the effect of pan thermal mass and coupling, it is possible to calibrate using one pan type and then use another pan type without incurring substantial errors. In summary, since the Tzero technology uses more information in the DSC measurement, it is more accurate under a wider range of conditions compared with the ordinary DSC without calibration under those specific conditions.

The thermal lag error is proportional to heat flow, heating rate, and to the mass of the sample/pan system. Hence, this error becomes greatest with fast scanning rates, large sample masses, massive sample pans, or sample specimens with an especially high heat capacity. The thermal lag error is also proportional to the thermal resistance between the sample and sensors so it is made worse by using pans made of poor thermal conductivity or pans making poor thermal contact. However, even in ordinary polymer samples, using optimally coupled aluminum pans, the error produced could be more than two Celsius degrees because of poor thermal conductivity; for other samples it could be several times larger. In summary, the Tzero technology has enabled better DSC experiments to be carried out.

References

19.1 W. Martienssen, H. Warlimont (eds): *Springer Handbook of Condensed Matter and Materials Data* (Springer, Berlin Heidelberg New York 2005)

19.2 O. Madelung: *Semiconductors: Data Handbook*, 3rd edn. (Springer, Berlin Heidelberg New York 2004)

19.3 S. Adachi: *Properties of Group IV, III–V and II–VI Semiconductors* (Wiley, Chichester 2005)

19.4 P. Debye: Ann. Phys. **39**, 789 (1912)

19.5 S.O. Kasap: *Principles of Electronic Materials and Devices*, 3rd edn. (McGraw-Hill, Boston 2005)

19.6 S. Elliott: *The Physics and Chemistry of Solids* (Wiley, Chichester 1998)

19.7 R.B. Stephens: Phys. Rev. B **8**, 2896 (1973)

19.8 J. De Launay: *Solid State Physics Vol. 2*, ed. by F. Seitz, D. Turnbull (Academic, New York 1956)

19.9 K. Ichikawa: J. Phys. C **18**, 4631 (1985)

19.10 D.W. Van Krevelen, P.J. Hoftyzer: *Properties of Polymer* (Elsevier, Amsterdam 1976)

19.11 M. Pyda, E. Nowak-Pyda, J. Mays, B. Wunderlich: J. Polymer Sci. B **42**, 4401 (2004)

19.12 B. Wunderlich: Thermochim. Acta **300**, 43 (1997) and references therein

19.13 D.E. Sharp, L.B. Ginther: J. Am. Ceram. Soc. **34**, 260 (1951)

19.14 S.A. Khalimovskaya-Churkina, A.I. Priven: Glass Phys. Chem., **26**, 531 (2000) and references therein

19.15 T. Wagner, S.O. Kasap: Philos. Mag. **74**, 667 (1996)

19.16 S. Inaba, S. Oda: J. Non-Cryst. Solids **325**, 258 (2003)

19.17 Y.P. Joshi, G.S. Verma: Phys. Rev. B **1**, 750 (1970)

19.18 C.J. Glassbrenner, G. Slack: Phys. Rev. **134**, A1058 (1964)

19.19 M.G. Holland: *The Proceedings of the 7th Int. Conf. Phys. Semicond., Paris* (Dunond, Paris 1964) p.1161. The data were extracted from 19.2 (Fig. 2.11.11) in which the original data were taken from this reference.

19.20 C.M. Bhandari, C.M. Rowe: *Thermal Conduction in Semiconductors* (Wiley, New Delhi 1988)

19.21 M.P. Zaitlin, A.C. Anderson: Phys. Rev. Lett. **33**, 1158 (1974)

19.22 C. Kittel: Phys. Rev. **75**, 972 (1949)

19.23 C. Kittel: *Introduction to Solid State Physics*, 8th edn. (Wiley, New York 2005)

19.24 P.B. Allen, J.L. Feldman: Phys. Rev. Lett. **62**, 645 (1989)

19.25 P.B. Allen, J.L. Feldman: Phys. Rev. B **48**, 12581 (1993)

19.26 A. Jagannathan, R. Orbach, O. Entin-Wohlman: Phys. Rev. B **30**, 13465 (1989)

19.27 C. Oligschleger, J.C. Schön: Phys. Rev. B **59**, 4125 (1999)

19.28 R.C. Zeller, R.O. Pohl: Phys. Rev. B **4**, 2029 (1971)

19.29 K. Eiermann: Kolloid Z. **201**, 3 (1965)

19.30 Y. Agari, A. Ueda, Y. Omura, S. Nagai: Polymer **38**, 801 (1997)

19.31 B. Weidenfeller, M. Höfer, F. Schilling: Composites A **33**, 1041 (2002)

19.32 B. Weidenfeller, M. Höfer, F.R. Schilling: Composites A **35**, 423 (2004)

19.33 R. Bube: *Electronic Properties of Crystalline Solids: An Introduction to Fundamentals* (Academic, New York 1974)

19.34 A. Jezowski, J. Mucha, G. Pompe: J. Phys. D **20**, 1500 (1987)

19.35 See Chapter 1 Selected Topic entitled "Thermal Expansion" in the CDROM *Principle of Electronic Materials and Devices*, 3rd Edition, McGraw-Hill, Boston, (2005)

19.36 Y. Okada, Y. Tokumaru: J. Appl. Phys. **56**, 314 (1984)

19.37 J.M. Hutchinson, P. Kumar: Thermochim. Acta **391**, 197 (2002)

19.38 R.C. Mackenzie: Thermochim. Acta **28**, 1 (1979)

19.39 W. W. Wedlandt: *Thermal Analysis*, 3 edn. (Wiley, New York 1986) p. 3
19.40 B. Wunderlich: *Thermal Analysis* (Academic, New York 1990)
19.41 E. F. Palermo, J. Chiu: Thermochim. Acta **14**, 1 (1976)
19.42 S. Sarig, J. Fuchs: Thermochim. Acta **148**, 325 (1989)
19.43 W. Y. Lin, K. K. Mishra, E. Mori, K. Rajeshwar: Anal. Chem. **62**, 821 (1990)
19.44 T. Ozawa: Thermochim. Acta **355**, 35 (2000)
19.45 J. Wong, C. A. Angell: *Glass, Structure by Spectroscopy* (Marcel Dekker, New York 1976) and references therein
19.46 J. Zaryzycki: *Glasses and the Vitreous State* (Cambridge University Press, Cambridge 1991)
19.47 J. Jäckle: Rep. Prog. Phys. **49**, 171 (1986)
19.48 C. A. Angell: J. Res. Natl. Inst. Stand. Technol. **102**, 171 (1997)
19.49 C. A. Angell, B. E. Richards, V. Velikov: J. Phys. Cond. Matter **11**, A75 (1999)
19.50 I. Gutzow, B. Petroff: J. Non-Cryst. Solids **345**, 528 (2004)
19.51 H. N. Ritland: J. Am. Ceram. Soc. **37**, 370 (1954)
19.52 C. T. Moynihan, A. J. Easteal, M. A. DeBolt, J. Tucker: J. Am. Cer. Soc. **59**, 12 (1976)
19.53 M. A. DeBolt, A. J. Easteal, P. B. Macedo, C. T. Moynihan: J. Am. Cer. Soc. **59**, 16 (1976)
19.54 H. Sasabe, C. Moynihan: J. Polym. Sci. **16**, 1447 (1978)
19.55 O. V. Mazurin: J. Non-Cryst. Solids **25**, 131 (1977)
19.56 C. T. Moynihan, A. J. Easteal: J. Am. Ceram. Soc. **54**, 491 (1971)
19.57 H. Sasabe, C. T. Moynihan: J. Polym. Sci. **16**, 1447 (1978)
19.58 I. Avramov, E. Grantscharova, I. Gutzow: J. Non-Cryst. Solids **91**, 386 (1987) and references therein
19.59 S. Yannacopoulos, S. O. Kasap: J. Mater. Res. **5**, 789 (1990)
19.60 S. O. Kasap, S. Yannacopoulos: Phys. Chem. Glasses **31**, 71 (1990)
19.61 J. Malek: Thermochim. Acta **311**, 183 (1998)
19.62 S. O. Kasap, D. Tonchev: J. Mater. Res. **16**, 2399 (2001)
19.63 Z. Cernosek, J. Holubova, E. Cernoskova, M. Liska: J. Optoelec. Adv. Mater. **4**, 489 (2002) and references therein
19.64 G. Williams, D. C. Watts: Trans. Faraday Soc. **66**, 80 (1970)
19.65 R. Bohmer, C. A. Angell: Phys. Rev. B **48**, 5857 (1993)
19.66 W. Kauzmann: Chem. Rev. **43**, 219 (1948)
19.67 R. F. Boyer: J. Appl. Phys. **25**, 825 (1954)
19.68 G. M. Bartenev, I. A. Lukianov: Zh. Fiz. Khim **29**, 1486 (1955)
19.69 T. G. Fox, P. J. Flory: J. Polym. Sci. **14**, 315 (1954)
19.70 T. G. Fox, S. Loshaek: J. Polym. Sci. **15**, 371 (1955)
19.71 K. Tanaka: Solid State Commun. **54**, 867 (1985)
19.72 I. Avramov, T. Vassilev, I. Penkov: J. Non-Cryst. Solids **351**, 472 (2005)
19.73 I. W. Donald: J. Non-Cryst. Solids **345**, 120 (2004)
19.74 T. Ozawa: J. Therm. Anal. **2**, 301 (1970)
19.75 T. Ozawa: J. Therm. Anal. **7**, 601 (1975)
19.76 T. Ozawa: J. Therm. Anal. **9**, 369 (1976)
19.77 S. O. Kasap, C. Juhasz: J. Chem. Soc. Faraday Trans. II **81**, 811 (1985) and references therein
19.78 T. Kemeny, J. Sestak: Thermochim. Acta **110**, 113 (1987) and references therein
19.79 S. Yannacopoulos, S. O. Kasap, A. Hedayat, A. Verma: Can. Metall. Q. **33**, 51 (1994)
19.80 A. W. Coats, J. P. Redfern: Nature **201**, 68 (1964)
19.81 J. Sestak: Thermochim. Acta **3**, 150 (1971)
19.82 H. E. Kissinger: J. Res. Natl. Bur. Stand. **57**, 217 (1956)
19.83 H. E. Kissinger: Anal. Chem. **29**, 1702 (1957)
19.84 S. de la Parra, L. C. Torres-Gonzalez, L. M. Torres-Martínez, E. Sanchez: J. Non-Cryst. Solids **329**, 104 (2003)
19.85 I. W. Donald, B. L. Metcalfe: J. Non-Cryst. Solids **348**, 118 (2004)
19.86 W. Luo, Y. Wang, F. Bao, L. Zhou, X. Wang: J. Non-Cryst. Solids **347**, 31 (2004)
19.87 J. Vazquez, D. Garcia-G. Barreda, P. L. Lopez-Alemany, P. Villares, R. Jimenez-Garay: J. Non-Cryst. Solids **345**, 142 (2004) and references therein
19.88 A. Pratap, K. N. Lad, T. L. S. Rao, P. Majmudar, N. S. Saxena: J. Non-Cryst. Solids **345**, 178 (2004)
19.89 K. Matusita, S. Sakka: Bull. Inst. Chem. Res. **59**, 159 (1981)
19.90 K. Matusita, T. Komatsu, R. Yokota: J. Mater. Sci. **19**, 291 (1984)
19.91 J. A. Augis, J. W. E. Bennett: J. Therm. Anal. **13**, 283 (1978)
19.92 H. J. Borchardt, F. Daniels: J. Am. Ceram. Soc. **78**, 41 (1957)
19.93 G. O. Pilonyan, I. D. Ryabchikov, O. S. Novikova: Nature **212**, 1229 (1966)
19.94 T. Ozawa: Polymer **12**, 150 (1971)
19.95 H. S. Chen: J. Non-Cryst. Solids **27**, 257 (1978)
19.96 M. Reading, D. Elliott, V. L. Hill: J. Therm. Anal. **40**, 949 (1993)
19.97 M. Reading: Trends Polym. Sci. **1**, 248 (1993)
19.98 M. Reading, A. Luget, R. Wilson: Thermochim. Acta **238**, 295 (1994)
19.99 E. Verdonck, K. Schaap, L. C. Thomas: Int. J. Pharm. **192**, 3 (1999)
19.100 C. M. A. Lopes, M. I. Felisberti: Polym. Test. **23**, 637 (2004)
19.101 T. Wagner, M. Frumar, S. O. Kasap: J. Non-Cryst. Solids **256**, 160 (1999)
19.102 P. Boolchand, D. G. Georgiev, M. Micoulaut: J. Optoelectron. Adv. Mater. **4**, 823 (2002) and references therein
19.103 K. J. Jones, I. Kinshott, M. Reading, A. A. Lacey, C. Nikopoulos, H. M. Pollosk: Thermochim. Acta **305**, 187 (1997)
19.104 Z. Jiang, C. T. Imrie, J. M. Hutchinson: Thermochim. Acta **315**, 1 (1998)
19.105 B. Wunderlich: Thermochim. Acta **355**, 43 (2000)

19.106 H. Huth, M. Beiner, S. Weyer, M. Merzlyakov, C. Schick, E. Donth: Thermochim. Acta **377**, 113 (2001)
19.107 Z. Jiang, C. T. Imrie, J. M. Hutchinson: Thermochim. Acta **387**, 75 (2002)
19.108 T. Wagner, S. O. Kasap, K. Maeda: J. Mater. Res. **12**, 1892 (1997)
19.109 I. Okazaki, B. Wunderlich: J. Polym. Sci. **34**, 2941 (1996)
19.110 L. Thomas, A. Boller, I. Okazaki, B. Wunderlich: Thermochim. Acta **291**, 85 (1997)
19.111 L. Thomas: NATAS Notes (North American Thermal Analysis Society, Sacramento, CA,USA) **26**, 48 (1995)
19.112 B. Hassel: NATAS Notes (North American Thermal Analysis Society, Sacramento, CA, USA) **26**, 54 (1995)
19.113 J. M. Hutchinson, S. Montserrat: J. Therm. Anal. **47**, 103 (1996)
19.114 J. M. Hutchinson, S. Montserrat: Thermochim. Acta **305**, 257 (1997)
19.115 J. M. Hutchinson: Thermochim. Acta **324**, 165 (1998)
19.116 J. M. Hutchinson, S. Montserrat: J. Therm. Anal. **377**, 63 (2001) and references therein
19.117 A. Boller, C. Schick, B. Wunderlich: Thermochim. Acta **266**, 97 (1995)
19.118 J. M. Hutchinson, A. B. Tong, Z. Jiang: Thermochim. Acta **335**, 27 (1999)
19.119 D. Tonchev, S. O. Kasap: Mater. Sci. Eng. **A328**, 62 (2002)
19.120 J. E. K. Schawe: Thermochim. Acta **271**, 127 (1996)
19.121 P. Kamasa, M. Pyda, A. Buzin, B. Wunderlich: Thermochim. Acta **396**, 109 (2003)
19.122 D. Tonchev, S. O. Kasap: Thermal Characterization of Glasses and Polymers by Temperature Modulated Differential Scanning Calorimetry: Glass Transition Temperature. In: *High Performance Structures and Materials II*, ed. by C. A. Brebbia, W. P. De Wilde (WIT, Southampton, UK 2004) pp. 223–232
19.123 S. Weyer, M. Merzlyakov, C. Schick: Thermochim. Acta **377**, 85 (2001)
19.124 L. E. Waguespack, R. L. Blaine: Design of a new DSC cell with Tzero technology. In: *Proceedings of the 29th North American Thermal Analysis Society, St. Louis, September 24–26*, ed. by K. J. Kociba (NATAS, Sacramento 2001) pp. 722–727
19.125 R. L. Danley: Thermochim. Acta **395**, 201 (2003)

20. Electrical Characterization of Semiconductor Materials and Devices

Semiconductor materials and devices continue to occupy a preeminent technological position due to their importance when building integrated electronic systems used in a wide range of applications from computers, cell-phones, personal digital assistants, digital cameras and electronic entertainment systems, to electronic instrumentation for medical diagnositics and environmental monitoring. Key ingredients of this technological dominance have been the rapid advances made in the quality and processing of materials – semiconductors, conductors and dielectrics – which have given metal oxide semiconductor device technology its important characteristics of negligible standby power dissipation, good input–output isolation, surface potential control and reliable operation. However, when assessing material quality and device reliability, it is important to have fast, nondestructive, accurate and easy-to-use electrical characterization techniques available, so that important parameters such as carrier doping density, type and mobility of carriers, interface quality, oxide trap density, semiconductor bulk defect density, contact and other parasitic resistances and oxide electrical integrity can be determined. This chapter describes some of the more widely employed and popular techniques that are used to determine these important parameters. The techniques presented in this chapter range in both complexity and test structure requirements from simple current-voltage measurements to more sophisticated low-frequency noise, charge pumping and deep-level transient spectroscopy techniques.

The continued evolution of semiconductor devices to smaller dimensions in order to improve performance – speed, functionality, integration density and reduced cost – requires layers or films of semiconductors, insulators and metals with increasingly high quality that are well-characterized and that can be deposited and patterned to very high precision. However, it is not always the case that improvements in the quality of materials have kept pace with the evolution of integrated circuit down-scaling. An important aspect of assessing the material quality and device reliability is the development and use of fast, nondestructive and accurate electrical characterization techniques to determine important parameters such as carrier doping density, type and mobility of carriers, interface quality, oxide trap density, semiconductor bulk defect density, contact and other parasitic resistances and oxide electrical integrity. This chapter will discuss several techniques that are used to determine these important parameters. However, it is not an extensive compilation of the electrical techniques currently used by the research and development community; rather, it presents a discussion of some of the more widely used and popular ones [20.1–4].

An important aspect of electrical characterization is the availability of appropriate test components [20.1–4]. In this chapter, we concentrate on discussing techniques that use standard test devices and structures. In addition, we will use the MOSFET whenever possible because they are widely available on test chips. This is also motivated by the fact that MOSFETs continue to dominate the semiconductor industry for a wide range of applications from memories and microprocessors to signal and imaging processing systems [20.5]. A key reason for this dominance is the excellent quality of the silicon wafers and the silicon–silicon dioxide interface, both of which play critical roles in the performance and reliability of the device. For example, if the interface has many defects or interface states, or it is rough, then the device's carrier mobility decreases, low-frequency noise increases and its performance and reliability degrades. In particular, it is not only the interface that is important, but also the quality of the oxide; good-quality oxide prevents currents from flowing between the gate and substrate electrodes through the gate oxide. Both interface and oxide quality allows for excellent isolation between the input and output terminals of the MOSFETs, causing it to behave as an almost ideal switch. Therefore, it is important to have good experimental tools to study the interface properties and the quality of the gate dielectric.

Electrical characterization of semiconductors and the semiconductor–dielectric interface is important for a variety of reasons. For example, the defects at and in the interfacial oxide layer in silicon–silicon dioxide ($Si–SiO_2$) systems and in the bulk semiconductor play critical roles in their low-frequency noise, independent of whether the device is surface-controlled such as a MOSFET, or a bulk transport device such as a polysilicon emitter bipolar junction transistor (PE BJT). These defects can affect the charge transfer efficiency in charge coupled devices (CCDs), p–n photodiodes or CMOS imagers, and can be the initiation point of catastrophic failure of oxides. Interface and bulk states can act as scattering centers to reduce the mobility in MOSFETs, thus affecting their performance parameters such as switching speed, transconductance and noise.

This chapter is devoted to the electrical characterization of semiconductors, insulators and interfaces. In the first part (Sects. 20.1 and 20.2), the basic electrical properties of materials (such as resistivity, concentration and mobility of carriers) are studied. The main measurement techniques used to determine these electrical parameters are presented. Due to its increasing importance in modern ultrasmall geometry devices, electrical contacts are also studied. All of the characterization techniques presented in this first part are associated with specially designed test structures. In the second part (Sects. 20.4 to 20.7), we use active components such as capacitors, diodes and transistors (mainly MOSFETs) in order to determine more specific electrical parameters such as traps, oxide quality and noise level that are associated with material or devices. Of course this involves specific measurement techniques that are often more sophisticated than those discussed in the previous two sections.

20.1 Resistivity

Resistivity is one of the most important electrical parameters of semiconductors [20.1–4]. First, we present the basic physical relations concerning the bulk resistivity. The main electrical measurement techniques are then described: the two oldest ones that are still relevant today – the *four-point-probe* technique and the *van der Pauw* technique – and then the *spreading resistance* technique. Second, because it is closely linked with bulk resistivity measurement techniques and it is increasingly important in modern ultrasmall geometry devices, contact resistivity will be presented. Special attention will be given to *Kelvin contact resistance* (*KCR*) *measurement* and the *transmission line measurement* (*TLM*) techniques.

20.1.1 Bulk Resistivity

Physical Approach, Background and Basics

The bulk resistivity ρ is an intrinsic electrical property related to carrier drift in materials such as metals and semiconductors [20.6]. From a macroscopic point of view, the resistivity ρ can be viewed as the normalization of the bulk resistance (R) by its geometrical dimensions

– the cross-sectional area ($A = Wt$) through which the current flows, and the distance between the two ideal contacts L, as shown in Fig. 20.1. The resistivity is given by

$$\rho = \frac{RA}{L} \text{ in } \Omega\text{m or commonly } \Omega\text{cm}. \tag{20.1}$$

For thin semiconductor layers, the sheet resistivity ρ_s is often used instead of the bulk resistivity ρ. The sheet resistivity ρ_s is the bulk resistivity divided by the sample's thickness t. This normalized parameter is related to the resistance of a square of side L. For this particular geometry in Fig. 20.1, since $A = Wt$, then $\rho_s = R_\square$, the sheet resistance. The unit of sheet resistance is Ω/square or $\Omega/\square$. The parameter $R_\square$ is convenient for integrated circuit designers because it allows them to quickly design the geometry for a specific value of resistance using very thin implanted or diffused semiconductor regions or polycrystalline layers. Resistivity (or its inverse, the conductivity σ in $\Omega^{-1}\text{cm}^{-1}$ or S/cm) and its variation with temperature is often used to classify material into metals, semiconductors and insulators.

Since different semiconductors can have the same resistivity, and also different values of resistivity can be found for a given semiconductor, depending on how it is processed for example, then resistivity is not a fundamental material parameter. From solid state theory, in the case of homogeneous semiconductor materials, the resistivity expresses the proportionality between the applied electric field E and the drift current density J; that is, $J = (1/\rho)\,E$. It can be defined by the microscopic relation:

$$\rho = \frac{1}{q(n\mu_n + p\mu_p)}, \tag{20.2}$$

where q is the electronic charge, n and p are the free electron and hole concentrations, and μ_n and μ_p are the electron and hole drift mobilities, respectively. In this way, the resistivity is related to fundamental semiconductor parameters: the number of free carriers, and their

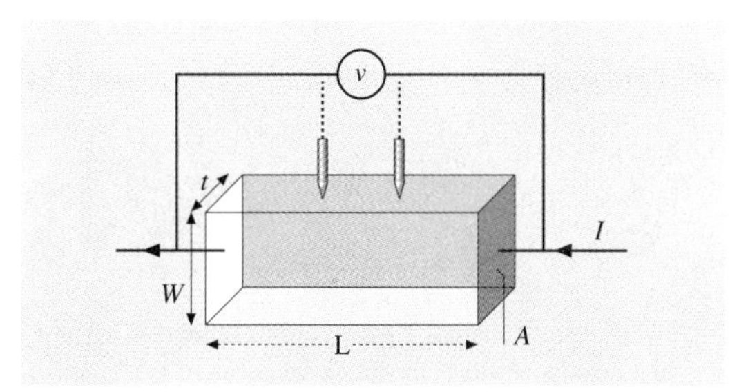

Fig. 20.1 Bulk resistance and its geometrical dimensions

ability to move in the lattice when an electric field is applied.

In n-type or donor (N_D)-doped, or p-type or acceptor (N_A)-doped semiconductors, the free carrier densities are determined by the ionized impurities (N_D or $N_A \gg$ the intrinsic carrier concentration n_i), then (20.2) can be simplified to:

$$\rho \approx \frac{1}{qn\mu_n}, \text{ for an n-type semiconductor}, \tag{20.3a}$$

$$\text{and } \rho \approx \frac{1}{qp\mu_p}, \text{ for an p-type semiconductor}. \tag{20.3b}$$

In the following sections, only single-type semiconductors will be studied. This corresponds to most semiconductor materials used in electronic and optoelectronic devices because either $N_D \gg N_A$ or $N_A \gg N_D$ in a typical semiconductor layer.

Measurement Techniques

The simplest way to determine bulk resistivity is to measure the voltage drop along a uniform semiconductor bar through which a DC current I flows, as shown in Fig. 20.1. Thus, the measured resistance and knowledge of the geometrical dimensions can lead to an estimate for the bulk resistivity according to (20.1). Unfortunately the measured resistance (R_{mea}) includes the unexpected contact resistance ($2R_c$), which can be significant for small-geometry samples because R_c is strongly dependent on the metal-semiconductor structure. Therefore, special processing technologies are used to minimize the influence of R_c (see Sect. 20.1.2). Now, the measured resistance is expressed as

$$R_{mea} = R + 2R_c. \tag{20.4}$$

If probes are used instead of large metal-semiconductor contacts, then the spreading resistance (R_{sp}) under the two probes must also be added, as shown in Fig. 20.2. In this case, (20.4) becomes

$$R_{mea} = R + 2R_{sp} + 2R_c, \tag{20.5}$$

where R_{sp} for a cylindrical contact of radius r, and for a semi-infinite sample, it can be expressed by

$$R_{sp} = \frac{\rho}{4r}. \tag{20.6}$$

For a hemispherical contact of radius r, R_{sp} is given by

$$R_{sp} = \frac{\rho}{2\pi r}. \tag{20.7}$$

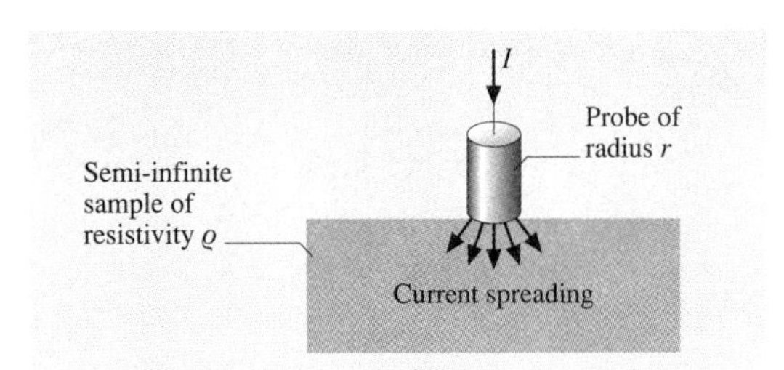

Fig. 20.2 Spreading resistance associated with a probe contact

In both cases, it is very difficult to provide a direct measurement or an accurate model of the contact resistance. So determining the bulk resistivity by this approach is not recommended, except when the spreading resistance is the dominant term in (20.5) and when (20.6) or (20.7) can be applied. In this case, the resistivity is determined by the *spreading resistance* technique measurement. Nevertheless, despite the lack of accuracy of the two contact techniques, it can be sufficient for monitoring some process steps and it is often used in the semiconductor industry as a process monitor.

Four-Point Probe Technique. In order to eliminate or at least minimize the contact contribution to the measured resistance value, techniques based on separate current injection and voltage drop measurements have been developed. First, the two-probe technique can be used, as reported in Fig. 20.1. This measurement is very simple, but it is affected by several parameters: lateral contact geometry, probe spacing, and minority carrier injection near the lateral contacts. The main disadvantage of this technique is the need for lateral contacts. This requirement is overcome with the four-point probe technique, where two probes are used for current injection and the other two probes are used to measure the voltage drop. The more usual probe geometry configuration is when the four probes are placed in a line, as shown in Fig. 20.3.

The voltage at probe 2, V_2, induced by the current flowing from probe 1 to probe 4 is given by:

$$V_2 = \frac{\rho I}{2\pi} \cdot \left(\frac{1}{s_1} - \frac{1}{s_2 + s_3} \right) . \tag{20.8a}$$

The voltage at probe 3 is:

$$V_3 = \frac{\rho I}{2\pi} \cdot \left(\frac{1}{s_1 + s_2} - \frac{1}{s_3} \right) . \tag{20.8b}$$

Then, by measuring $V = V_2 - V_3$, the voltage drop between probes 2 and 3, and the current I through probes 1 and 4, the resistivity can be determined using (20.8a) and (20.8b) as

$$\rho = \frac{2\pi \mathrm{V/I}}{\left(\frac{1}{s_1} + \frac{1}{s_2} - \frac{1}{s_2+s_3} - \frac{1}{s_1+s_2} \right)} \tag{20.9}$$

Thus, a direct measurement of the resistivity can be made using a high-impedance voltmeter and a current source. When the probe spacings are equal ($s_1 = s_2 = s_3 = s$), which is the most practical case, then (20.9) becomes

$$\rho = 2\pi s \cdot \frac{V}{I} . \tag{20.10}$$

Equations (20.9) and (20.10) are valid only for semi-infinite samples; that is, when both t and the sample surface are very large ($\to \infty$), and the probes' locations must be far from any boundary. Because these relations can be applied only to large ingots, then in many cases a correction factor f must be introduced in order to take into account the finite thickness and surface of the sample and its boundary effects. Further, for epitaxial layers, f must also consider the nature of the substrate – whether it is a conductor or an insulator. Thus, (20.10) becomes

$$\rho = 2\pi s \cdot \frac{V}{I} \cdot f . \tag{20.11}$$

For a thin semiconductor wafer or thin semiconducting layer deposited on an insulating substrate, and for the condition $t < s/2$, which represents most practical cases because the probe spacing s is usually on the order of a millimeter, then the correction factor due to the thickness is

$$f = \frac{(t/s)}{2\ln 2} \text{ so that } \rho = 4.532 t \frac{V}{I} . \tag{20.12}$$

The noninfinite sample surface must be corrected if the ratio of the wafer diameter to the probe spacing is not

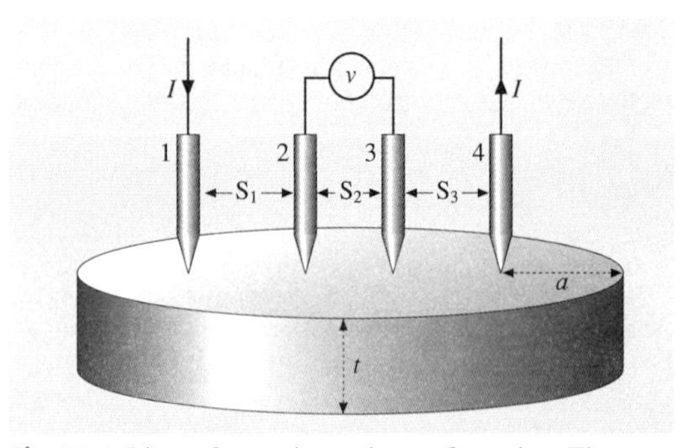

Fig. 20.3 Linear four-point probe configuration. The sample thickness is t and a is the distance from the edge or boundary of the sample

greater than 40, otherwise a correction factor of less than unity has to be introduced [20.3].

If the probe header is too close to any boundary, then (20.12) is no longer valid and another correction factor must be introduced. This correction factor is close to 1 until the ratio a/s is greater than 2, where a is the distance from the edge of the sample that is shown in Fig. 20.3. A study of various 8 inch n- and p-type silicon wafers have demonstrated that the edge exclusion limit is 5 mm [20.7].

In the case of a different arrangement of probes, for instance a square array or when a different measurement configuration of the four-point collinear probes is used, such as current injection between probes 1 and 3, other specific correction factors are required. Here, rather than detail all the different correction factors, complementary information can be found in Chap. 4 of [20.4] and Chap. 1 of [20.3].

Taking into account the appropriate correction factors as well as some specific material parameters such as hardness or surface oxidation, it is possible to map the resistivities of different types of semiconductor wafers or deposited semiconductor layers with an accuracy better than 1% over a large range of resistivity values using commercial equipment and appropriate computational techniques.

Van der Pauw Technique. Based on the same basic principle of separating the current injection and voltage measurement, the *van der Pauw* [20.8] measurement technique allows for the determination of resistivity on a sample of arbitrary shape using four small contacts placed on the periphery, as shown in Fig. 20.4. Then, the resistivity of a uniform sample of thickness t is given by

$$\rho = \frac{\pi t}{\ln 2} \frac{(R_A + R_B)}{2} f \,. \tag{20.13}$$

Here, R_A and R_B are resistances measured by injecting current on two adjacent contacts and by measuring the voltage drop on the two remaining ones. With the notation in Fig. 20.4, one can define

$$R_A = \frac{V_3 - V_4}{I_{1,2}} \,, \quad R_A = \frac{V_4 - V_1}{I_{2,3}} \,,$$
$$f \text{ is a correction factor that is} \tag{20.14}$$
$$\text{a function of the ratio} \quad R_f = \frac{R_A}{R_B} \,.$$

with R_f obtained from

$$\frac{(R_f - 1)}{(R_f + 1)} = \frac{f}{\ln 2} \cdot \operatorname{arccosh}\left(\frac{\exp(\ln 2/f)}{2}\right) \,. \tag{20.15}$$

In the case of samples with symmetrical geometries, and when the contacts are also symmetrical, as shown in Fig. 20.5, then $R_A = R_B$, $R_f = 1$ and $f = 1$, and (20.13) becomes

$$\rho = \frac{\pi t}{\ln 2} R_A = 4.532 t R_A \tag{20.16}$$

In order to minimize errors caused by the finite dimensions of the contacts (since ideally the contact area should be zero) and the finite thickness of the sample, then the distance between the contacts must be larger than both the diameter and the thickness of the contact. Also, the cloverleaf configuration in Fig. 20.5d is recommended to prevent contact misalignment, but this configuration requires a more complicated patterning technology.

The main advantage of the van der Pauw technique compared to the four-point probe technique is its use of a smaller area for the test structure. Therefore, this measurement technique is often used in integrated circuit technology. Also, because of its simple structure, the Greek cross configuration in Fig. 20.5b is

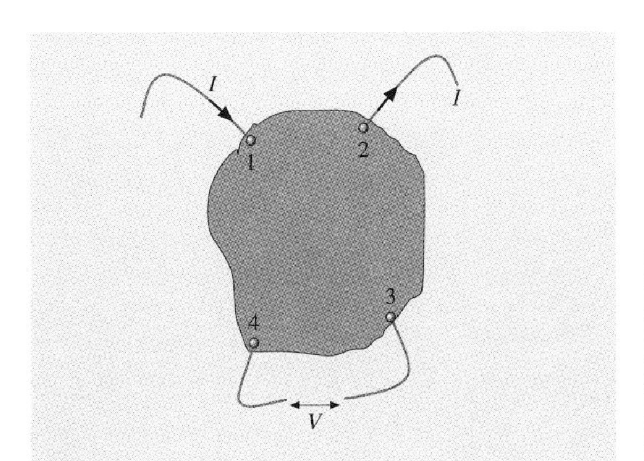

Fig. 20.4 van der Pauw method for an arbitrarily shaped sample

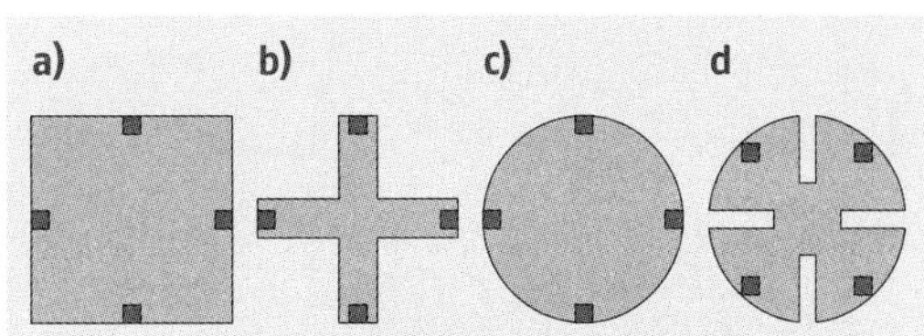

Fig. 20.5a–d Symmetrical van der Pauw structures: (**a**) square, (**b**) Greek cross, (**c**) circle and (**d**) cloverleaf

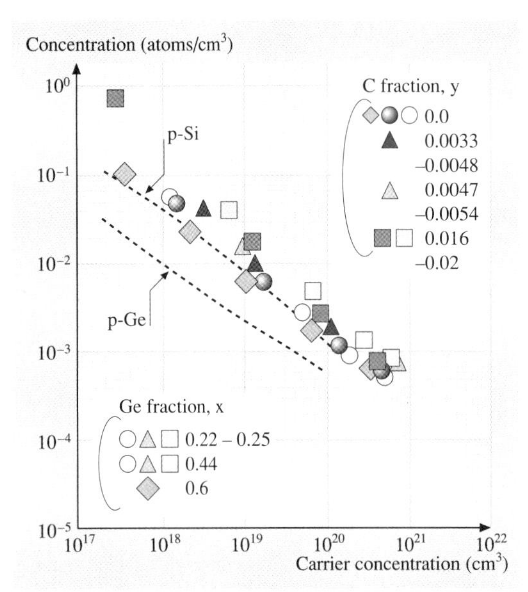

Fig. 20.6 Resistivity versus carrier concentration in $Si_{1-x-y}Ge_xC_y$ films obtained using a cloverleaf van der Pauw structure. These results are from [20.10]

widely used (experimental results obtained on SiGeC epitaxial layers are reported in Fig. 20.6 as an example). However, when narrows arms are used, current crowding at the corners may have a significant influence, and in this case a different Greek cross layout can be considered to reduce this current crowding effect [20.9].

Spreading Resistance Technique. The spreading resistance technique is based on the modeling of current spreading from a probe tip or a small metallic contact and flowing into a bulk semiconductor, as shown previously in Fig. 20.2. Equations (20.6) and (20.7) presented above are for cylindrical probes and hemispherical probes, respectively. Basically, the principle of this method is opposite to the previous four-contact techniques where the separation of the current injection from the measured voltage drop was used to avoid the spreading resistance. Here, the spreading resistance is expected to be the dominant term in (20.5). Only two contacts are needed: two closely aligned probes, a small top contact probe or a metallic contact and a large bottom contact. In the first case, surface mapping can be performed, but the main use of this compact probe configuration is for resistivity profiling using a bevelled sample [20.3]. The second configuration has been used to measure the substrate resistivity of silicon integrated circuits where simple test structures – for example the square top contact of 25 μm × 25 μm and 50 μm × 50 μm shown in [20.12] – have been included on a test chip.

More recently, semiconductor resistivity has been nanocharacterized using scanning spreading resistance microscopy (SSRM) with a standard atomic force microscope (AFM) of lateral resolution of 10 to 20 nm. A SSRM image of a 0.5 μm nMOSFET is given in

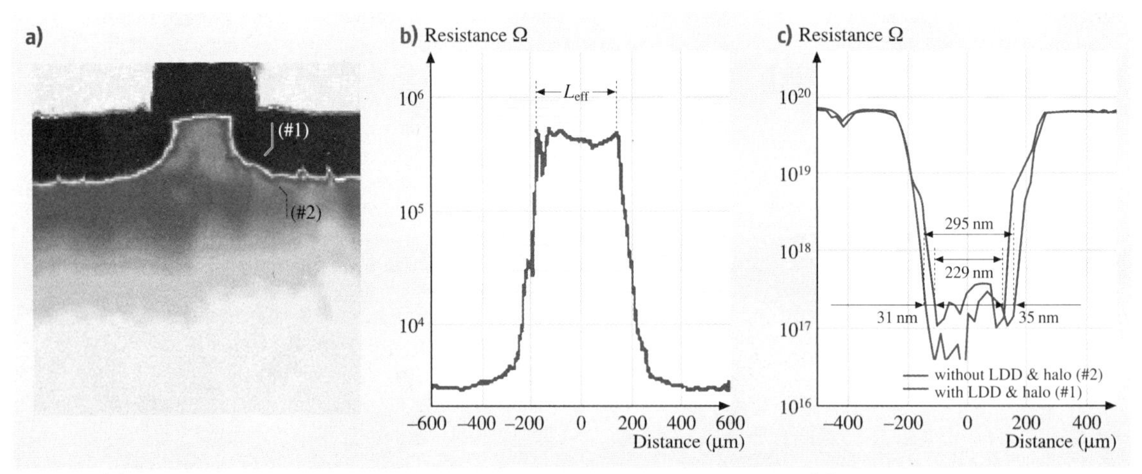

Fig. 20.7 (a) SSRM resistance image (scan size: 1.5 μm × 1.5 μm) of a 0.5 μm nMOSFET; (b) lateral section taken 10 nm under the gate oxide of the same transistor; (c) lateral carrier concentration profiles measured with SSRM 10 nm under the gate oxide for two 0.35 μm nMOSFETs (with and without halo and LDD process). After [20.11]

Fig. 20.7a [20.11]. The resistance is low in the highly doped regions (dark) and high in the lower doped regions (bright): source, drain, gate and well regions are clearly observed in Fig. 20.7a or the resistance profile in Fig. 20.7b. With such a high resolution, scanning the lateral and vertical diffusion of dopants in active regions of submicron transistors is possible. An example is shown in Fig. 20.7c where the extra implantations [halo and lightly doped drain (LDD)] in a 0.35 μm nMOSFET process are clearly visible and result in a change of L_{eff} from 295 nm without extra implantations to 229 nm with the extra implantations.

20.1.2 Contact Resistivity

The contact resistance of an active device and interconnection becomes larger as the dimensions are scaled down. As a consequence, the performance of single transistors as well as integrated circuits can be seriously limited by increasing *RC* time constants and power consumption. This is of major interest for the semiconductor industry, as reported by the International Technology Roadmap for Semiconductors, ITRS 2001 [20.5], and in [20.13].

Contact Resistance Elements

Basically, the contact resistance R_c is the resistance localized from a contact pad, a probe or from the metallization process to an active region. However, it does not include all of the access resistances between these two regions, as shown in Fig. 20.8a for a horizontal contact and Fig. 20.8b for a vertical contact.

Starting from the contact pad (Fig. 20.8a and Fig. 20.9), the contact resistance includes the resistance of the metal R_m, the interfacial metal-semiconductor resistance R_i, and the resistance associated with the semiconductor just below the contact in the contact region R_{sc}. Thus, the contact resistance can be expressed

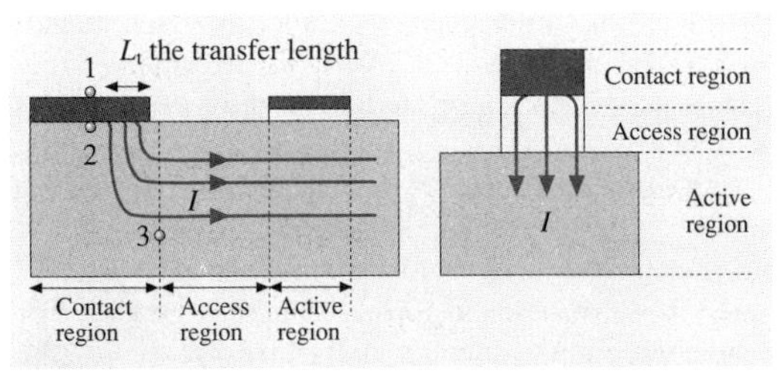

Fig. 20.8 (**a**) Horizontal contact and (**b**) vertical contact. *Black* indicates the metallic conductor, *white* the semiconductor material or an insulator

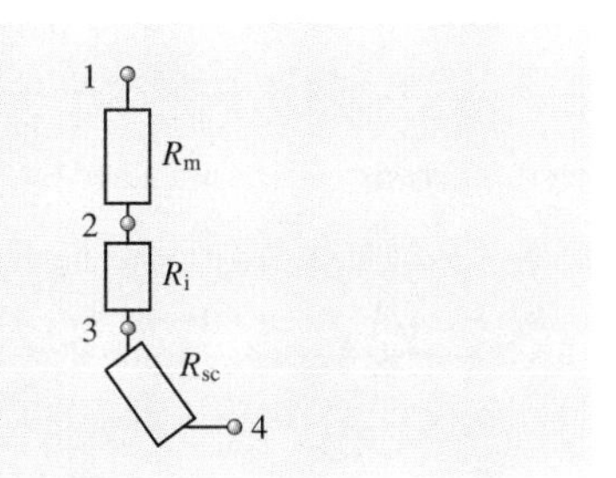

Fig. 20.9 Different components of the contact resistance

as

$$R_c = R_m + R_i + R_{sc} \,. \tag{20.17}$$

The last component R_{sc} cannot be accurately defined because the boundary between the contact and access regions is very difficult to determine due to (for example) interdiffusion of metal and semiconductor atoms, and because the current flow into this region is not homogeneous due to current spreading and lateral or vertical current crowding at the periphery of the contact. The relative importance of each component of R_c is strongly dependent on different parameters of the process itself – annealing temperature, doping density and the geometry used (lateral or vertical).

When comparing different contact technologies and different contact areas, the most convenient parameter to use is the contact resistivity ρ_c, which is referred to as the specific contact resistance in $\Omega\,cm^2$, and ρ_c is given by

$$\rho_c = R_c A_{ceff} \,, \tag{20.18}$$

where A_{ceff} is the effective contact area; that is, the current injection area. The concept of an effective contact area can be approximated by the contact geometry in the case of a vertical contact in Fig. 20.8b. However, A_{ceff} is more difficult to specify for a lateral contact, where a transfer length L_T, representing the length where the current flow transfers from the contact into the semiconductor just underneath, must be introduced, as shown in Fig. 20.8a. L_T is defined as the length over which the voltage drops to e^{-1} of its value at the beginning of the contact [20.3], and is given by

$$L_T = \sqrt{\frac{\rho_c}{\rho_{sc}}} \,, \tag{20.19}$$

where ρ_{sc} is the sheet resistivity of the semiconductor below the contact.

Because of its various components, it is difficult to accurately model the contact resistivity. Nevertheless, a theoretical approach to the interfacial resistivity [see R_i in (20.17)], ρ_i, can be determined from the well-known Schottky theory of metal–semiconductor contacts. The interfacial resistivity ρ_i is defined by

$$\rho_i = \left.\frac{\partial V}{\partial J}\right|_{V=0} . \quad (20.20)$$

This metal–semiconductor structure is equivalent to an abrupt p-n junction. According to the Schottky theory (for more details see Chap. 5 of [20.14]), the J–V characteristic of a metal–semiconductor contact in the case of a low-doped semiconductor is given by

$$J = A^* T^2 \exp\left(-\frac{q\phi_B}{kT}\right)\left[\exp\left(\frac{qV}{kT}\right) - 1\right] , \quad (20.21)$$

where A^* is Richardson's constant, and T the absolute temperature. ϕ_B is the barrier height formed at the metal–semiconductor interface – the difference between the vacuum level and the Fermi level of the metal and of the semiconductor materials respectively, and ϕ_B is given by

$$\phi_B = \phi_M - \chi \quad (20.22)$$

where ϕ_M is the metal work function and χ the semiconductor electron affinity.

The energy band diagram for a low-doped n-type semiconductor–metal contact is shown in Fig. 20.10. In this case, the current transport is dominated by the thermionic emission current, resulting in a rectifying contact.

Thus, when the conduction mechanism is controlled by the thermionic emission, the interfacial resistivity in (20.20) is simply obtained from the derivative of (20.21), and $\rho_{i,TE}$ is

$$\rho_{i,TE} = \frac{k}{qA^*T} \exp\left(\frac{q\phi_B}{kT}\right) \quad (20.23)$$

Due to the presence of surface states, the barrier height ϕ_B is positive and weakly dependent on the metal–semiconductor material. ϕ_B is $\approx 2E_g/3$ for an n-type

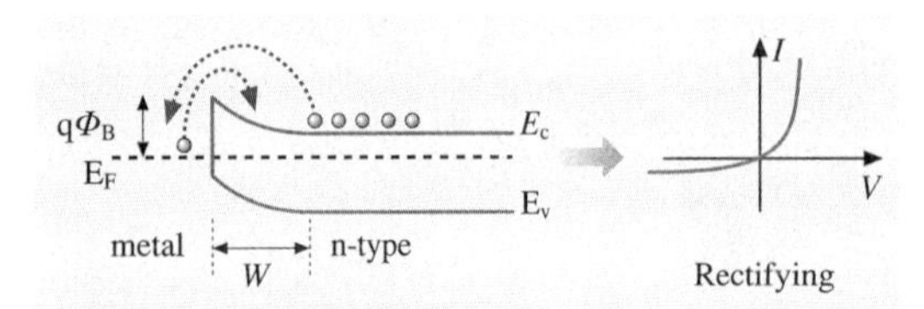

Fig. 20.10 Energy-band diagram of an n-type semiconductor–metal contact and related rectifying contact. W is the width of the depletion layer

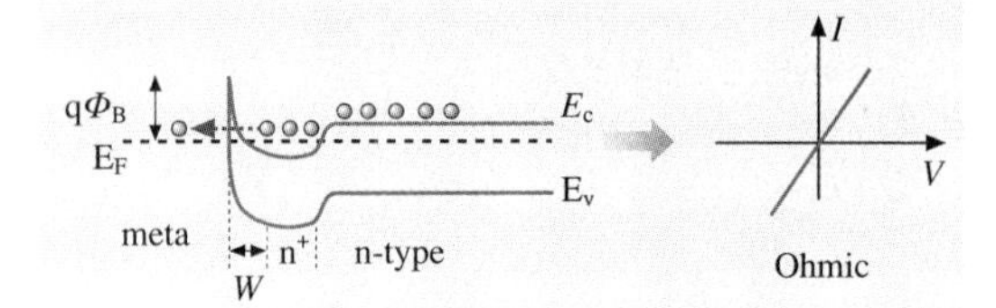

Fig. 20.11 Energy-band diagram of an n^+-n semiconductor–metal structure and related ohmic contact

semiconductor and $\approx E_g/3$ for a p-type semiconductor. Therefore, high values of interfacial resistivity $\rho_{i,TE}$ are usually obtained except when narrow bandgap semiconductors are used.

The way to fabricate ohmic contacts with low contact resistivity values is to process the metal on a heavily doped semiconductor layer. In this case, the depletion width decreases ($W \approx N_D^{-1/2}$) and the probability of carrier tunneling through the barrier increases. Thus, the conduction mechanism is dominated by tunneling, as shown in Fig. 20.11.

The electron tunneling current is expressed as

$$J_{tun} \approx \exp\left(\frac{q\phi_B}{E_{00}}\right) , \quad (20.24)$$

where

$$E_{00} = \frac{q\hbar}{2}\sqrt{\frac{N_D}{\varepsilon_s m_n^*}} , \quad (20.25)$$

ε_s is the permittivity of the semiconductor and m_n^* is the effective mass of the electron.

From (20.20), (20.24) and (20.25), the interfacial resistivity $\rho_{i,T}$ is found to be

$$\rho_{i,T} \propto \frac{2\sqrt{\varepsilon_s m_n^*}}{\hbar} \cdot \frac{\phi_B}{\sqrt{N_D}} . \quad (20.26)$$

Comparing $\rho_{i,TE}$ from (20.23) to $\rho_{i,T}$ from (20.26), we see that a highly doped layer can significantly reduce the interfacial resistivity. For $N_D \geq 10^{19}\,\mathrm{cm}^{-3}$, the tunneling process dominates the interfacial resistivity, while for $N_D \leq 10^{17}\,\mathrm{cm}^{-3}$, the thermionic emission current is dominant.

As most semiconductors such as Si, SiGe, GaAs, InP are of relatively wide bandgap, the deposition of a heavily doped layer before the metallization is commonly used in order to form a tunneling contact. For compound semiconductor manufacturing processes, the contact layer is generally formed from the same semiconductor material, or at least from the same material as the substrate. For silicon and related materials such as SiGe alloys or polysilicon, silicidation techniques are commonly used to make the contact layer with very thin silicide layers such as $CoSi_2$ or $TiSi_2$ layers.

Measurement Techniques

As mentioned above, it is difficult to accurately model the contact resistance, so direct measurements of the contact resistance or of the contact resistivity are of great importance. The two main test structures used to determine contact characteristics will now be discussed: the *cross Kelvin resistor* (*CKR*) *test structure* and the *transmission line model* (*TLM*) structure.

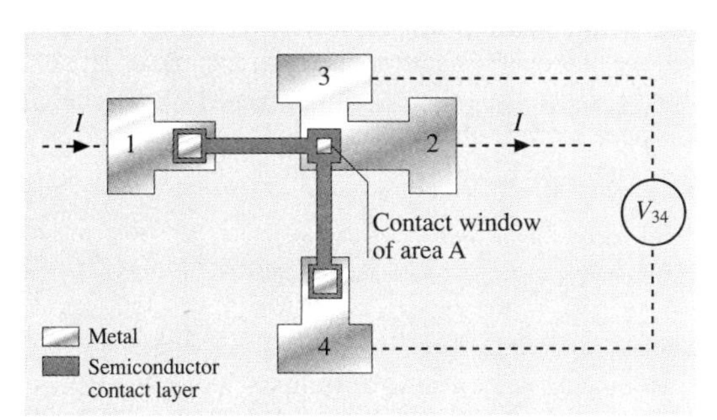

Fig. 20.12 Cross Kelvin resistor test structure

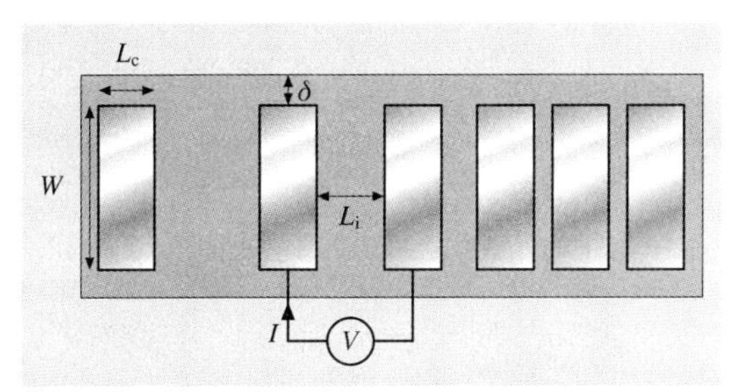

Fig. 20.13 Transmission line model (TLM) test structure

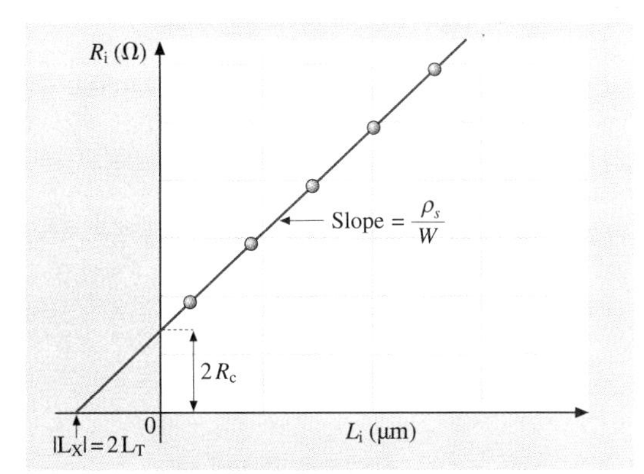

Fig. 20.14 Determination of the sheet resistivity and characterization of the contact using a TLM test structure

Kelvin Test Structure. The Kelvin test structure, also referred to as the cross Kelvin resistor (CKR) test structure, is shown in Fig. 20.12. The contact resistance R_c is determined from the potential drop in the contact window (V_{34}) when a current I is forced through the contact window from contact pad 1 to pad 2, and R_c is

$$R_c = \frac{V_{34}}{I} . \tag{20.27}$$

Therefore, a measure of R_c and knowledge of the contact area A allows for direct extraction of the contact resistivity ρ_c, given by

$$\rho_c = R_c A . \tag{20.28}$$

This basic approach is not valid when parasitic effects are present. One of the main problems is current crowding around the contact. In order to extract accurate values for the contact resistivity using Kelvin test structures, it is necessary to take into account the two-dimensional current-crowding effect. This is achieved using the results from numerical simulations [20.15]. Nevertheless, the development of ohmic contacts with very low values of contact resistivities require complex technology with different materials and usually with several interfaces. In this case, a large discrepancy between the extracted and the measured contact parameters can be found [20.15, 16]. To improve the accuracy, three-dimensional models are now used to take into account the different interfacial and vertical parasitic effects [20.17].

Transmission Line Model Test Structures. The transmission line model test structure (TLM) consists of depositing a metal grid pattern of unequal spacing L_i between the contacts. This leads to a scaled planar resistor structure. Each resistor changes only by its distance L_i between two adjacent contacts, as shown in Fig. 20.13, and it can be expressed by

$$R_i = \frac{\rho_s L_i}{W} + 2R_c \tag{20.29}$$

Then, by plotting the measured resistances as a function of the contact spacing L_i, and according to (20.29), the layer sheet resistivity ρ_s and the contact resistance R_c can be deduced from the slope and from the intercept at

$L_i = 0$ respectively, as shown in Fig. 20.14:

$$\text{Slope} = \frac{\rho_s}{W}\ ;\ R_i(\text{intercept}) = 2R_c\ ;\quad \text{and } |L_i(\text{intercept})| = 2L_T \tag{20.30}$$

As discussed in Sect. 20.1.2, the most suitable parameter for characterizing a contact is its contact resistivity (ρ_c) or the specific contact resistance ($R_c A_{ceff}$), given by

$$\rho_c = R_c A_{ceff} = R_c W L_T \tag{20.31}$$

As shown in Fig. 20.8a, for a planar resistor, the effective contact area requires the notion of the transfer length L_T. According to (20.19), and assuming that the sheet resistance under the contact ρ_{sc} is equal to the sheet resistance between the contacts ρ_s, then L_T can be expressed by (20.19).

Therefore, the substitution of R_c into (20.31) in (20.29) leads to

$$R_i = \frac{\rho_s}{W} L_i + \frac{\rho_s}{W} 2L_T \tag{20.32}$$

Now, extrapolation to $R_i = 0$ allows us to determine the value of L_T. The main advantage of the *TLM* method is its ability to give two main electrical parameters, the resistivity of the semiconductor contact layer ρ_s and the contact resistance R_c. However, this is done at the expense of a questionable assumption that the sheet resistance under the contact must be equal to the sheet resistance between the contacts. More on this technique can be found in [20.3].

20.2 Hall Effect

As mentioned before, the resistivity of a semiconductor is not a fundamental material parameter. One can consider the carrier density (n or p) or the carrier mobility (μ_n or μ_p) to be fundamental or microscopic parameters. For a semiconductor material, the resistivity is related to these two parameters (density and mobility) by (20.2). The strength of the Hall effect is to directly determine the sheet carrier density by measuring the voltage generated transversely to the current flow direction in a semiconductor sample when a magnetic field is applied perpendicularly, as shown in Fig. 20.15a. Together with a resistivity measurement technique such as the four-point probe or the van der Pauw technique, Hall measurements can be used to determine the mobility of a semiconductor sample.

In modern semiconductor components and circuits, knowledge of these two fundamental parameters n/p and μ_n/μ_p is critical. Currently, Hall effect measurements are one of the most commonly used characterization tools in the semiconductor industry and research laboratories. This is not just because of the parameters that can be extracted for use in device modeling or materials characterization, but also because of the quantum Hall effect (QHE) in condensed matter physics [20.18]. Moreover, in the applied electronics domain, one should note the development of different sensors based on the physical principle of the Hall effect, such as commercial CMOS Hall sensors.

As is very often the case, the development of a characterization technique is related to its cost, simplicity

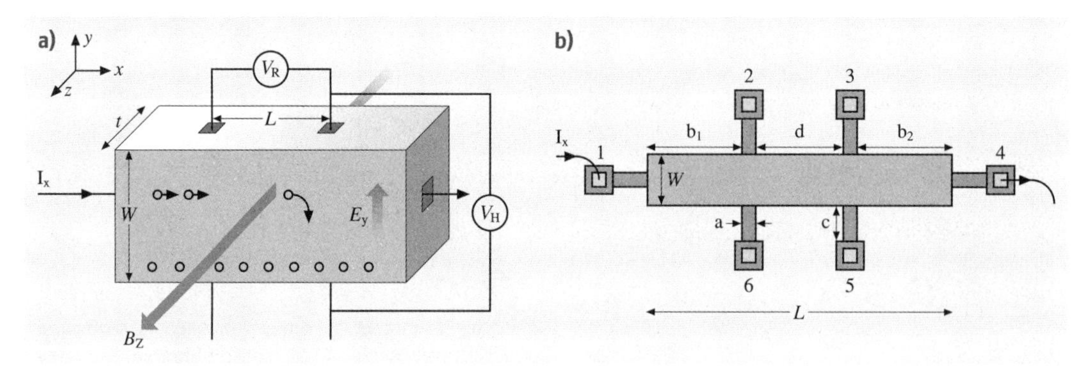

Fig. 20.15 (a) Representation of the Hall effect in an p-type bar-shaped semiconductor. (b) Practical sample geometry: a six-terminal Hall-bar geometry

of implementation and ease of use. Since these practical characteristics are satisfied even when specially shaped samples are required, then the Hall effect measurement technique has become a very popular method of characterizing materials.

In this section, we will first present the physical principle of the Hall effect. Then we will show how it can be used to determine the carrier density and mobility. Finally, the influence of the Hall scattering factor will be presented, followed by some practical issues about the implementation of the Hall effect method.

20.2.1 Physical Principles

The Hall effect was discovered by *Hall* in 1879 [20.19] during an experiment on current transport in a thin metal strip. A small voltage was generated transversely when a magnetic field was applied perpendicularly to the conductor.

The basic principle of this Hall phenomenon is the deviation of some carriers from the current line due to the Lorentz force induced by the presence of a transverse magnetic field. As a consequence, a voltage drop V_H is induced transversely to the current flow. This is shown in Fig. 20.15a for a p-type bar-shaped semiconductor, where a constant current flow I_x in the x-direction and a magnetic field in the z-direction results in a Lorentz force on the holes. If both holes and electrons are present, they deviate towards the same direction. Thus, the directions of electrical and magnetic fields must be accurately specified.

The Lorentz force is given by the vector relation

$$F_L = q(\boldsymbol{v}\times\boldsymbol{B}) = -qv_x B_z\,, \tag{20.33}$$

where v_x is the carrier velocity in the x-direction. Assuming a homogeneous p-type semiconductor

$$v_x = \frac{I}{qtWp}\,. \tag{20.34}$$

As a consequence, an excess surface electrical charge appears on one side of the sample, and this gives rise to an electric field in the y-direction E_y. When the magnetic force F_L is balanced by the electric force F_{EL}, then the Hall voltage V_H is established, and from a balance between F_L and F_{EL}, we get

$$F = F_L + F_{EL} = -qv_x B_z + qE_y = 0\,, \tag{20.35}$$

$$\text{so } E_y = \frac{BI}{qtWp}\,. \tag{20.36}$$

Also, the Hall voltage V_H is given by

$$V_H = V_y = E_y W = \frac{BI}{qtp}\,. \tag{20.37}$$

So if the magnetic field B and the current I are known, then the measurement of the Hall voltage gives the hole sheet concentration p_s from

$$p_s = pt = \frac{BI}{qV_H}\,. \tag{20.38}$$

If the conducting layer thickness t is known, then the bulk hole concentration can be determined [see (20.40)] and expressed as a function of the Hall coefficient R_H, defined as

$$R_H = \frac{tV_H}{BI} \tag{20.39}$$

$$\text{and } p = \frac{1}{qR_H} \tag{20.40}$$

Using the same approach for an n-type homogeneous semiconductor material leads to

$$R_H = -\frac{tV_H}{BI}\,, \tag{20.41}$$

$$\text{and } n = -\frac{1}{qR_H} \tag{20.42}$$

Now, if the bulk resistivity ρ is known or can be measured at the same time using a known sample such as a Hall bar or van der Pauw structure geometry in zero magnetic field, then the carrier drift mobility can be obtained from

$$\mu = \frac{|R_H|}{\rho} \tag{20.43}$$

There are two main sample geometries commonly used in Hall effect measurements in order to determine either the carrier sheet density or the carrier concentration if the sample thickness is known, and the mobility. The first one is the van der Pauw structure presented in Sect. 20.1.1. The second one is the Hall bar structure shown in Fig. 20.15b, where the Hall voltage is measured between contacts 2 and 5, and the resistivity is measured using the four-point probes technique presented in Sect. 20.1.1 (contacts 1, 2, 3 and 4). Additional information about the shapes and sizes of Hall structures can be found in [20.3, 4, 20].

Whatever the geometry used for Hall measurements, one of the most important issues is related to the offset voltage induced by the nonsymmetric positions of the contact. This problem, and also those due to spurious voltages, can be controlled by two sets of measurements, one for a magnetic field in on direction and another for a magnetic field in the opposite direction.

The Hall effect has also been investigated on specific structures, and an interesting example can be found in reference [20.21], where a Hall bar structure was combined with a double-gate n-SOI MOSFET. This was done in order to understand the mobility behavior in ultra-thin devices and to validate the classical drift mobility extraction method based on current–voltage measurements.

In the Hall effect experiment, the measurement of the Hall coefficient R_H leads to the direct determination of the carrier concentration and mobility. Moreover, the sign of R_H can be used to determine the type of conductivity of the semiconductor sample. If various types of carriers are present, then the expression for R_H becomes more complex and approximations in the limit of low and high magnetic field are necessary (see Chap. 8 of [20.3]).

We have so far discussed the Hall effect on a uniformly doped substrate or single semiconductor layer deposited on an insulating or semi-insulating substrate. In the case of a semiconductor layer deposited on a semiconducting substrate of opposite doping type, Hall effect measurements can be performed if the space charge region can act as an insulator. In the case of multilayers, the problem is more difficult, but an approximation for transport experiments has been developed for two-layer structures [20.23] and applied to different MESFET structures, for instance [20.24].

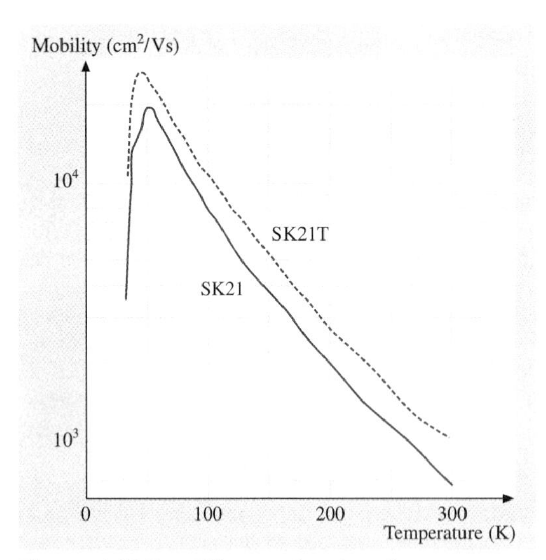

Fig. 20.16 Hall mobility as a function of temperature on two SOI films. After [20.22]

20.2.2 Hall Scattering Factor

The relations presented above are based on an energy-independent scattering mechanism. With this assumption made, the Hall carrier concentration and the Hall mobility are equal to the carrier concentration and the carrier drift mobility. When this assumption is no longer valid, these electrical parameters are different and the Hall scattering factor r_H must be taken into account. In this case (20.40), (20.42) and (20.43) must be modified as follows:

$$p_H = \frac{r_H}{qR_H} = r_H p\,, \tag{20.44}$$

$$n_H = -\frac{r_H}{qR_H} = r_H n\,, \tag{20.45}$$

and

$$\mu_H = r_H \frac{|R_H|}{\rho} = r_H \mu\,. \tag{20.46}$$

The Hall scattering factor [20.25] is related to the energy dependence of the mean free time between carrier collisions $\tau(E)$, and r_H is given by:

$$r_H = \frac{\langle \tau^2 \rangle}{\langle \tau \rangle^2}\,. \tag{20.47}$$

According to theory [20.3], the Hall scattering factor tends to unity in the limit of high magnetic field. Therefore, r_H at low magnetic fields can be determined by measuring the Hall coefficient in the limit of both high and low magnetic fields [20.25] using:

$$r_H = \frac{R_H(B)}{R_H(\infty)} \tag{20.48}$$

Depending on the scattering mechanism involved (lattice, ionized or neutral impurity, electron, or phonon scattering), r_H is found to vary between 0.6 and 2 [20.26]. However, due to valence band distortion effects, values as low as 0.26 have been found in strained p-type SiGe epilayers [20.27]. Therefore, the Hall carrier concentration and especially the Hall mobility must be distinguished from carrier concentration and carrier drift mobility.

As the different scattering mechanisms have different temperature (T) dependences, then the Hall mobility as function of temperature is often used to separate the different scattering processes. An example is given in Fig. 20.16 for silicon-on-insulator (SOI) films [20.22]. The increase in the mobility between 4–45 K, which is given by $\mu \propto T^{2.95}$, is related to the ionized donor scattering mechanism. The decrease in mobility between 46–120 K given by $\mu \propto T^{-1.55}$ is associated with lattice scattering. However, after 150 K, the rapid decrease in mobility observed, where $\mu \propto T^{-2.37}$, suggests that other scattering mechanisms as well as the lattice scattering mechanism, such as electron or phonon scattering, must be taken into account.

20.3 Capacitance–Voltage Measurements

Capacitance–voltage (C–V) measurements are normally made on metal–oxide semiconductor (MOS) or metal–semiconductor (MS) structures in order to determine important physical and defect information about the insulator and semiconductor materials. For example, high-frequency (HF) and low-frequency (LF) or quasi-static C–V measurements in these structures are used to determine process and material parameters – insulator thickness, doping concentration and profile, density of interface states, oxide charge density, and work function or barrier height. In this section, we describe various C–V measurements and how they can be used to provide process parameters as well as valuable information about the quality of the materials. A typical C–V curve for a MOS capacitor with an n-type semiconductor is shown in Fig. 20.17. For a MOS capacitor with a p-type substrate, the C–V curve be similar to that in Fig. 20.17, but reflected about the y-axis.

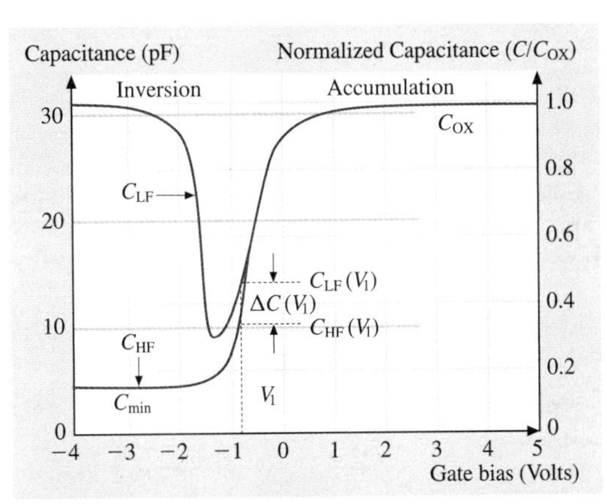

Fig. 20.17 Typical C–V curve for a MOS capacitor on an n-type substrate. After [20.28]

20.3.1 Average Doping Density by Maximum–Minimum High-Frequency Capacitance Method

The maximum–minimum high-frequency (HF) capacitance method uses the HF capacitance under strong accumulation (C_{OX}) and strong inversion ($C_{HF,min}$) to determine the average doping density (see [20.29][pp. 406–408]). Note that under strong inversion and at high frequencies, the interface trap capacitance is negligible ($C_{it} \approx 0$). Under strong inversion, the depletion width (w_{max}) is a maximum and so the high frequency capacitance per unit area $C_{HF,min}$ is a minimum, since the minority carriers cannot respond to the high-frequency signal. Since the inversion layer is very thin compared to the depletion layer, then

$$w_{max} = \varepsilon_{Si}\left(\frac{1}{C_{HF,min}} - \frac{1}{C_{OX}}\right) , \tag{20.49}$$

where ε_{Si} is the permittivity of silicon and C_{OX} is the gate oxide capacitance per unit area.

At the conditions for w_{max}, the band bending ψ_{max} is a maximum, and it is

$$\begin{aligned}\psi_{max} &= 2\phi_B + \frac{kT}{q}\ln\left(2\frac{q}{kT}\phi_B - 1\right)\\ &= 2\frac{kT}{q}\left\{\ln\left(\frac{n}{n_i}\right) + \frac{1}{2}\ln\left[2\ln\left(\frac{n}{n_i}\right) - 1\right]\right\} ,\end{aligned} \tag{20.50}$$

where $\phi_B = (k_B T/q) \times \ln(n/n_i)$ is the shift of the Fermi level from the intrinsic Fermi level $\phi_i = (E_c - E_v)/2q$ in the bulk of the silicon in the MOS structure due to the doping concentration n, and n_i is the thermally generated carrier concentration in silicon. For a uniformly doped sample,

$$w_{max}^2 = \frac{2\varepsilon_{Si}\psi_{max}}{qn} \tag{20.51}$$

and from (20.49) and (20.51), a relation between the doping concentration n and the measured capacitance

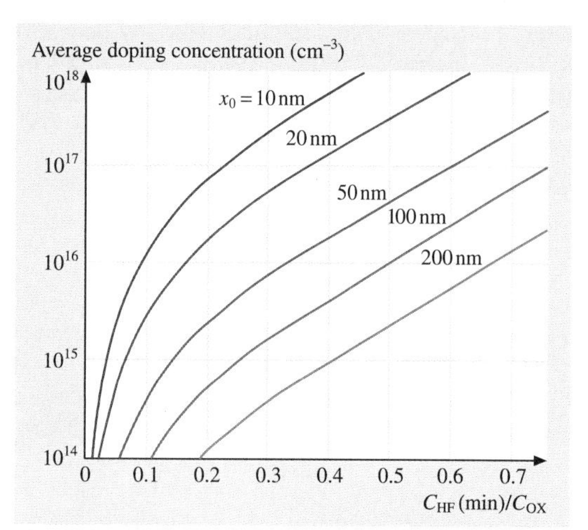

Fig. 20.18 Doping concentration n as function of $C_{HF,min}/C_{OX}$ with oxide thickness, based on (20.52). After [20.29]

can be established ([20.29], p. 407) as

$$\frac{n}{\ln\left(\frac{n}{n_i}\right)+\frac{1}{2}\ln\left[2\ln\left(\frac{n}{n_i}\right)-1\right]} = \frac{4kT}{q^2\varepsilon_{Si}}\frac{C_{OX}^2}{\left(\frac{C_{OX}}{C_{HF,min}}-1\right)^2}\,. \tag{20.52}$$

Equation (20.52) is a transcendental equation in average doping concentration n that can be solved numerically by iteration. Figure 20.18 shows the solutions as function of $C_{HF,min}/C_{OX}$ with oxide thickness, and this can be used to obtain the average doping n graphically. Equation (20.52) can be further simplified by neglecting the term $0.5\ln[2\ln(n/n_i)-1]$, and assuming $C_{OX} = C_{HF,max}$ [20.30]. Also, an approximation of (20.52) for the average doping concentration n in unit cm^{-3} is obtained in [20.4] and [20.31] for silicon MOS structures at room temperature, and this is given by

$$\begin{aligned}\log_{10}(n) &= 30.38759+1.68278\\ &\quad\times\log_{10}(C_{DM}-0.03177\\ &\quad\times[\log_{10}(C_{DM})]^2\,,\end{aligned} \tag{20.53}$$

where the depletion capacitance (per cm^2 of area) C_{DM} is defined as

$$C_{DM} = \frac{C_{HF,min}C_{OX}}{C_{OX}-C_{HF,min}}\,, \tag{20.54}$$

where all capacitances are in units of F/cm^2.

20.3.2 Doping Profile by High-Frequency and High–Low Frequency Capacitance Methods

The doping profile in the depletion layer can be obtained ([20.29], Sect. 9.4) by assuming that the depletion capacitance per unit area C_D and the oxide capacitance per unit area C_{OX} are connected in series; that is, that the measured high frequency capacitance C_{HF} is given by

$$\frac{1}{C_{HF}} = \frac{1}{C_D}+\frac{1}{C_{OX}} \Rightarrow \frac{1}{C_D} = \frac{1}{C_{HF}}-\frac{1}{C_{OX}}\,. \tag{20.55}$$

For a particular gate biasing V_G of the MOS structure, the depletion thickness $w(V_G)$ is obtained from C_D as

$$w(V_G) = \varepsilon_{Si}\left(\frac{1}{C_{HF}}-\frac{1}{C_{OX}}\right)\,. \tag{20.56}$$

The doping concentration $n(V_G)$ is given by the slope of the $(1/C_{HF})^2$ versus V_G characteristic, given by

$$n(w) = \frac{-2}{q\varepsilon_{Si}\frac{\partial}{\partial V_G}\left(\frac{1}{C_{HF}^2}\right)}\,. \tag{20.57}$$

Note that a plot of $1/C_{HF}^2$ versus V_G (Fig. 20.19) can yield important information about the doping profile. The average n is related to the reciprocal of the slope in the linear part of the $1/C_{HF}^2$ versus V_G curve, and the intercept with V_G at a value of $1/C_{OX}^2$ is equal to

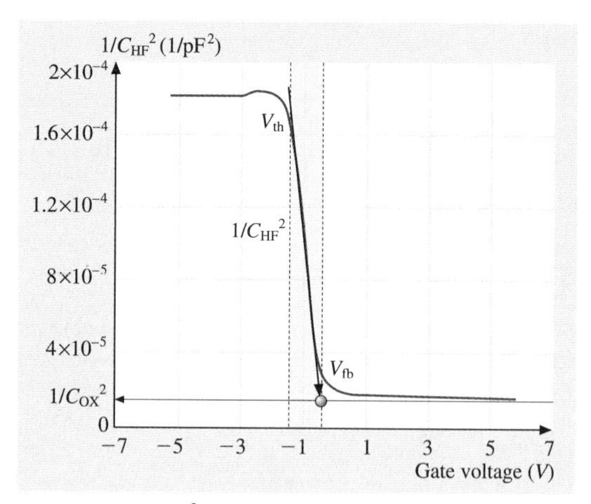

Fig. 20.19 A $1/C_{HF}^2$ versus V_G plot [20.30]. The slope of the *fitted arrow line* is proportional to the average doping, and the *arrow* points to the flat-band voltage V_{FB}, obtained at the V intercept with $1/C_{OX}^2$, shown with the *second horizontal arrow*

the flat-band voltage V_{FB} caused by the fixed surface charge Q_{SS} and the gate–semiconductor work function ψ_{MS} [20.3, 30].

Equation (20.57) does not take into account the impact of interface traps, which cause the C–V curve to stretch. The traps are slow and do not respond to the high frequency of the test signal, but they do follow the changes in the gate bias. Therefore, ∂V_G must be replaced with ∂V_{G0} in (20.57), with ∂V_{G0} representing the case when no interface traps are present.

The value of ∂V_{G0} can be obtained by comparing high- and low-frequency (quasi-static) C–V curves for a MOS structure at the same gate biases V_G. Therefore, the ratio $\partial V_{G0}/\partial V_G$ can be found at any gate bias V_G, since the band-bending is the same for both HF and LF capacitances. In [20.29] (Sect. 9.4), it is shown that

$$\frac{\partial V_{G0}}{\partial V_G}=\frac{C_{OX}+C_D}{C_{OX}+C_D+C_{IT}}=\frac{1-C_{LF}/C_{OX}}{1-C_{HF}/C_{OX}}\,. \qquad (20.58)$$

In this case (20.57) is modified to

$$n(w)=\frac{-2}{q\varepsilon_{Si}\frac{\partial}{\partial V_G}\left(\frac{1}{C_{HF}^2}\right)}\,\frac{1-C_{LF}/C_{OX}}{1-C_{HF}/C_{OX}} \qquad (20.59)$$

as originally proposed in [20.32] and illustrated in Fig. 20.20. As seen from Fig. 20.20, the stretching of the C–V curves due to the interface states induced by stress in Fig. 20.20a causes a disparity in the doping profile in Fig. 20.20b if only the high frequency capacitance is used. The disparity is well-suppressed in Fig. 20.20c by the high–low frequency capacitance measurement, taking into account the stretching of the C–V characteristics. Provided that the depletion layer capacitance is measured at a high frequency, the depletion layer width w is still obtained by (20.56).

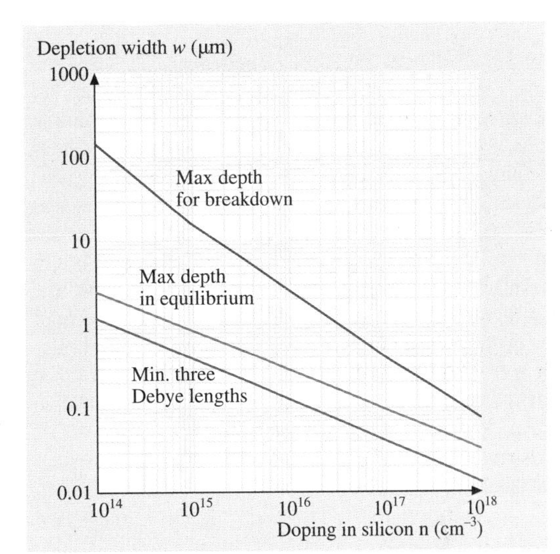

Fig. 20.21 Limitations on the depth achievable when profiling the doping of silicon MOS structures via C–V measurements at room temperature (After [20.4] p. 86)

Note that the maximum depth w_{max} (20.51) and the resolution Δ_w of the doping profile by means of C–V measurements is limited by the maximum band-bending ψ_{max} and the extrinsic Debye length λ_{Debye}, given by (20.50) and (20.60), respectively, and λ_{Debye} is

$$\lambda_{Debye}=\sqrt{\frac{\varepsilon_{Si}kT}{q^2 n}}\,. \qquad (20.60)$$

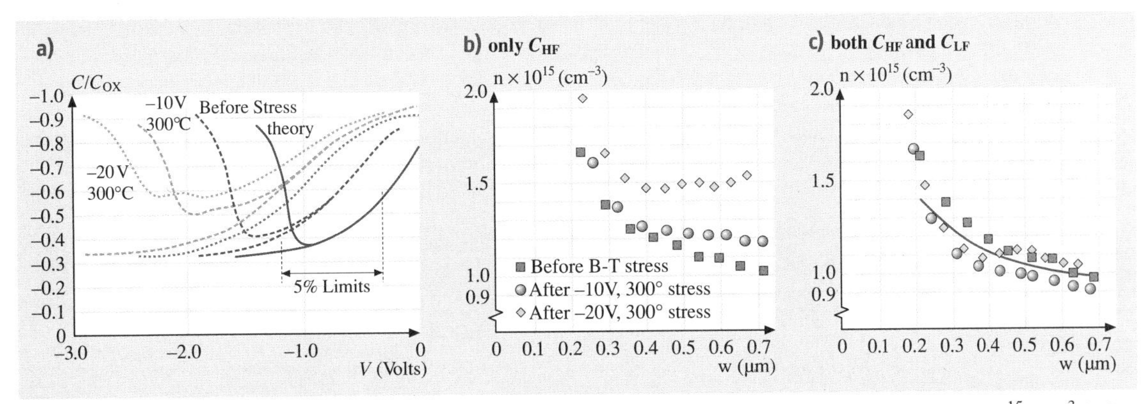

Fig. 20.20a–c C–V curves and doping profiles of a MOS structure with 145 nm oxide and uniform doping of 10^{15} cm^{-3} before and after bias temperature stress [20.32]. (**a**) Theoretical and (stretched) measured C–V curves. (**b**) Doping profile deduced from only C_{HF} (see (20.57)); (**c**) Doping profile deduced from both C_{HF} and C_{LF} (see (20.59)). After [20.32]

The doping profile obtained in this way is reliable for depths w of between $3\lambda_{\text{Debye}}$ and $w_{\max}/2$, when the MOS structure is in depletion and weak inversion, but not in accumulation. That is, $C_{\text{LF}} < 0.7C_{\text{OX}}$ as a simple rule. As illustrated in Fig. 20.21, the range of w values between $3\lambda_{\text{Debye}}$ and equilibrium, obtained via quasi-static C–V measurements, cover about half-a-decade. With proper corrections, the lower distance decreases to one Debye length [20.30]. Using nonequilibrium (transient) C–V measurements in deep depletion, the profiling can be extended to higher distances by about an order of magnitude, but further limitations can appear due to the high-frequency response of the interface charge, measurement errors, avalanche breakdown in deep depletion, or charge tunneling in highly doped substrates and thin oxides. More details are presented in [20.29].

20.3.3 Density of Interface States

Interface traps change their charge state depending on whether they are filled or empty. Because interface trap occupancy varies with the slow gate bias, stretching of the C–V curves occurs, as illustrated in Fig. 20.20. A quantitative treatment of this "stretch-out" can be obtained from Gauss' law as

$$C_{\text{OX}}(V_{\text{G}} - \psi_{\text{S}}) = -Q_{\text{S}} - Q_{\text{IT}} = -Q_{\text{T}} , \quad (20.61)$$

where Q_{S} and Q_{IT} are the surface and interface trap charges (per unit area), which are both dependent on the surface band-bending ψ_{S}, $Q_{\text{T}} = Q_{\text{S}} + Q_{\text{IT}}$ is the total charge in the MOS structure, C_{OX} is the gate capacitance (per unit area), and V_{G} is the bias applied at the gate of the MOS structure. For simplicity, the (gate metal)-to-(semiconductor bulk) potential ψ_{MS} is omitted in (20.61), but in a real structure the constant ψ_{MS} must be subtracted from V_{G}. As follows from (20.61), small changes ∂V_{G} in gate bias cause changes $\partial\psi_{\text{S}}$ in the surface potential bending, and the surface C_{S} and interface trap C_{IT} capacitances (both per unit area) can represent Q_{S} and Q_{IT}, given by

$$C_{\text{OX}}\partial V_{\text{G}} = (C_{\text{OX}} + C_{\text{S}} + C_{\text{IT}}) \cdot \partial\psi_{\text{S}} . \quad (20.62)$$

C_{S} and C_{IT} are in parallel and in series with the C_{OX}, respectively. Therefore, the measured low-frequency capacitance C_{LF} (per unit area) of the MOS structure becomes

$$C_{\text{LF}} = \frac{\partial Q_{\text{T}}}{\partial V_{\text{G}}} = \frac{\partial Q_{\text{T}}}{\partial \psi_{\text{S}}} \cdot \frac{\partial \psi_{\text{S}}}{\partial V_{\text{G}}} = \frac{C_{\text{OX}}\,(C_{\text{S}} + C_{\text{IT}})}{C_{\text{OX}} + C_{\text{S}} + C_{\text{IT}}} . \quad (20.63)$$

Equation (20.63) shows that stretch-out in the C–V curve can arise due to a non-zero value of C_{IT}, which deviates from the ideal case of $C_{\text{IT}} = 0$.

According to [20.29] (p. 142), D_{IT} is the density of interface states per unit area (cm^2) and per unit energy (1 eV) in units of $\text{cm}^{-2}\text{eV}^{-1}$. Since the occupancy of the interface states has a Fermi–Dirac distribution, then upon integrating over the silicon band-gap, the relation between C_{IT} and D_{IT} is

$$C_{\text{IT}}(\psi_{\text{S}}) = qD_{\text{IT}}(\phi_{\text{B}} + \psi_{\text{S}}) , \quad (20.64)$$

where $\phi_{\text{B}} = (k_{\text{B}}T/q)\ln(n/n_i)$ is the shift of the Fermi level from the intrinsic level $\phi_{\text{i}} = (E_{\text{c}} - E_{\text{v}})/2q$ in the silicon bulk of the MOS structure due to the doping concentration n, and n_i is the thermally generated carrier concentration in silicon. Since the derivative of the Fermi–Dirac distribution is a sharply peaking function, then $C_{\text{IT}}(\psi_{\text{S}})$ at particular ψ_{S} probes $D_{\text{IT}}(\phi_{\text{B}} + \psi_{\text{S}})$ over a narrow energy range of $k_{\text{B}}T/q$, in which D_{IT} can be assumed to be constant and zero outside this interval. Thus, varying the gate bias V_{G}, and therefore ψ_{S}, (20.64) can be used to obtain the density of states D_{IT} at a particular energy shift $q(\phi_{\text{B}} + \psi_{\text{S}})$ from the silicon intrinsic (mid-gap) energy E_{i}.

It is evident from (20.63) and (20.64) that the experimental values for D_{IT} can be obtained only when C_{IT}, and ψ_{S} are determined from C–V measurements. The simplest way to determine ϕ_{B} is to get the average doping density n using the maximum–minimum high-frequency capacitance method (see (20.52) and Fig. 20.18), or to use the values of n from doping profiles at $0.9w_{\max}$ - see (20.58) [20.30]. Either the high-frequency or the low-frequency $C - V$ measurement can be used to obtain C_{IT}, but it is necessary to calculate C_{S} as function of ψ_{S}, which makes it difficult to process the experimental data.

The most suitable technique for experimentally determining D_{IT} is the combined high–low frequency capacitance method ([20.29], Sect. 8.2.4, p. 332). The interface traps respond to the measurement of low–frequency capacitance C_{LF}, whereas they do not respond to the measurement of the high-frequency measurement C_{HF}. Therefore, C_{IT} can be obtained from measurements by "subtracting" C_{HF} from C_{LF}, given by

$$C_{\text{IT}} = \left(\frac{1}{C_{\text{LF}}} - \frac{1}{C_{\text{OX}}}\right)^{-1} - \left(\frac{1}{C_{\text{HF}}} + \frac{1}{C_{\text{OX}}}\right)^{-1} . \quad (20.65)$$

Denoting $\Delta C = C_{\text{LF}} - C_{\text{HF}}$, the substitution of (20.65) into (20.64) provides a direct estimate of D_{IT} from C–V

measurements (see also [20.3], p. 371) as

$$D_{\mathrm{IT}} = \frac{\Delta C}{q}\left(1 - \frac{C_{\mathrm{LF}}}{C_{\mathrm{OX}}}\right)^{-1}\left(1 - \frac{C_{\mathrm{HF}}}{C_{\mathrm{OX}}}\right)^{-1} \quad (20.66)$$

Note that the combined high–low frequency capacitance method provides C_{IT} and D_{IT} as function of gate bias V_{G}. However, if D_{IT} needs to be plotted as a function of the position in the energy band-gap, the surface band-bending ψ_{S} must also be determined as function of gate bias V_{G}, as follows from (20.64).

There are several ways to obtain the relation between ψ_{S} and V_{G}. One way is to create a theoretical plot of C_{HF} versus ψ_{S} and then, for any choice of C_{HF}, a pair of values for ψ_{S} and V_{G} is found (see [20.29] p. 327). This method is relatively simple if the doping concentration n in the silicon is uniform and known, because the high-frequency silicon surface capacitance C_{S} under depletion and accumulation is a simple function of the band-bending ψ_{S}, and the flat-band capacitance C_{FB} ([20.29] pp. 84, 97, 164) is given by

$$C_{\mathrm{FB}} = \frac{\varepsilon_{\mathrm{Si}}}{\lambda_{\mathrm{Debye}}} = \sqrt{\frac{\varepsilon_{\mathrm{Si}} q^2 n}{k_{\mathrm{B}} T}} \quad (20.67)$$

$$C_{\mathrm{S}}(\psi_{\mathrm{s}}) \approx \begin{cases} \frac{C_{\mathrm{FB}}}{\sqrt{2}} \frac{\exp\left(\frac{q\psi_{\mathrm{S}}}{k_{\mathrm{B}}T}\right)-1}{\sqrt{\exp\left(\frac{q\psi_{\mathrm{S}}}{k_{\mathrm{B}}t}\right)-\frac{q\psi_{\mathrm{S}}}{k_{\mathrm{B}}T}-1}}, \\ \quad \psi_{\mathrm{S}} > 0 \text{ in accumulation} \\ C_{\mathrm{FB}},\ \psi_{\mathrm{S}} = 0 \text{ at flat band,} \\ \frac{C_{\mathrm{FB}}}{\sqrt{2}} \frac{1-\exp\left(\frac{q\psi_{\mathrm{S}}}{k_{\mathrm{B}}T}\right)}{\sqrt{\exp\left(\frac{q\psi_{\mathrm{S}}}{k_{\mathrm{B}}t}\right)-\frac{q\psi_{\mathrm{S}}}{k_{\mathrm{B}}T}-1}}, \\ \quad \psi_{\mathrm{S}} < 0 \text{ in depletion}. \end{cases} \quad (20.68)$$

Since C_{S} is in series with C_{OX}, then the theoretical C_{HF} is obtained as a function of the band-bending ψ_{S} by

$$\frac{1}{C_{\mathrm{HF}}(\psi_{\mathrm{S}})} = \frac{1}{C_{\mathrm{OX}}} + \frac{1}{C_{\mathrm{S}}(\psi_{\mathrm{S}})} \quad (20.69)$$

For a uniformly doped silicon with SiO_2 as the insulator, the ratio $C_{\mathrm{HF}}(V_{\mathrm{FB}})/C_{\mathrm{OX}}$ at gate bias for flat-band conditions is given ([20.3], p. 349) by

$$\frac{C_{\mathrm{HF}}(V_{\mathrm{FB}} \text{ or } \psi_{\mathrm{S}} = 0)}{C_{\mathrm{OX}}} = \frac{1}{1 + \frac{136\sqrt{T/300}}{t_{\mathrm{ox}}\sqrt{n}}}, \quad (20.70)$$

where t_{ox} is the oxide thickness (cm), n is the doping (cm^{-3}), and the T is the temperature (K).

It was demonstrated in [20.29] that the method of using a theoretical plot to obtain the relation between ψ_{S} and V_{G} works well in the case of uniformly doped silicon even if only high-frequency C–V measurements are used to obtain the density of states; that is, the $1/C_{\mathrm{HF}}^2$ versus V_{G} plot is almost a straight line. However, with substrates that are not uniformly doped, the method is inconvenient because the corrections in (20.67) and (20.68) are difficult to implement. Therefore, in practice, a method based on low-frequency C–V measurement is preferred [20.30].

Low-frequency C–V measurement was first used to obtain the relation between ψ_{S} and V_{G} [20.34]. This method is based on the integration of (20.65) from an initial gate bias V_{G0}, arbitrarily chosen either under strong accumulation or strong inversion, to the desired V_{G} at which the band-bending $\psi_{\mathrm{S}}(V_{\mathrm{G}})$ is to be obtained. Since C_{IT} is part of (20.65), then the low-frequency C–V curve ([20.29] p. 93) is integrated as

$$\psi_{\mathrm{S}}(V_{\mathrm{G}}) = \psi_{\mathrm{S0}} + \int_{V_{\mathrm{G0}}}^{V_{\mathrm{G}}} \left(1 + \frac{C_{\mathrm{LF}}(V_{\mathrm{G}})}{C_{\mathrm{OX}}}\right) \mathrm{d}V_{\mathrm{G}}\,. \quad (20.71)$$

The value of ψ_{S0} is selected such that $\psi_{\mathrm{S}}(V_{\mathrm{FB}}) = 0$ when integrating from V_{G0} to the flat-band gate voltage V_{FB}. In this case, V_{FB} is usually obtained beforehand from the point of V-intercept with $1/C_{\mathrm{OX}}^2$ when extrapolating

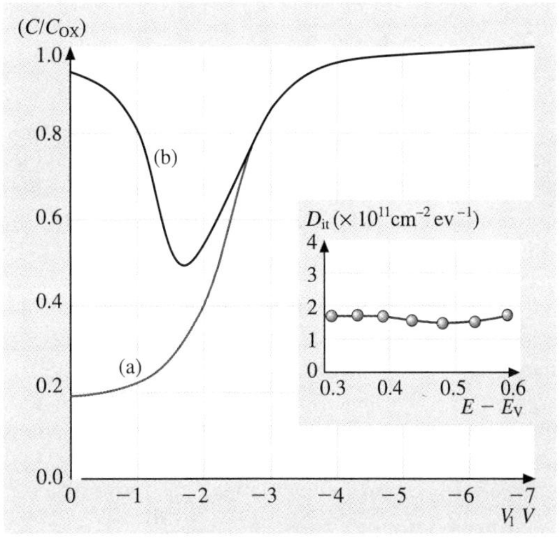

Fig. 20.22a,b Results from the combined high–low frequency capacitance method [20.33]. (**a**) High-frequency C–V curve; (**b**) low-frequency C–V curve. The energy profile for the density of interface states D_{IT} is shown in the *inset*, as calculated by (20.64), (20.66) and (20.71)

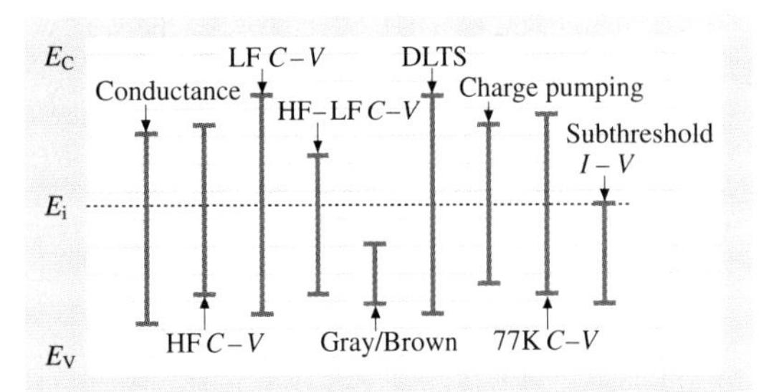

Fig. 20.23 Energy ranges in the silicon band-gap of a p-type substrate over which the density of interface traps can be determined using various measurement methods and characterization techniques. (After [20.3], p. 104)

the linear part of the $1/C_{HF}^2$ versus V_G curve toward the V_G axis (Fig. 20.19). After determining ψ_{S0}, (20.71) provides the relation between ψ_S and V_G. Thus, the density of states D_{IT} obtained from (20.66) as function of V_G using the combined high–low frequency capacitance method can be plotted against the position of D_{IT} in the silicon band gap, as given by (20.64). High and low frequency $C-V$ measurements can therefore be used to plot the data, as illustrated in the insert of Fig. 20.22.

Overall, many different techniques are used to determine the density of states D_{IT} (please see Fig. 20.13). For some of these techniques, the ability to sense the energy position of D_{IT} in the band-gap of silicon is summarized in [20.3]. Most of them use $C-V$ measurements, but others are based on $I-V$ measurements taken during the subthreshold operation of MOS transistors, deep-level transient spectroscopy (DLTS), charge pumping (CP) in a three-terminal MOS structure, cryogenic temperature measurements, and so on. Each technique has its strengths and weaknesses, which are discussed in [20.3, 35].

In the methods discussed above, it has been assumed that the gate bias V_G varies slowly with time, 20–50 mV/s, and that the MOS structure is in equilibrium; that is, the minority carriers are generated and the inversion layer is readily formed in the MOS structure when V_G is above the threshold. However, the time constant for minority carrier generation is high in silicon (≈ 0.1 s or more), and it is possible to use nonequilibrium high-frequency $C-V$ measurements to further analyze the properties of the MOS structure. Some applications of these methods are presented later.

20.4 Current–Voltage Measurements

20.4.1 $I-V$ Measurements on a Simple Diode

Current–voltage measurements of mainstream semiconductor devices are perhaps the simplest and most routine measurements performed, and they can provide valuable information about the quality of materials used. For example, if we consider the $I-V$ characteristics of a p–n diode structure, the source–substrate or drain–substrate junctions can provide useful information on the quality of the junction, such as whether defects are present (they give rise to generation–recombination currents or large parasitic resistances for the contacts at the source, drain or substrate terminals). This is easily seen from the current–voltage relation given by the sum of the diffusion (I_{DIFF}) and recombination (I_{GR}) currents:

$$I_D = I_{DIFF} + I_{GR} = I_{D0}\left[\exp\left(\frac{eV_D}{nk_BT}\right) - 1\right] + I_{GR0}\cdot\exp\left(\frac{eV_D}{2k_BT}\right) \quad (20.72)$$

where I_{D0} and I_{GR0} are the zero-bias diffusion and recombination currents respectively, n is an ideality factor (typically 1), and V_D is the voltage across the intrinsic diode, which is given by

$$V_D = V_{applied} - I_D R_{parasitic}\,. \quad (20.73)$$

For (20.72), a plot of $\ln(I_D)$ versus V_D allows us to separate out the diffusion and the recombination current components. From (20.72) and (20.73), we can also use the diode's $I-V$ characteristics to determine the parasitic resistance in series with the intrinsic diode, as described in detail in [20.36]. In most cases, this $R_{parasitic}$ is dominated by the contact resistance.

20.4.2 $I-V$ Measurements on a Simple MOSFET

Simple current–voltage measurements – drain current versus gate voltage ($I_{DS}-V_{GS}$), and I_{DS} versus drain voltage (V_{DS}) – are routinely taken on MOSFETs in order to study their electrical characteristics; however, these can also be used to obtain useful information

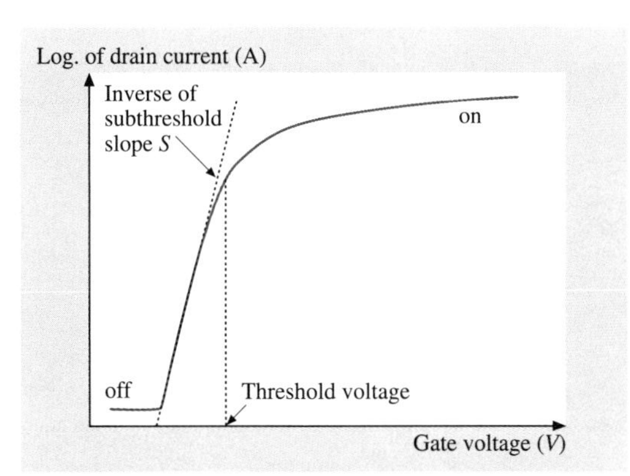

Fig. 20.24 Typical subthreshold characteristics of a MOSFET. The interface state density can be extracted from S

on the quality of the semiconductor, contacts, oxide and semiconductor–oxide interface. For example, the $I_{DS} - V_{GS}$ characteristics at very small V_{DS} biases (linear region of operation) for a set of test transistors of fixed channel width and different channel lengths is often used to extract parameters such as the threshold voltage (V_T), the transconductance (g_m), the intrinsic mobility (μ_o) and the mobility degradation coefficients θ_0 and η, the parasitic source (R_S) and the drain resistances (R_D) in series with the intrinsic channel, the channel length reduction ΔL, the output conductance (g_{DS}) and the subthreshold slope (S) [20.37].

These parameters are required for modeling and they directly impact the device's performance. However, some of these parameters can also be used to assess the quality of the silicon–silicon dioxide (Si–SiO_2) interface [20.38, 39]. For example, interface states at the Si–SiO_2 interface can change the threshold voltage, the subthreshold slope and the mobility, all of which will impact on the drain–source (I_{DS}) current flowing through the device. Here, we look at one parameter in more detail – the subthreshold slope S in mV (of V_{GS})/decade (of I_{DS}).

First, the interface trap density (D_{IT}) can be determined from a semi-log plot of $I_{DS} - V_{GS}$ characteristics at very low drain biases, as shown in Fig. 20.24. We start with the expression for the subthreshold slope

$$S = \frac{k_B T}{q} \ln(10) \left(1 + \frac{C_D + C_{IT}}{C_{OX}}\right) \quad (20.74)$$

in which

$$C_D = \frac{q \varepsilon_{Si} N_A}{\sqrt{2\psi + |V_B| - k_B T/q}} \quad (20.75)$$

D_{IT} can then be calculated from

$$C_{IT} = q D_{IT} \quad (20.76)$$

once C_{IT} is determined from (20.74). In fact, a recent comparison in [20.40] of the interface trap densities extracted from capacitance, subthreshold and charge pumping measurements produced similar results, demonstrating that simple and fast I–V measurements based on the subthreshold technique can provide useful information on the Si–SiO_2 quality.

20.4.3 Floating Gate Measurements

The floating gate technique is another simple I–V measurement in which the evolution of the drain current I_{DS} is monitored after the gate bias has been removed. It has been shown to be particularly useful when monitoring early-mode hot-carrier activity in MOS transistors [20.40, 41]. In this measurement, we first check the oxide quality by biasing the transistor in the strong linear region (very low V_{DS} and a V_{GS} well above V_T), and then lift the gate voltage probe so that $V_{GS} = 0$ V and measure the evolution of I_{DS} with time. For a high-quality gate and spacer oxide, I_{DS} remains constant for a long time, indicating that there is negligible carrier injection across the gate oxide through Fowler–Nordheim tunneling or other leakage mechanisms.

A second precaution is to have a dry or inert gas such as nitrogen flowing over the chip to reduce the possibility of other leakage mechanisms such as that from water vapor. The measurement set-up for this experiment is shown in Fig. 20.25. The

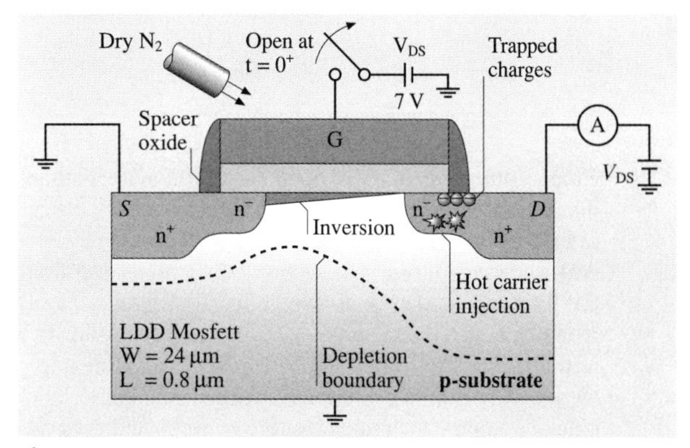

Fig. 20.25 Schematic diagram showing the set-up used for floating gate measurements. The area where charge is trapped after hot electrons are applied is shown

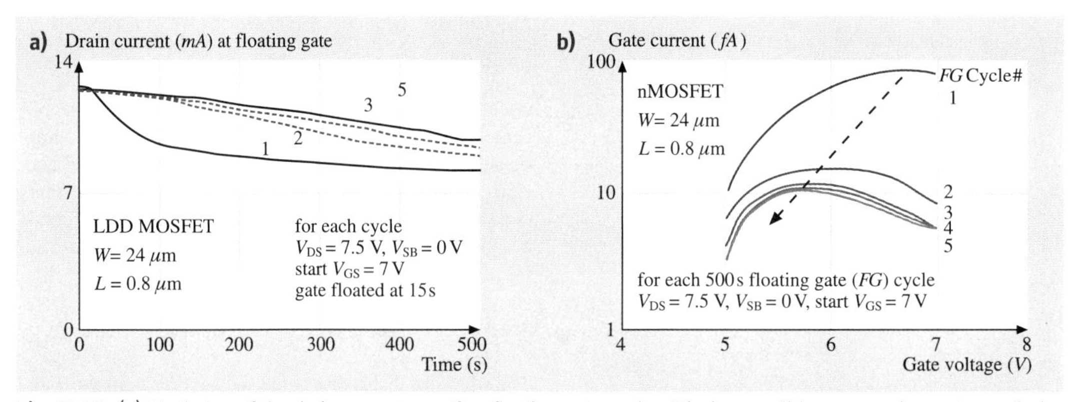

Fig. 20.26 (**a**) Evolution of the drain current over five floating gate cycles. Biasing conditions were chosen to maximize hot-electron gate currents. Note that the maximum drop in drain current occurs after the first floating gate cycle. (**b**) Extracted gate currents using (20.77) and the measurements in (a). As with the drain current, the maximum change in gate current occurs between floating gate cycles 1 and 2. The shift in the peak of the gate current is explained in [20.38,41].

evolution of the floating gate current over several cycles with the gate voltage applied and then removed is shown in Fig. 20.26. The biasing voltages and time at which the gate is floated are also given in Fig. 20.26a. For this experiment, a biasing condition of $V_{GS} \approx V_{DS}$ was chosen for a high-impact ionization-induced gate current, but a lower-than-maximum electron injection situation was used for the initial biasing condition.

From the evolution of I_{DS} and the $I_{DS} - V_{GS}$ characteristics of a virgin (not intentionally stressed) transistor at the same V_{DS} as the floating gate measurements, and from measurements of the total capacitance associated with the gate (C_G), the gate current (I_G) evolution after each floating gate cycle can be determined using

$$I(V_{GS}) = C_G \frac{dV_{GS}}{dt} \tag{20.77}$$

This $I_G - V_{GS}$ evolution is shown in Fig. 20.26b. An ancillary benefit of the floating gate technique is that very small gate currents (in the fA range or even smaller) can be easily determined by measuring much larger drain currents using, for example, a semiconductor parameter analyzer. The reason for this is that the gate current is not directly measured in this technique – it is determined from $I_{DS} - V_{GS}$ and (20.77). Also, the change in the floating gate current after the first few cycles can be used to monitor for early mode failure after statistical evaluation.

20.5 Charge Pumping

Charge pumping (CP) is another electrical technique that is well suited to studying semiconductor–insulator interfaces in MOSFETs [20.42–47]. There are several versions of the CP technique: spatial profiling CP [20.43–47], energy profiling CP [20.48], and, more recently, new CP techniques [20.49] that permit the determination of both interface states (N_{IT}) and oxide traps (N_{OT}) away from the interface and inside the oxide. The charge pumping technique is more complicated than either of the I–V or floating gate methods. However, it is a very powerful technique for assessing interface quality and it works well even with very small transistor geometries and very thin gate oxides, where tunneling can be a problem.

The charge pumping technique was first used in 1969 [20.50] to measure the interface traps at Si–SiO_2 interface. Since then, there have been numerous publications with enhancements, refinements and applications of the technique to a variety of semiconductor–insulator interfacial studies. In the basic charge pumping experiment, the gate of an NMOST (for example) is pulsed from a low value (V_L) when the device is in accumulation to a high value (V_H) when the device is in inversion. This results in the filling of traps between

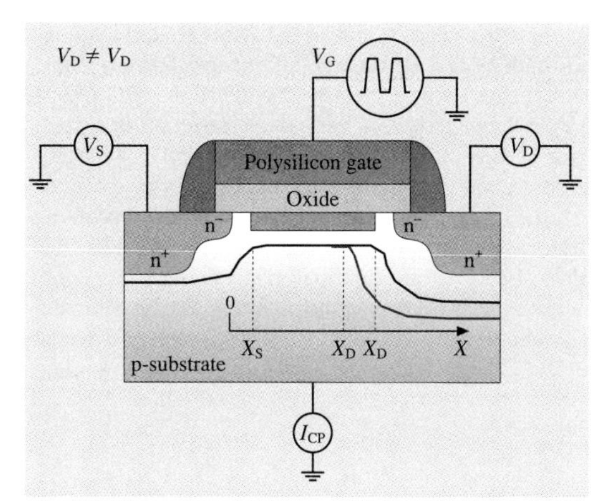

Fig. 20.27 Example of spatial profiling charge-pumping set-up used when the source and drain biases are slightly different

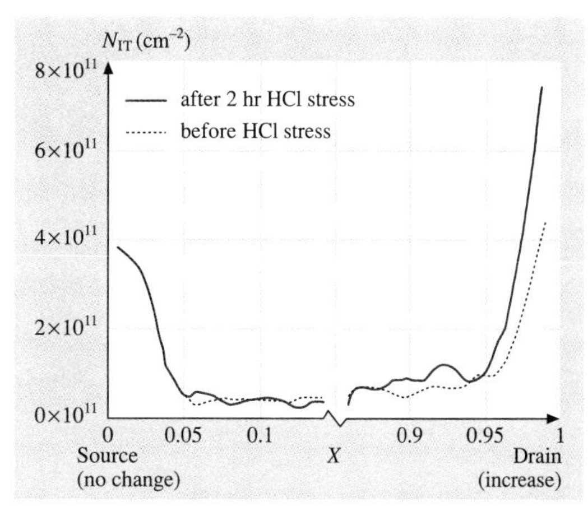

Fig. 20.29 Spatial interface state distribution over the channel in a 1 μm-long device. The stress was applied for 2 h at $V_{DS} = 5$ V and $V_{GS} = 2.4$ V

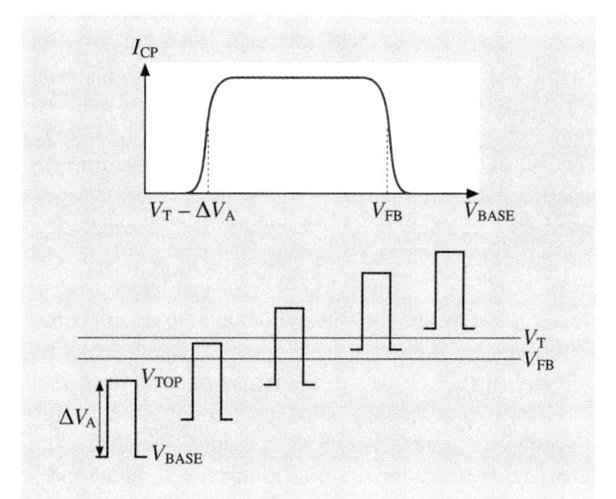

Fig. 20.28 Demonstration of how the CP curve is generated by varying the base level of the pulse so that the entire pulse is between V_{FB} and V_T

$E_{F,ACC}$ (corresponding to V_L) and $E_{F,INV}$ (corresponding to V_H) with holes and electrons, respectively. When pulsing the gate between accumulation at V_L and inversion at V_H, a current flows due to the repetitive recombination at the interface traps of minority carriers from the source and drain junctions with majority carriers from the substrate. This current is termed the charge pumping current, and it was found to be proportional to the frequency of the gate pulse, the gate area and the interface state density. Its sensitivity is better than 10^9 cm^{-2}eV^{-1}. In the traditional CP experiment shown in Fig. 20.27, but with $\Delta V_S = |V_D - V_S| = 0$ V, the gate G is connected to a pulse generator, a reverse bias V_R or no bias is applied to both sources S and drain D terminals, and the charge pumping current flowing in the substrate terminal, I_{CP}, is measured. To generate a typical charge pumping curve (as shown in the top part of Fig. 20.28), the base level of the pulse is varied, taking the transistor from below flat-band to above surface inversion conditions, as shown in the bottom part of Fig. 20.28.

In the traditional CP experiment, the charge pumping current I_{CP} is given by

$$I_{CP} = qA_{GATE}\overline{D_{IT}}\Delta E\,, \qquad (20.78)$$

where

$$\Delta E = \left(E_{F,INV} - E_{F,ACC}\right)\,. \qquad (20.79)$$

This expression assumes that the electrical and physical channel lengths are the same. However, for short channel devices, this assumption results in an error. Therefore, a more accurate expression is

$$I_{CP} = qfW\int_{x_s}^{x_d} N_{IT}(x)\cdot(q\Delta\psi_{SO}) \qquad (20.80)$$

$$x_d = L - \sqrt{\frac{2\varepsilon_{Si}}{qN}}\left(\sqrt{V_D+\psi_S} - \sqrt{\psi_S}\right)\,. \qquad (20.81)$$

The interface state density at the edge of the drain depletion region $[N_{IT}(x_d)]$ is given by

$$N_{IT}(x_d) = \frac{1}{qfWq\Delta\psi_{S0}} \left(\frac{dx_s}{dV_S}\right)^{-1} \cdot \left(\frac{dI_{CP}}{dV_S}\right)_{V_D=\text{constant}} \quad (20.82)$$

When performing spatial profiling CP experiments, some precautions are required. The first is that a voltage difference ΔV_S between V_D and V_S that is too small results in a difference in I_{CP} that is too small as well, and hence a large error in $N_{IT}(x)$, as indicated from (20.82). On the other hand, values of ΔV_S that are too large result in a I_D that is too high and hence more substrate current I_B. This current I_B can interfere with I_{CP} if ΔV_S is large or if L is very short, resulting in a large error in $N_{IT}(x)$. The range 50–100 mV for ΔV_S seems to be a good compromise for the devices investigated in [20.43, 44]. Experimental results for spatial profiling CP measurements indicate that $N_{IT}(x)$ peaks near S/D edges. However, after normal mode stress, $N_{IT}(x)$ only increases near D. This is shown in Fig. 20.29. More details about charge pumping can be found in a recent review [20.51].

20.6 Low-Frequency Noise

20.6.1 Introduction

Low-frequency noise (LFN) spectroscopy requires very good experimental skills in the use of low-noise instrumentation as well as grounding and shielding techniques. Other special considerations are also required, which are discussed later. Although it is time-consuming to perform, it has been widely used to probe microscopic electrical transport in semiconductors and metals. LFN is very sensitive to defects in materials and devices, and large differences in LFN characteristics can be observed in devices with identical electrical current–voltage characteristics. This is mainly because electrical I–V measurements only probe the average or macroscopic transport in devices and so are not as sensitive to defects as LFN. Due to its sensitivity to defects, traps or generation–recombination centers, LFN has been proposed as a good tool for predicting device reliability. For example, LFN has been used to predict the reliabilities of metal films [20.52], and has been used in processing steps that produce photodetectors with better performance [20.53, 54]. LFN noise is sensitive to both bulk and surface defects or contaminants of a material.

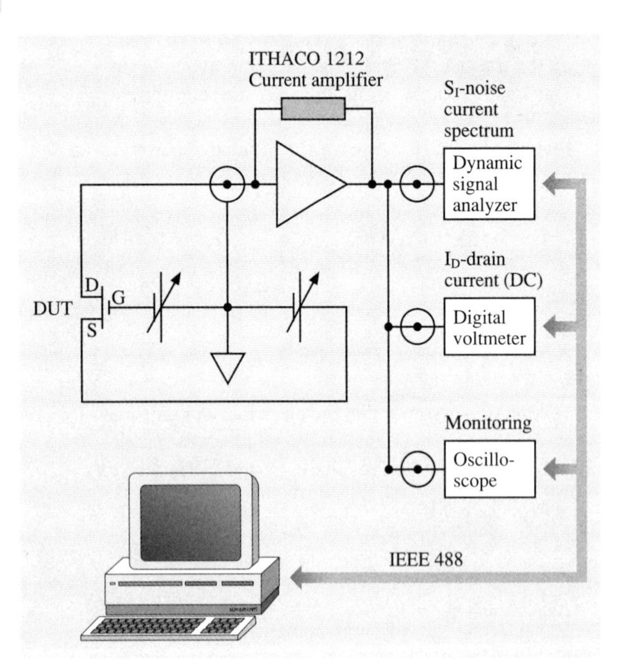

Fig. 20.30 System for measuring low-frequency noise. In this diagram, the device under test is any field-effect or thin-film transistor

Using low-frequency noise spectroscopy and biasing the transistor in saturation, we can spatially profile the defect density near the drain and source terminals for devices in normal and reverse modes of operation [20.55]. Low-frequency noise in the linear region also allows us to extract the average defect density over the entire channel region at the silicon–silicon dioxide interface [20.56, 57]. Noise experiments were performed on small-geometry polysilicon emitter bipolar transistors to investigate the number of interface states in the thin interfacial oxide layer between the monocrystalline and polycrystalline silicon [20.58–68]. Recent experiments using body or substrate bias (V_B) in a MOS transistor allowed us to look at the contribution of bulk defects (defects away from the silicon–silicon dioxide interface) and their contribution to device noise [20.69–71]. This is important since substrate biasing has been proposed as a means to cleverly manage power dissipation and speed in emerging circuits and systems [20.72].

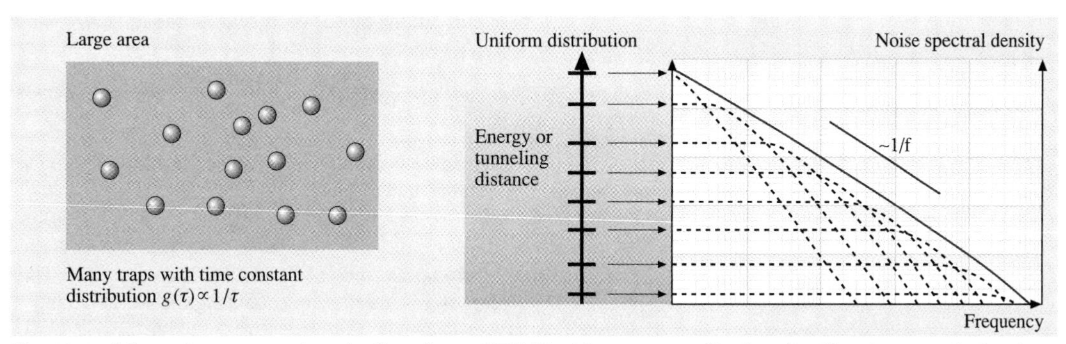

Fig. 20.31 Schematic representation of a "large"-area PE-BJT with many traps distributed uniformly across the band gap and the emitter area, and with a $g(\tau) = 1/\tau$ distribution for the time constant. The resulting spectrum is $1/f$ noise

We will discuss how low-frequency noise (LFN) spectroscopy can be applied to the interfacial oxide layer between the mono-silicon and polysilicon emitter in bipolar junction transistors (BJTs) here. The experimental system shown in Fig. 20.30 is used for LFN measurements of FETs; the same system can also be used for BJTs.

As mentioned before, special attention must be paid to grounding and shielding in LFN measurements, as this is crucial to minimizing the effects of experimental and environmental noise sources on the device under test (DUT). Because electric power supplies are noisy, especially at 60 Hz (in North America) and its harmonics, and this noise can dominate the noise of the DUT, batteries are often used to supply the voltage. Metal film resistors are the preferred means of changing the biasing conditions, because of their better low-noise characteristics compared to carbon resistors, for example.

With these experimental precautions taken, the noise signal from the transistor might still be too low to be directly measured using a spectrum or signal analyzer. Therefore, a low-noise voltage or current amplifier, whose input noise sources are lower than that of the noise signal, is used to boost the noise signal. In addition, other instruments might be used to measure currents or voltages, or to display the waveforms (as shown in Fig. 20.30). An example of a low-frequency noise characterization system that we have used to study the noise in thin film polymer transistors is shown in Fig. 20.30. Note that LFN measurements are time intensive because a large number of averages are required for smooth spectra. Also, in noise measurements, the power spectrum densities S_V and S_I for the noise voltages and currents are measured, in units V^2/Hz and I^2/Hz, respectively.

20.6.2 Noise from the Interfacial Oxide Layer

Here we present some sample results and show how low-frequency noise spectra in ultrasmall devices can be used to estimate the oxide trap density. Generally, the low-

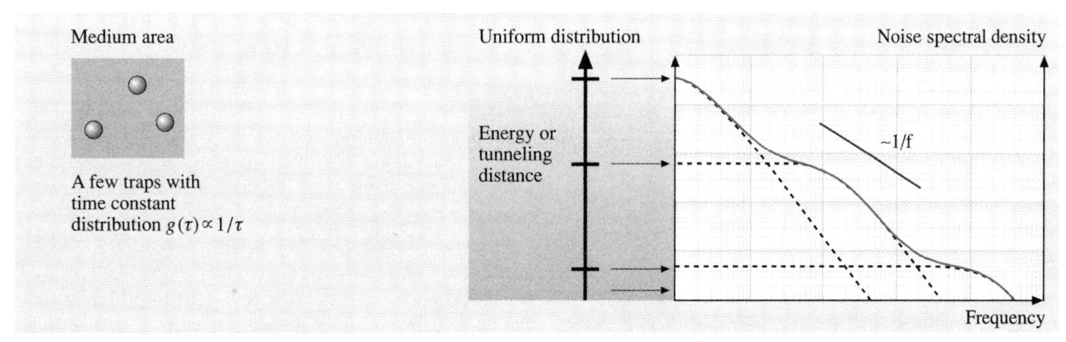

Fig. 20.32 Schematic representation of a "medium"-area PE-BJT ($\approx 0.5\,\mu m^2$) with a few traps. Note that g–r bumps appear for each trap since there are only a few traps

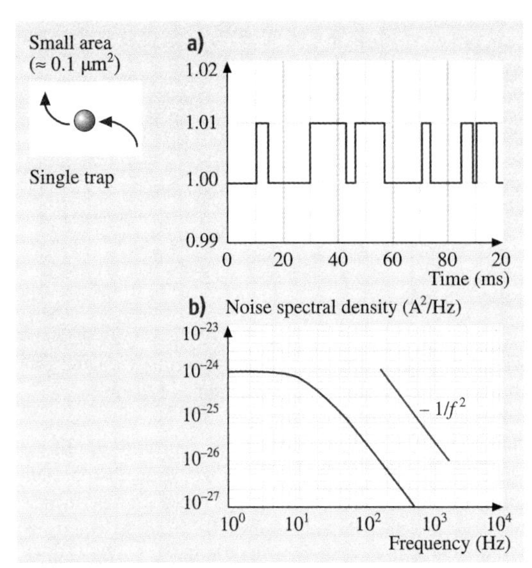

Fig. 20.33 Schematic representation of a "small"-area PE-BJT ($\approx 0.1\,\mu\text{m}^2$) with one trap. Note that a single RTS and g–r spectrum appear because there is only one trap

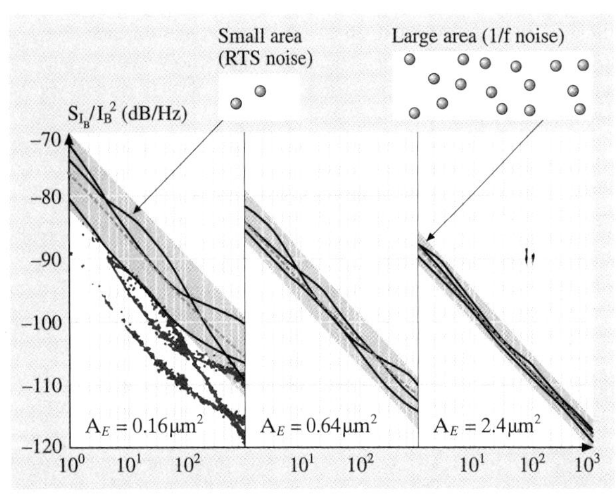

Fig. 20.34 Experimental results for low-frequency noise spectra from sets of large-, medium- and small-area PE-BJTs. In all cases, the average spectrum is $1/f$ noise and the relative magnitude of the $1/f$ noise is the same; that is, the area of $K_{\mathrm{F}}\times$ is the same for the three sets of transistors. After [20.59–62]

frequency noise spectra of polysilicon emitter (PE) BJTs are made up of $1/f$ noise, generation–recombination (g–r) noise and shot noise sources. In the case of the base current, the noise spectra can be modeled as

$$S_{I_{\mathrm{B}}} = \frac{K_{\mathrm{F}} \cdot I_{\mathrm{B}}^{A_{\mathrm{F}}}}{f} + \sum_{i=1}^{n} \frac{B_i \tau_i}{1+(2\pi f \tau_i)^2} + 2qI_{\mathrm{B}}\,, \quad (20.83)$$

where the symbols have their usual meanings, see [20.68] for example.

As described in [20.63–68], the LFN in PE BJTs originates from the thin layer of oxide between the monocrystalline and polycrystalline silicon emitter. The defects at this interface may be dangling oxygen bonds, oxygen vacancies, interface states or oxide traps [20.63]. Devices with large emitter areas have many traps, and these produce generation–recombination noise which produce $1/f$ noise when added. This is schematically shown in Fig. 20.31.

As the device area is reduced, and assuming a constant trap density (which is normally true for devices on the same wafer), then there are fewer traps in the interfacial oxide layer for smaller area devices. In this case, the spectral density of the noise changes and it gains characteristic "bumps" associated with resolvable g–r noise components. This is schematically shown in Fig. 20.32.

In very small devices with only a single trap, for example, the noise spectrum changes dramatically; only g–r noise is observed in the frequency domain along with a random telegraph signal (RTS) in the time domain. This is schematically shown in Fig. 20.33. Real experimental results are shown for three sizes of transistors (2.4, 0.64 and $0.16\,\mu\text{m}^2$) in Fig. 20.34. Here, one can see how $1/f$ noise is made up of g–r spectra as the emitter geometries are scaled to smaller and smaller values. For the PE BJT with an emitter area of $0.16\,\mu\text{m}^2$, a lower bound of $\approx 10^9/\text{cm}^2$ can be approximated for the oxide trap density. Similar results have been obtained for MOSFETs [20.73].

20.6.3 Impedance Considerations During Noise Measurement

Two basic circuits can be employed when measuring the low-frequency noise (LFN) in a device. These configurations are sketched in Fig. 20.35. In voltage noise measurement (Fig. 20.35a), a low-noise preamplifier senses the voltage across the device, and this signal is sent to a spectrum analyzer or a fast Fourier transform (FFT) analyzer. In current noise measurements (Fig. 20.35b), the low-noise preamplifier senses the current through the device, converts it into a voltage, and forwards the voltage to a FFT or spectrum analyzer.

In principle, both configurations can be used for LFN measurement, but the impact of the nonideality of the amplifier (such as the input impedance, noise voltage and current) changes when the device impedance changes. Also, the noise from the bias source varies with each measurement set-up.

The noise equivalent circuit used for voltage measurement is shown in Fig. 20.36. The noise from the amplifier is represented by the input-referred noise voltage ($S_{V\mathrm{n}}$) and noise current ($S_{I\mathrm{n}}$) sources. The noise voltage from the bias is represented by S_{V0}. The impedance of the bias source is R_0, whereas the input impedance of the amplifier is neglected, since it is usually very high compared to R_0. The impedance of DUT is r_d. The noise current $S_{I\mathrm{d}}$ of the device that can be measured, assuming that the noise voltage at the input of the amplifier $S_{V\mathrm{m}} = S_{I\mathrm{d}} \times r_\mathrm{d}^2$. However, the amplifier sees a different level of $S_{V\mathrm{m}}$, given by

$$S_{V\mathrm{m}} = \frac{S_\mathrm{OUT}}{A^2} = S_{V0}\left(\frac{r_\mathrm{d}}{r_\mathrm{d}+R_0}\right)^2 + S_{V\mathrm{n}} + (S_{II} + S_{II})Z^2 \tag{20.84}$$

where

$$Z = (r_\mathrm{d}//R_0) = \frac{r_\mathrm{d}\cdot R_0}{r_\mathrm{d}+R_0} \tag{20.85}$$

and A is the voltage gain of the amplifier. Therefore, the estimated value for $S_{I\mathrm{d}}$ is

$$S_{I\mathrm{d}} = \frac{S_{V\mathrm{m}} - S_{V\mathrm{n}}}{Z^2} - \frac{S_{V0}}{R_0^2} - S_{I\mathrm{n}}\,. \tag{20.86}$$

The uncertainty in (20.86) is

$$\frac{\Delta S_{I\mathrm{d}}}{S_{I\mathrm{d}}} = \frac{\Delta S_{V\mathrm{n}}}{S_{V\mathrm{m}}} + \frac{\Delta S_{V0}}{S_{V\mathrm{m}}}\left(\frac{Z}{R_0}\right)^2 + \frac{\Delta S_{I\mathrm{n}}}{S_{V\mathrm{m}}}Z^2 \tag{20.87}$$

where $\Delta S \le S$ denotes the uncertaintie in each noise source. As seen from (20.87), the impact of the bias source noise ΔS_{V0} and the input current noise ΔS_{In} can be reduced if the impedance of the measurement circuit Z is low and the ratio r_d/R_0 is kept much less than 1; in other words, the *voltage noise measurement is more appropriate for low-impedance devices, such as diodes at forward biasing*, and the noise floor of the measurement is limited by the input-referred voltage noise S_{V_n} of the amplifier.

For the other (dual) case, current noise measurement, the noise equivalent circuit is shown in Fig. 20.37. The corresponding equations for the measured noise current $S_{I\mathrm{m}}$, Z, the device noise $S_{I\mathrm{d}}$, and the uncertainty, respectively, are given by (20.88–20.91) below

$$S_{I\mathrm{m}} = \frac{S_\mathrm{OUT}}{R^2} = \frac{S_{V0}}{(r_\mathrm{d}+R_0)^2}S_{I\mathrm{n}} + \frac{S_{V\mathrm{n}}}{Z^2} + S_{I\mathrm{d}}\left(\frac{r_\mathrm{d}}{r_\mathrm{d}+R_0}\right)^2 \tag{20.88}$$

$$Z = (r_\mathrm{d}+R_0)//R = \left(\frac{1}{R}+\frac{1}{r_\mathrm{d}+R_0}\right)^{-1} \tag{20.89}$$

$$S_{I\mathrm{d}} = (S_{I\mathrm{m}} - S_{I\mathrm{n}})\left(1+\frac{R_0}{r_\mathrm{d}}\right)^2 - \frac{S_{V0}}{r_\mathrm{d}^2} - \frac{S_{V\mathrm{n}}}{Z^2}\left(1+\frac{R_0}{r_\mathrm{d}}\right)^2 \tag{20.90}$$

$$\frac{\Delta S_{I\mathrm{d}}}{S_{I\mathrm{d}}} = \frac{\Delta S_{I\mathrm{n}}}{S_{I\mathrm{m}}} + \frac{\Delta S_{V0}}{S_{I\mathrm{m}}}\frac{1}{(r_\mathrm{d}+R_0)^2} + \frac{\Delta S_{V\mathrm{n}}}{Z^2 S_{I\mathrm{m}}} \tag{20.91}$$

As expected from duality, it is apparent from (20.91) that the impact of the bias source noise ΔS_{V0} and the input

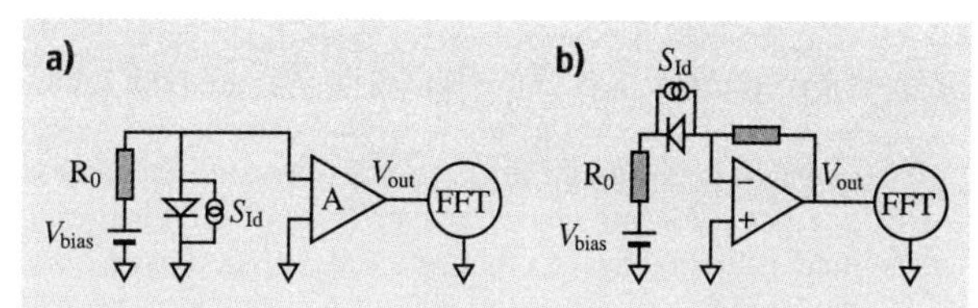

Fig. 20.35a,b Basic circuits used to measure the low-frequency noise (LFN) in a device (DUT). **(a)** Voltage noise measurement; **(b)** current noise measurement

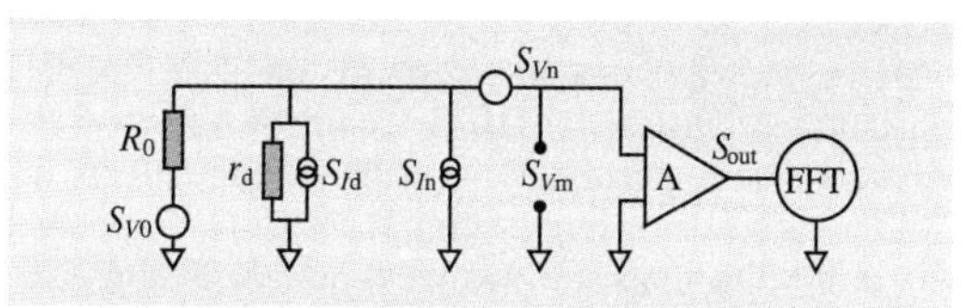

Fig. 20.36 Noise equivalent circuit for voltage noise measurements

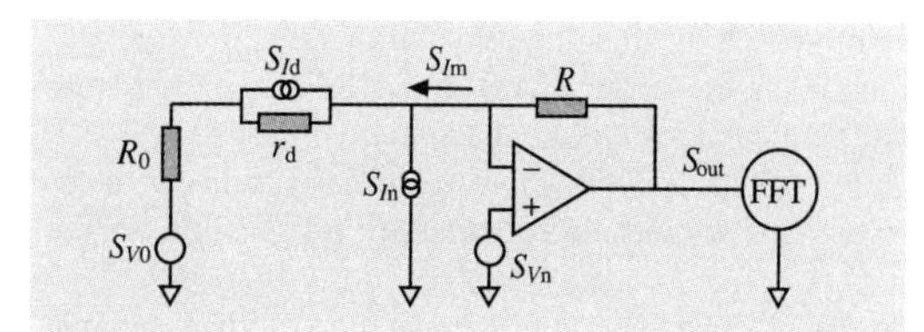

Fig. 20.37 Noise equivalent circuit of the current noise measurement

voltage noise $\Delta S_{I\text{n}}$ can be reduced if the impedance of the measurement circuit Z and $(r_\text{d} + R_0)$ are both high; in other words, the *current noise measurement is more appropriate for high-impedance devices, such as diodes at reverse biasing*, and the noise floor of the measurement is limited by the input-referred current noise S_{I_n} of the amplifier.

This analysis above demonstrates that the choice of the measurement configuration follows our expectation that voltage should be measured in low-impedance devices and current in high-impedance devices. Also, the noise floor limiting parameter of the preamplifier is of the same type as the type of measurement; that is, input-referred noise voltage for voltage noise measurement and input-referred noise current for current noise measurement. Note that there is a trade-off between the voltage and current noise in amplifiers, which implies that the measurement configuration – either voltage or current measurement – should also be carefully selected with respect to the impedance of the device under test. In addition, four-point connection can be used to measure the noise in very low impedance devices ($r_\text{d} < 100$). These and other considerations for low-frequency noise instrumentation are discussed in many papers, for example [20.74–76].

20.7 Deep-Level Transient Spectroscopy

Deep-level transient spectroscopy (DLTS) is a fairly complicated electrical characterization technique where the temperature is varied in large range from cryogenic temperatures (< 80 K) to well above room temperature (> 400 K). However, it is a powerful and versatile technique for investigating deep-level defects and it also gives accurate values for the capture cross-sections of defects. There are several DLTS techniques and [20.77, 78] provide recent reviews of the subject. In DLTS, the semiconductor device or junction is pulsed with an appropriate signal, and the resulting transient (such as capacitance, voltage or current) is monitored at different temperatures. Using these recorded transients at different temperatures, it is possible to generate a spectrum with peaks, each of which is associated with a deep level. The heights of the peaks are proportional to the defect density.

Here, we will focus on a new version of DLTS: the constant resistance (CR) DLTS technique [20.79–81]. We were able to accurately investigate bulk defects in a variety of test structures with CR-DLTS. Using body bias in a MOS transistor, we were able to distinguish interfacial and bulk defects that are important for different applications. For example, interfacial defects are important for electronic applications, and bulk defects are important for imaging or radiation detection applications. Examples of results from DLTS studies with and without body bias will be discussed.

CR-DLTS is well-suited to investigations of electrically active point defects that are responsible for the

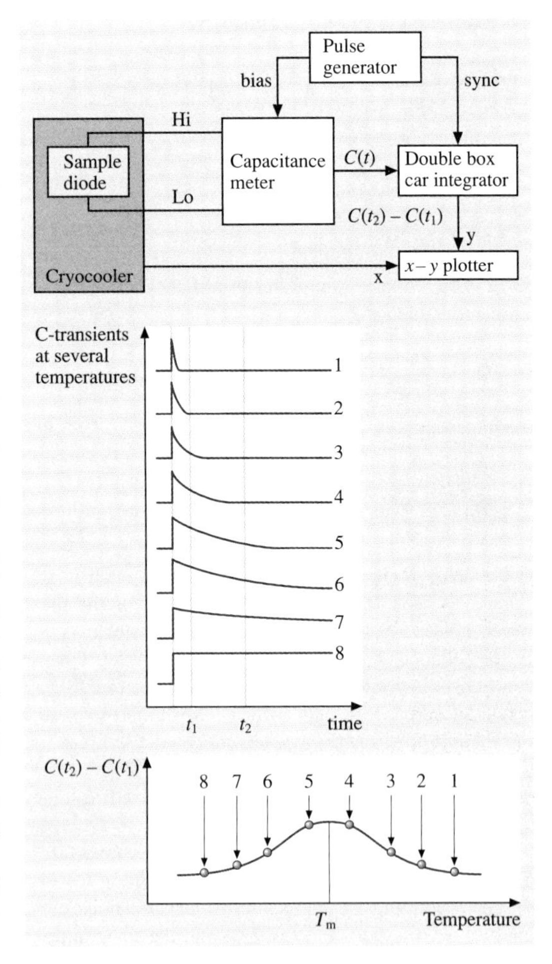

Fig. 20.38 Schematic representation of a conventional DLTS system. The time scans from which the DLTS temperature spectrum is obtained are shown on the *right*

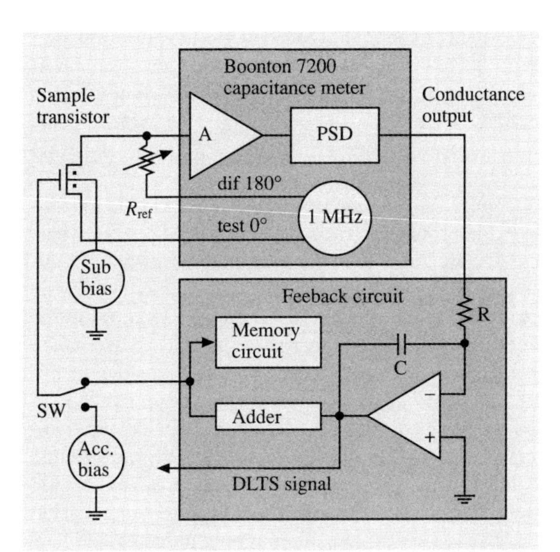

Fig. 20.39 Block diagram representation of the CR-DLTS system

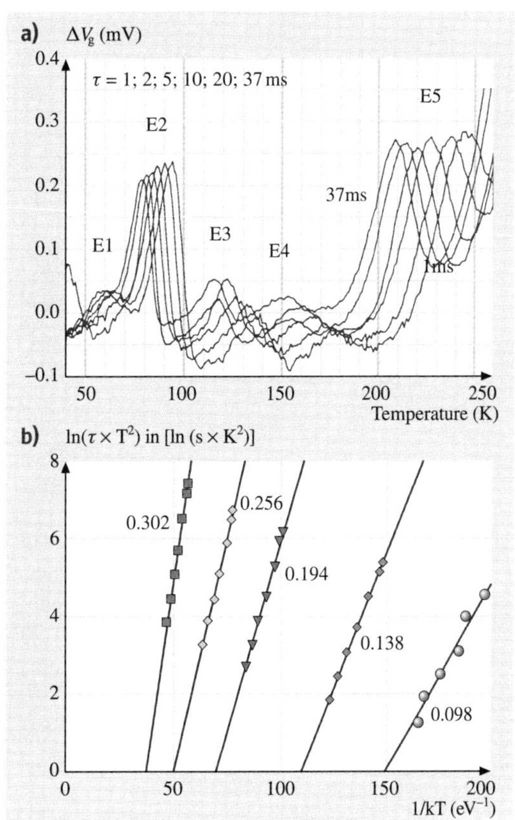

Fig. 20.40 (**a**) CR-DLTS spectra of a $50\,\mu m \times 20\,\mu m$ MOSFET damaged with 2.7×10^9 pronton/cm^2. (**b**) Arrhenius plot derived from the CR-DLTS spectra, showing the energies of the five traps E1–E5. After [20.80, 82]

creation of deep levels in the semiconductor band-gap. CR-DLTS can also be used to distinguish bulk traps and interface traps in MOSFETs.

A conventional DLTS system is shown schematically in Fig. 20.38. In DLTS, an excitation pulse is applied to the sample to fill all of the traps and then the pulsing is stopped. The next step is to detect the transient signal from the sample due to charge emission from the traps. The right side of Fig. 20.38 shows capacitance transients at eight different temperatures. By selecting a time window from t_1 to t_2, and then plotting $[C(t_1) - C(t_2)]$ as a function of temperature, a DLTS spectrum with a characteristic peak is obtained as shown in the bottom of Fig. 20.38.

This peak is a signature from a specific defect level. To determine the properties of the defect (its energy level and capture cross-section), the time window ($\tau = t_2 - t_1$) is changed. In this case, different DLTS spectra are obtained at different temperatures. Using the time difference τ and the temperatures at which the peaks occur, Arrhenius plots are constructed in order to determine the defect energy level and its capture cross-section. Examples of DLTS spectra and Arrhenius plots associateed with CR-DLTS are presented later (in Fig. 20.40).

A block representation of the CR-DLTS system is shown in Fig. 20.39. More details can be found in [20.79–83]. A discussion of the signal processing and averaging techniques used with this DLTS technique can be found in [20.83]. Here, the gate bias voltage of the field-effect transistor is adjusted using a feedback circuit so that the resistance corresponding to the source–drain conductance matches that of a reference resistor R_{ref}, which is typically around 1 MΩ. The voltage transient due to the change in occupancy of the traps appears as a compensation voltage on the gate. This voltage change can be regarded as a threshold voltage change because the flat-band voltage of the device changes when the occupancy of the traps change. More details on how this change in the threshold is related to the traps can be found in [20.77, 79, 81].

Some important advantages of the CR DLTS technique are that the surface mobility of the MOS transistors does not need to be high, and that it is theoretically independent of the gate area of the transistor. This is expected, since the small amount of charge trapped beneath

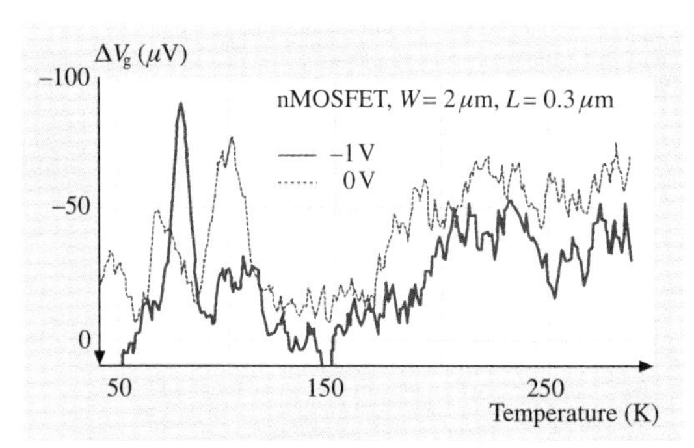

Fig. 20.41 Effect of body bias on CR-DLTS spectra. The body bias affects the surface and bulk traps in different ways

the gate must be balanced by a correction voltage applied across a relatively small gate–substrate capacitance.

Figure 20.40a shows six DLTS spectra for a JFET damaged with 2.7×10^9 protons/cm^2 [20.79, 81] with six selected rate windows. Using the temperatures at which the peaks occur and the rate windows, Arrhenius plots can be constructed as shown in Fig. 20.40b, where the energies of five electron trap levels below the conduction band are also indicated. For the five traps, the extracted capture cross sections were 4.6×10^{-15} cm^2 (E1), 6.3×10^{-15} cm^2 (E2), 1.2×10^{-16} cm^2 (E3), 8.5×10^{-16} cm^2 (E4) and 3.4×10^{-15} cm^2 (E5).

Figure 20.41 shows CR-DLTS spectra as the source–body bias voltage is varied. The scans with a body bias of −1 V are lower in magnitude than those without substrate bias, except for the peak associated with the hole trap at 0.13 eV above the valence band [20.81]. When the reverse substrate bias is increased, the gate control of the space charge region near the channel decreases, meaning that fewer interface traps participate in the capture and emission of charges. However, the increased reverse substrate bias results in an increased space charge region in the silicon beneath the gate, so more bulk deep levels can participate in the capture and emission processes. This explains the increased deep-level peak (below 75 K) when −1 V is applied to the body. These differences between the CR-DLTS spectra demonstrate the ability to distinguish bulk traps from surface traps when the substrate bias of the MOSFET is varied.

References

20.1 M.J. Deen: Proc. Sixth Symp. Silicon Nitride and Silicon Dioxide Thin Insulating Films, ed. by R.E. Sah, M.J. Deen, D. Landheer, K.B. Sundaram, W.D. Brown, D. Misra (Electrochem. Soc. Paris, Paris 2003) pp. 3–21

20.2 P. Rai-Choudhury, J. Benton, D. Schroder (Eds.): Proc. Symp. Diagnostic Techniques Semiconductor Materials and Devices, Proc. Vol. 97-12 (The Electrochemical Society Press, New Jersey 1997)

20.3 D. Schroder: *Semiconductor Material and Device Characterization*, 2nd edn. (Wiley, New York 1998)

20.4 W. Runyan, T Shaffner: *Semiconductor Measurements and Instrumentation*, 2nd edn. (McGraw Hill, New York 1997)

20.5 ITRS: *International Technology Roadmap for Semiconductors 2001 version*, URL: http://public.itrs.net/Files/2001ITRS/Home.htm (2001)

20.6 R. Pierret: *Advance Semiconductor Fundamentals*, Modular Ser. Solid State Dev. 6 (Addison-Wesley, Reading 1987)

20.7 W. Sawyer: Proc. 1998 Int. Conf. on Ion Implantation Technology, (IEEE Press, Piscataway, 1998)

20.8 L.J. van der Pauw: Philos. Res. Rev. **13**, 1–9 (1958)

20.9 M. Newsam, A. Walton, M. Fallon: Proc. 1996 Int. Conf. Microelectronic Test Structures (1996) pp. 247–252

20.10 T. Noda, D. Lee, H. Shim, M. Sakuraba, T. Matsuura, J. Murota: Thin Solid Films **380**, 57–60 (2000)

20.11 P. De Wolf, R. Stephenson, S. Biesemans, Ph. Jansen, G. Badenes, K. De Meyer, W. Vandervorst: Int. Electron Dev. Meeting (IEDM) Tech. Dig. (1998) pp. 559–562

20.12 J.H. Orchard-Webb, R. Coultier: Proc. IEEE Int. Conf. Microelectron. Test Structures, (1989) pp. 169–173

20.13 J. D. Plummer, P. B. Griffin: Proc. IEEE **89**(3) (2001) 240–258

20.14 S. Sze: *Physics of Semiconductor Devices*, 2nd edn. (Wiley, New York 1981)

20.15 W. M. Loh, S. E. Swirhun, T. A. Schreyer, R. M. Swanson, K. C. Saraswat: IEEE Trans. Electron Dev. **34**(3), 512–524 (1987)

20.16 S. Zhang, M. Östling, H. Norström, T. Arnborg: IEEE Trans. Electron Dev. **41**(8), 1414–1420 (1994)

20.17 A. S. Holland, G. K. Reeves: Microelectron. Reliab. **40**, 965–971 (2000)

20.18 Y. Qiu: *Introduction to the Quantum Hall Effect*, URL: http://www.pha.jhu.edu/~qiuym/qhe (1997)

20.19 E. H. Hall: Am. J. Math. **2**, 287–292 (1879)

20.20 P. Elias, S. Hasenohrl, J. Fedor, V. Cambel: Sensors Actuat. A **101**, 150–155 (2002)

20.21 A. Vandooren, S. Cristoloveanu, D. Flandre, J. P. Colinge: Solid-State Electron. **45**, 1793–1798 (2001)

20.22 D.T. Lu, H. Ryssel: Curr. Appl. Phys. **1**(3-5), 389–391 (2001)
20.23 R.L. Petriz: Phys. Rev. **110**, 1254–1262 (1958)
20.24 P. Terziyska, C. Blanc, J. Pernot, H. Peyre, S. Contreras, G. Bastide, J.L. Robert, J. Camassel, E. Morvan, C. Dua, C.C. Brylinski: Phys. Status Solidi A **195**(1), 243–247 (2003)
20.25 G. Rutsch, R.P. Devaty, D.W. Langer, L.B. Rowland, W.J. Choyke: Mat. Sci. Forum **264–268**, 517–520 (1998)
20.26 P. Blood, J.W. Orton: *The Electrical Characterization of Semiconductor: Majority Carriers and Electron States*, (Techniques of Physics, Vol. 14) (Academic, New York 1992)
20.27 Q. Lu, M.R. Sardela Jr., T.R. Bramblett, J.E. Greene: J. Appl. Phys. **80**, 4458–4466 (1996)
20.28 S. Wagner, C. Berglund: Rev. Sci. Instrum. **43**(12), 1775–1777 (1972)
20.29 E.H. Nicollian, J.R. Brews: *MOS (Metal Oxide Semiconductor) Physics and Technology* (Wiley, New York 1982)
20.30 *Model 82-DOS Simultaneous C-V Instruction Manual* (Keithley Instruments, Cleveland 1988)
20.31 W. Beadle, J. Tsai, R. Plummer: *Quick Reference Manual for Silicon Integrated Circuit Technology* (Wiley, New York 1985)
20.32 J. Brews: J. Appl. Phys., **44**(7), 3228–3231 (1973)
20.33 M. Kuhn: Solid-State Electron. **13**, 873–885 (1970)
20.34 C.N. Berglund: IEEE Trans. Electron Dev., **13**(10), 701–705 (1966)
20.35 S. Witczak, J. Schuele, M. Gaitan: Solid-State Electron. **35**, 345 (1992)
20.36 M.J. Deen: Electron. Lett. **28**(3), 1195–1997 (1992)
20.37 Z.P. Zuo, M.J. Deen, J. Wang: Proc. Canadian Conference on Electrical and Computer Engineering (IEEE Press, Piscataway, 1989) pp. 1038-1041
20.38 A. Raychaudhuri, M.J. Deen, M.I.H. King, W. Kwan: IEEE Trans. Electron Dev. **43**(7), 1114–1122 (1996)
20.39 W.S. Kwan, A. Raychaudhuri, M.J. Deen: Can. J. Phys. **74**, S167–S171 (1996)
20.40 T. Matsuda, R. Takezawa, K. Arakawa, M. Yasuda, T. Ohzone, T. Kameda, E. Kameda: Proc. International Conference on Microelectronic Test Structures (ICMTS 2001) (IEEE Press, Piscataway, 2001) pp. 65-70
20.41 A. Raychaudhuri, M.J. Deen, M.I.H. King, W. Kwan: IEEE Trans. Electron Dev. **43**(1), 110–115 (1996)
20.42 G. Groeseneken, H. Maes, N. Beltram, R. DeKeersmaker: IEEE Trans. Electron Dev. **31**, 42–53 (1984)
20.43 X. Li, M.J. Deen: Solid-State Electron. **35**(8), 1059–1063 (1992)
20.44 X.M. Li, M.J. Deen: IEEE International Electron Devices Meeting (IEDM) (IEEE Press, Piscataway, 1990) pp. 85-87
20.45 D.S. Ang, C.H. Ling: IEEE Electron Dev. Lett. **19**(1), 23–25 (1998)
20.46 H. Uchida, K. Fukuda, H. Tanaka, N. Hirashita: International Electron Devices Meeting (1995) pp. 41-44
20.47 N.S. Saks, M.G. Ancona: IEEE Trans. Electron Dev. **37**(4), 1057–1063 (1990)
20.48 S. Mahapatra, C.D. Parikh, V.R. Rao, C.R. Viswanathan, J. Vasi: IEEE Trans. Electron Dev. **47**(4), 789–796 (2000)
20.49 Y.-L. Chu, D.-W. Lin, C.-Y. Wu: IEEE Trans. Electron Dev. **47**(2), 348–353 (2000)
20.50 J.S. Bruglers, P.G. Jespers: IEEE Trans. Electron Dev. **16**, 297 (1969)
20.51 D. Bauza: J. Appl. Phys. **94**(5), 3239–3248 (2003)
20.52 L.M. Head, B. Le, T.M. Chen, L. Swiatkowski: Proceedings 30th Annual International Reliability Physics Symposium (1992) pp. 228-231
20.53 S. An, M.J. Deen: IEEE Trans. Electron Dev. **47**(3), 537–543 (2000)
20.54 S. An, M.J. Deen, A.S. Vetter, W.R. Clark, J.-P. Noel, F.R. Shepherd: IEEE J. Quantum Elect. **35**(8), 1196–1202 (1999)
20.55 M.J. Deen, C. Quon: 7th Biennial European Conference - Insulating Films on Semiconductors (INFOS 91), ed. by W. Eccleston, M. Uren (IOP Publishing Ltd., Liverpool U.K., 1991) pp. 295-298
20.56 Z. Celik-Butler: IEE P.-Circ. Dev. Syst. **149**(1), 23–32 (2002)
20.57 J. Chen, A. Lee, P. Fang, R. Solomon, T. Chan, P. Ko, C. Hu: Proceedings IEEE International SOI Conference (1991) pp. 100-101
20.58 J. Sikula: *Proceedings of the 17th International Conference on Noise in Physical Systems and 1/f Fluctuations* (*ICNF 2003*), (CNRL,Prague, 2003)
20.59 M.J. Deen, Z. Celik-Butler, M.E. Levinhstein (Eds.): SPIE Proc. Noise Dev. Circ. **5113** (2003)
20.60 C.R. Doering, L.B. Kish, M. Shlesinger: *Proceedings of the First International Conference on Unsolved Problems of Noise*, (World Scientific Publishing, Singapore, 1997)
20.61 D. Abbot, L.B. Kish: *Proceedings of the Second International Conference on Unsolved Problems of Noise and Fluctuations*, (American Institute of Physics Conference Proceedings: 511, Melville, New York 2000)
20.62 S.M. Bezrukov: *Proceedings of the Third International Conference on Unsolved Problems of Noise and Fluctuations*, Washington, DC (American Institute of Physics Conference Proceedings: 665, Melville, New York 2000)
20.63 M.J. Deen, S.L. Rumyantsev, M. Schroter: J. Appl. Phys. **85**(2), 1192–1195 (1999)
20.64 M. Sanden, O. Marinov, M. Jamal Deen, M. Ostling: IEEE Electron Dev. Lett., **22**(5), 242–244 (2001)
20.65 M. Sanden, O. Marinov, M. Jamal Deen, M. Ostling: IEEE Trans. Electron Dev. **49**(3), 514–520 (2002)
20.66 M.J. Deen, J.I. Ilowski, P. Yang: J. Appl. Phys. **77**(12), 6278–6288 (1995)

20.67 M. J. Deen, E. Simoen: IEE P.-Circ. Dev. Syst. **49**(1), 40–50 (2002)
20.68 M. J. Deen: IEE Proceedings - Circuits, Devices and Systems – Special Issue on Noise in Devices and Circuits **151**(2) (2004)
20.69 M. J. Deen, O. Marinov: IEEE Trans. Electron Dev. **49**(3), 409–414 (2002)
20.70 O. Marinov, M. J. Deen, J. Yu, G. Vamvounis, S. Holdcroft, W. Woods: Instability of the Noise Level in Polymer Field Effect Transistors with Non-Stationary Electrical Characteristics, Third International Conference on Unsolved Problems of Noise and Fluctuations (UPON 02), Washington, DC (AIP Press, Melville, 2002)
20.71 M. Marin, M. J. Deen, M. de Murcia, P. Llinares, J. C. Vildeuil: IEE P.-Circ. Dev. Syst. **151**(2), 95–101 (2004)
20.72 A. Chandrakasan: Proceedings European Solid-State Circuits Conference (ESSCIRC 2002), (AIP Press, Melville, 2002) pp. 47–54
20.73 R. Brederlow, W. Weber, D. Schmitt-Landsiedel, R. Thewes: IEDM Technical Digest (1999) pp. 159–162
20.74 L. Chaar, A. van Rheenen: IEEE Trans. Instrum. Meas. **43**, 658–660 (1994)
20.75 C.-Y. Chen, C.-H. Kuan: IEEE Trans. Instrum. Meas. **49**, 77–82 (2000)
20.76 C. Ciofi, F. Crupi, C. Pace, G. Scandurra: IEEE Trans. Instrum. Meas. **52**, 1533–1536 (2003)
20.77 P. Kolev, M. J. Deen: Development and Applications of a New DLTS Method and New Averaging Techniques. In: *Adv. Imag. Electr. Phys.*, ed. by P. Hawkes (Academic, New York 1999)
20.78 P. McLarty: Deep Level Transient Spectroscopy (DLTS). In: *Characterization Methods for Submicron MOSFETs*, ed. by H. Haddara (Kluwer, Boston 1996) pp. 109– 126
20.79 P. V. Kolev, M. J. Deen: J. Appl. Phys. **83**(2), 820–825 (1998)
20.80 P. Kolev, M. J. Deen, T. Hardy, R. Murowinski: J. Electrochem. Soc. **145**(9), 3258–3264 (1998)
20.81 P. Kolev, M. J. Deen, J. Kierstead, M. Citterio: IEEE Trans. Electron Dev. **46**(1), 204–213 (1999)
20.82 P. Kolev, M. J. Deen: Proceedings of the Fourth Symposium on Low Temperature Electronics and High Temperature Superconductivity, **97**-2, ed. by C. Claeys, S. I. Raider, M. J. Deen, W. D. Brown, R. K. Kirschman (The Electrochemical Society Press, New Jersey, 1997) pp. 147–158
20.83 P. V. Kolev, M. J. Deen, N. Alberding: Rev. Sci. Instrum. **69**(6), 2464–2474 (1998)